Parallel Processing in a Control Systems Environment

Prentice Hall International
Series in Systems and Control Engineering

M. J. Grimble, Series Editor

BENNETT, S., *Real-time Computer Control: an Introduction*
BITMEAD, R. R., GEVERS, M. and WERTZ, V., *Adaptive Optimal Control*
BUTLER, H., *Model Reference Adaptive Control*
ISERMANN, R., LACHMANN, K. H. and MATKO, D., *Adaptive Control Systems*
KUCERA, V., *Analysis and Design of Discrete Linear Control Systems*
LUNZE, J., *Feedback Control of Large-Scale Systems*
McLEAN, D., *Automatic Flight Control Systems*
OLSSON, G. and PIANI, G., *Computer Systems for Automation and Control*
PARKS, P. C. and HAHN, V., *Stability Theory*
PATTON, R., CLARK, R. N. and FRANK, P. M. (editors), *Fault Diagnosis in Dynamic Systems*
PETKOV, P. H., CHRISTOV, N. D. and KONSTANTINOV, M. M., *Computational Methods for Linear Control Systems*
ROGERS, E. and LI, Y., *Parallel Processing in a Control Systems Environment*
SÖDERSTROM, T. and STOICA, P., *System Identification*
SOETERBOEK, A. R. M., *Predictive Control: A unified approach*
STOORVOGEL, A., *The H_∞ Control Problem*
WATANABE, K., *Adaptive Estimation and Control*
WILLIAMSON, D., *Digital Control and Instrumentation*

Parallel Processing in a Control Systems Environment

Eric Rogers

Advanced Systems Research Group, Department of Aeronautics and Astronautics, University of Southampton

Yun Li

Department of Electronics and Electrical Engineering, The University, Glasgow

Prentice Hall

New York London Toronto Sydney Tokyo Singapore

First published 1993 by
Prentice Hall International (UK) Ltd
Campus 400, Maylands Avenue
Hemel Hempstead
Hertfordshire, HP2 7EZ
A division of
Simon & Schuster International Group

Typeset in 10/12 pt Times
by Mathematical Composition Setters Ltd, Salisbury, Wiltshire

Printed and bound in Great Britain
at the University Press, Cambridge

Library of Congress Cataloging-in-Publication Data

Parallel processing in a control systems environment / [edited] by Eric Rogers, Yun Li.
p. cm. -- (Prentice Hall international series in systems and control)
Includes bibliographical references and index.
ISBN 0–13–651530–4
1. Automatic control. 2. Parallel processing (Computer science) I. Rogers, Eric, II. Li, Yun. III. Series.
TJ213.P298 1993
629.8′9--dc20 92–36364
CIP

British Library Cataloguing in Publication Data

A catalogue record for this book is available from the British Library

ISBN 0–13–651530–4 (hbk)

1 2 3 4 5 97 96 95 94 93

Contents

Contributors

D. P. ATHERTON, School of Engineering and Applied Sciences, University of Sussex, Brighton BN1 9QT, UK.

M. I. BARLOW, Department of Marine Technology, University of Newcastle-upon-Tyne, Newcastle-upon-Tyne NE1 7RU, UK.

M. BROWN, Advanced Systems Research Group, Department of Aeronautics and Astronautics, University of Southampton, Southampton SO9 5NH, UK.

S. E. BURGE, Department of Engineering, University of Lancaster, LA1 4YR, UK.

L. CHISCI, Dipartimento di Sistemi E Informatica, Universita di Firenze, 50139, Firenze, Italy.

C. J. HARRIS, Advanced Systems Research Group, Department of Aeronautics and Astronautics, University of Southampton, Southampton SO9 5NH, UK.

K. J. HUNT, Daimler-Benz AG ALT-Moabit 916, D-1000 Berlin 21, Germany.

D. M. A. HUSSAIN, School of Engineering and Applied Sciences, University of Sussex, Brighton BN1 9QT, UK.

Y. LI, Department of Electronics and Electrical Engineering, University of Glasgow, Glasgow G12 8QQ, UK.

D. A. LINKENS, Department of Automatic Control and Systems Engineering, University of Sheffield, S1 3JD, UK.

G. A. MONTAGUE, Department of Chemical and Process Engineering, University of Newcastle-upon-Tyne, Newcastle-upon-Tyne NE1 7RU, UK.

A. J. MORRIS, Department of Chemical and Process Engineering, University of Newcastle-upon-Tyne, Newcastle-upon-Tyne NE1 7RU, UK.

E. ROGERS, Advanced Systems Research Group, Department of Aeronautics and Astronautics, University of Southampton, Southampton SO9 5NH, UK.

A. P. ROSKILLY, Department of Marine Technology, University of Newcastle-upon-Tyne, Newcastle-upon-Tyne NE1 7RU, UK.

R. W. STEWART, Signal Processing Division, Department of Electronic and Electrical Engineering, University of Strathclyde, Glasgow G1 1XW, UK.

M. T. THAM, Department of Chemical and Process Engineering, University of Newcastle-upon-Tyne, Newcastle-upon-Tyne NE1 7RU, UK.

A. M. TYRRELL, Department of Electronics, University of York, York YO1 5DD, UK.

M. J. WILLIS, Department of Chemical and Process Engineering, University of Newcastle-upon-Tyne, Newcastle-upon-Tyne NE1 7RU, UK.

G. ZAPPA, Dipartimento di Sistemi E Informatica, Universita di Firenze, 50139, Firenze, Italy.

Preface

The last two to three decades, in particular, have seen rapid and far-reaching developments in the design and implementation of control systems. This progress has, of course, been critically dependent on the 'enabling technology' resulting from the equally rapid and massive developments over the same period in computing power and flexibility. In particular, the development of general purpose mini- and microcomputers with adequate power and flexibility to service both the computing loads generated and related tasks, such as database management and input/output facilities, has played a very significant role.

These developments in the general subject area have naturally led to (realistic) consideration of systems with greatly increased size and complexity. For example, there is increasing interest in the control of complex distributed systems such as large space structures or interconnected power systems networks. In terms of existing controller design algorithms, such as those arising in optimal control, the major problem here is that the dynamics of such systems typically extend over 'wide' frequency ranges or spatial domains and hence model order reduction techniques are often not appropriate. Consequently, successful application of these algorithms to problems with this general characteristic will only be achieved if it is possible to allow for an increase in the model dimension of at least one order of magnitude with negligible reduction in reliability. Hence in the case of linear, state-space or transfer-function matrix based design, this means the reformulation, sometimes termed algorithmic engineering, of essentially matrix based computations.

Consider now the digital implementation of a control system with a prescribed sampling rate or period. Then (typically) the hardware/software used must complete real-time control and identification functions, together with a number of support functions such as data logging and checking, within this period. In a large number of the (successful) applications to date, the use of general-purpose mini- and microcomputers for this task has proved adequate. This advantage rapidly disappears, however, with increasing control requirements to meet demands such as increased speed and complexity, improved dynamic range and accuracy, and lower cost/increased flexibility.

The origins of these increased control requirements are generally application-specific. Consider, for example, motor control. In this application the control requirements are often expressed in terms of short (or 'fast') sampling times which, in turn, can lead to constraints on the execution time of the control algorithm used, which are incompatible with the available implementation medium. Alternatively, controller complexity may increase owing to the quantity and quality of the information to be processed, such as that from the sensors in vision based systems. Similarly, implementation of state-space based control schemes, such as those employing a Kalman filter, can easily place an unacceptable burden on the proposed computing system.

As these difficulties became apparent, the control community began to explore alternative hardware solutions. For example, special-purpose control computers which make use of array processing power to exploit matrix-vector operations which dominate a large number of algorithms have been developed. Further, the dramatic advances in parallel processing have attracted interest as a means of providing more general-purpose solutions.

One general area where all forms of parallel processing is well established, to the extent that devices based on this approach are at the commercial viability stage, is digital signal processing. Further, the existence of strong structural links between large numbers of algorithms and architectures used in digital signal processing and control problems is a standard fact. These mean that, appropriately exploited, parallel processing offers the possibility of very significant breakthroughs across a very broad spread of control oriented computations. In particular, it is possible to provide efficient solutions to many classes of currently intractable high-order problems and real-time computations.

In the case of conventional (or traditional) control systems, the most obvious benefit of parallel processing is increased computational speed. The importance of this area has been supported by its inclusion in a number of strategy documents for future research and development on control systems produced by various organizations, such as the Society for Industrial and Applied Mathematics. These have identified, in addition to complex distributed systems, robotics and adaptive control as examples of areas which should benefit significantly from appropriately targeted research in the short to medium term. To this list can be added, amongst others, fault tolerance, i.e. by using a parallel processing based system, computation operations can be organized in a distributed sense and hence an operational failure results in performance degradation rather than complete controller failure.

The attributes of conventional control systems are well known and documented, but there still remains a large class of dynamic systems of sufficient complexity to prevent adequate analytical representation or modelling – the essential first step of the conventional approach. One alternative in such cases is to consider the use of so-called intelligent control techniques which are currently undergoing rapid and far-reaching developments, both in terms of basic research and applications. Further, a survey of the progress to date, plus anticipated developments, clearly reveals the underlying role of various forms of parallel processing in this exciting area.

It is clear that the various forms of parallel processing offer the very real prospect of major and far-reaching breakthroughs in the computations associated with the broadest possible classification of control/automation systems. This applies equally to the following:

1. Extending the effective operating range of conventional approaches.
2. The development and appropriate application of new approaches such as intelligent control.

The research challenges posed to the control community at large by this enabling technology are immense, ranging from theoretic algorithm development (in the broadest sense) through to problems associated with the fabrication of special-purpose chips such as systolic arrays. This text is motivated both by the rapid developments in parallel processing as a subject in its own right and in the acceleration in interest in parallel processing techniques amongst the general control systems community. The following are its major aims.

1. To provide a 'state of the art' report on the main areas of progress to date and hence to demonstrate beyond doubt the potential of this general approach.
2. To define and set general objectives for further research and development.

This book consists of twelve self-contained chapters, each written by leading researchers. They are arranged in the following four parts to provide a coherent and comprehensive coverage:

Part I Systolic arrays.
Part II Architectures for intelligent control.
Part III Transputer networks.
Part IV Parallel machines and algorithms.

Each part is preceded by a short introduction to 'set the scene'. In Part I, various systolic and wavefront arrays are developed which are targeted towards realization by very large scale/wafer scale integration (VLSI/WSI). The architectures developed here are for the problems associated with a generic class of real-time embedded control computations – including performance assessment, adaptive and self-tuning schemes, predictive control, system identification and Kalman filtering.

Part II presents massively parallel and artificial intelligence type architectures based on neural networks. The neural network concept is not limited to hardware realizations and can also be used to develop software schemes which can be realized in existing parallel machines. Applications covered here are modelling, identification, stochastic estimation, prediction, nonlinear learning control, self-organizing and fuzzy logic based control.

In Part III the emphasis is on the use of commercially available processors and, in particular, transputers and A100 parallel signal processors as building blocks in the reconfiguration of user-tailored systems for control applications. Subjects considered here are a generic class of digital controllers in the form of an infinite impulse response filter, adaptive and self-tuning schemes, robot control systems and multiple target tracking problems. Further, task partitioning for parallel realizations, load balancing and processor clustering issues are also considered, together with software and hardware techniques for fault-tolerant implementations.

Part IV gives a brief overview of the use of commercially available parallel machines for control applications. In particular, it is emphasized that, in this context, such machines are best suited to the development of parallel algorithms for control systems design. This feature is illustrated by algorithms for multivariable frequency response computation and the algebraic Riccati equation based on a hypercube machine.

This division into parts is mainly based on the granularity, i.e. a measure which matches computational power to interprocessor communications overheads, of the parallel architectures, starting with system-specific architectures targeted towards VLSI realization through to commercially available general-purpose machines. These parts should not, however, be considered as isolated from each other. For example, in the development stages of a 'fine-grain' architecture, such as a systolic array, existing coarse grain machines, such as a transputer network, can be used to assist in the design and simulation of the architecture. Similarly, algorithms and architectures developed using a 'coarse-grained' machine can be mapped onto a 'fine-grain' processor array and hence, by fabrication, a more compact system. Further, given the level of research in this general area, it is to be expected that there will be even greater interactions between different architectures in both hardware and software design, leading to the availability of even more parallel architectures suitable for applications encountered by control engineers.

ACKNOWLEDGEMENTS

The ideas for this book first surfaced when the authors were with the Division of Dynamics and Control, University of Strathclyde. We are extremely grateful for the advice and encouragement provided in those early days by, in particular, David Owens (now at the University of Exeter). Allison King from Prentice Hall provided the 'necessary input' to keep going during times when, for various reasons, 'progress was slow'. Finally, we must thank all the contributors for their patience and support which greatly eased the production process.

August 1992
Eric Rogers
Yun Li

Part I

Systolic Arrays

The first generation, still widely used, of digital control systems implementations are based on the single general-purpose processor architecture arising from the von Neumann model. Further, the development of these systems, measured in terms of increased speed and improved performance, has been historically linked to developments in computing technology. This has led to the use of special-purpose digital signal processors and, subsequently, control-dedicated 'digital control processors', both of which use the uniprocessor architecture.

Rapid developments in very large scale integration (VLSI) circuit technology now permit the fabrication of multiple processing elements (PEs) onto a single silicon chip. This, in turn, has led to the development of architectures in the form of systolic and wavefront arrays which have been the subject of intensive research, motivated by a number of computationally intensive applications areas. For example, in digital signal processing, an area which shares strong structural links with control, these arrays have been applied to problems such as correlation and convolution, infinite impulse response filters, fast Fourier transform, orthogonal triangularization and LU decomposition.

In effect, a systolic array is a network of processors which rhythmically computes and passes data through the network. Its major features are (a) modularity and regularity in terms of VLSI implementation, (b) linear-rate pipelinability, (c) spatial and temporal locality, (d) data input to the array only pipelined through boundary PEs, and (e) synchrony of data flow and computation. A wavefront array has the same architecture but here the control of data flow and computation is self-timed and data-driven, rather than controlled by a synchronized global clock. At word-level, these two types of architecture directly map an algorithm onto the hardware PE level and in this section they are applied to control problems.

After a brief review of previous work and the relevant features of systolic/wavefront arrays, Chapter 1 considers the design of these architectures for a generic class of real-time feedback control schemes. A crucial point stressed here is that, in contrast to other areas of application, both processing speed (or throughput rate) and processing delay (or system latency) must be included in measuring the performance of a parallel architecture for real-time feedback control schemes. Chapter 2 then gives a systematic treatment of the design of systolic arrays for adaptive control schemes and includes architectures for least-squares estimation and predictive control. Finally, Chapter 3 develops systolic architectures for the Kalman filter whose various forms have found wide applications in control/digital signal processing. Hence this chapter should be seen as complementing the already existing designs.

1

VLSI Systolic/Wavefront Array Processing for Recursive Filtering and Compensation

Y. Li and E. Rogers

1.1 INTRODUCTION

Early implementations of real-time digital feedback control schemes were almost exclusively in the process industries. The major reason for this was that the process dynamics typically encountered were 'slow' and hence compatible with the processing power of the computers then available. Other contributory factors were the relatively large size of these machines, which prevented their effective application to the real-time control of moving objects, and their cost. Since these early days, the development of real-time digital control systems and computer technology has been largely complementary.

This chapter develops parallel architectures for the implementation of feedback control schemes in a general, rather than an applications-specific, setting, with the aim of answering fundamental questions as an (obvious) starting point for detailed development studies. In particular, the use of systolic/wavefront array architectures will be considered. The objective of this work, therefore, is the exploitation of these well-established, and continually evolving, devices to develop dedicated architectures for a (reasonably) generic class of feedback control schemes.

The main body of this chapter begins in Section 1.2 with the necessary preliminaries which consist of a survey of related work, the required background on systolic/wavefront arrays, and the specification of the scheme to be considered. Section 1.3 then details basic architectural manipulations that lead on to the systolic/wavefront arrays of Section 1.4. Finally, Section 1.5 critically reviews the progress achieved in the work reported here and details some immediate and longer term future research/development goals.

1.2 PRELIMINARIES

1.2.1 Background and Requirements

A familiar real-time digital control scheme with unity negative feedback is shown schematically in Figure 1.1 where, owing to the data pipeline, processing of the data sampled at the plant output must be completed before the beginning of the next sample. The processing (computing) speed determines the sampling rate of the system. Further, 'high' accuracy performance from such a system often requires a 'high'-order model coupled with a relatively 'short' sampling period. It is well known that the higher the sampling rate, the smaller the relative error between the actual system response and its sampled version. For example, Åström and Wittenmark (1984) have shown that using a standard digital to analog (D/A) converter (typically a zero-order hold in digital control systems) to reconstruct a sine wave gives a 30% relative error at a sampling rate 10 times higher than the Nyquist rate and a 1% error if a 300 times higher rate is used. In general terms, reducing this rate can often result in a degradation of the implemented system performance (Moroney, 1983). Further, a 'low' sampling rate also prolongs the system response time. Increasing the sampling rate, however, increases the load on the computer used.

Clearly, the processing speed of the digital control system is dictated by the speed of the device used to implement the controller (or compensator). Over the past 30 years, in particular, there has been a very rapid development in circuit and computer technology, which has yielded a 'large' increase in processing speed. This has led to many successful implementations of real-time digital control schemes which, in turn, have stimulated theoretical developments. As a result, for example, the design and synthesis of advanced control schemes based on numerous well-known techniques is now a very mature area. In the case of adaptive/ self-tuning control (see also Chapters 2 and 9) it is assumed here that the

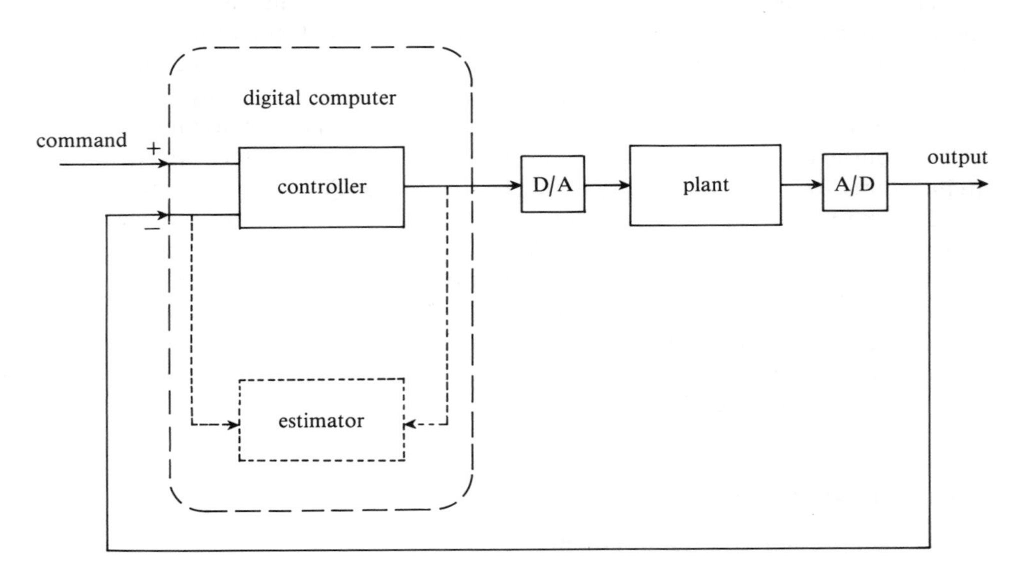

Figure 1.1 Block diagram of a typical digital feedback control scheme

identification/estimation phase, represented in Figure 1.1 by the inner set of dotted lines, is completed using any one of the numerous well-established techniques.

Such algorithms offer, in theory, the promise of high performance but, in implementation terms, typically require a 'high' sampling rate and/or are often computationally intensive and hence require 'high-speed' computation for real-time operations (Åström and Wittenmark, 1989; Middleton and Goodwin, 1990). The solution often adopted in such cases is to reduce the algorithm complexity and/or the sampling rate with a consequent possible compromise in terms of performance. This approach is also one explanation for the popularity in routine applications of simple structure controllers such as proportional plus integral plus derivative (PID) controllers and phase lead/lag.

A number of important factors exist which restrict the options for increasing computing speed. These include device operation or logic circuit speed and the well-known memory–processor communication bottleneck of the von Neumann architecture and hence low processor utilization. Note also that the increase in computing speed which has been achieved has largely been as a result of increasing circuit speed which, of course, provides only limited scope.

Conventional computers are based on the von Neumann model, which in effect consists of a single processing unit separated from the memory and this architecture limits the speed increase possible in two major respects. Firstly, since it is a single processor system, the speed of this computer cannot be increased beyond the limit imposed on one processor. Secondly, for every operation or data sample, the von Neumann architecture requires the following:

1. Fetching of the instructions and operands from memory serially.
2. Decoding of these by the central processing unit (CPU).
3. Intermediate storage of the result.

Consequently, when the processor speed is increased, the passage of data for this sequence of operations through the bus system forms a communication 'bottleneck' and hence a barrier to improving the speed of the overall system. Hence, the only feasible means of providing a further increase in speed is to update the computer architecture. A concurrent, or parallel/pipelined, processing methodology has long been recognized as one feasible option for achieving this transition.

The last decade, in particular, has seen a very dramatic advance in very large scale integration (VLSI) circuit technology and associated computer-aided design (CAD) tools. This removes a basic constraint on the design of computer architectures, which is limited by the cost of processing elements (PEs), and makes concurrent architectures feasible. Note also that this modern circuit technology offers the potential of a circuit density in excess of the device speed itself. Correspondingly, high-speed operation can be obtained, at a relatively low cost, by connecting together a number of simple PEs to form a processing array with every PE running in parallel. Further, one of many other benefits of this high-degree circuit density is the ability to offer a longer word length and hence higher accuracy at a relatively low cost. This facility also improves speed by avoiding the use of multibyte operations which would be the case if 'short' word length processors were used.

A combination of demand and manufacturing capability has led to the emergence of various concurrent systems, both mainframe and microprocessor based, which provide supercomputing power at a relatively 'low' cost. Particular examples include multiprocessor, vector processor, pipelined, hypercube, data flow, neural network and transputer based systems. These have been making an increasing impact in large areas of engineering and science based computing and, in particular, the general area of real-time digital signal processing (DSP). For example, VLSI oriented systolic/wavefront array architectures have been proposed for a large number of DSP problems and some of these have reached the stage of commercial availability (Kung, 1988).

In a feedback control scheme, such as Figure 1.1, the processed data must be fed back to the plant at the beginning of the next sample, i.e. the system latency must be one cycle. A processing delay which occupies several (or indeed thousands) of cycles is of little real consequence in digital filtering, or open-loop applications, but is a critical factor in feedback control schemes. This is because a pure time delay introduces a shifted or distorted signal input to the plant (Åström and Wittenmark, 1984) with consequent problems in terms of system stability and performance. At worst, an unstable implementation could result. It follows immediately, therefore, that particular attention must be given to latency considerations.

From the above discussion, the essential requirements of a 'good' parallel architecture for feedback control can be summarized as follows:

1. High processing throughput.
2. Low system latency (measured in processing cycles).

Other important factors include reliability, efficiency and speedup (Li, 1990). The existence of a strong duality between large numbers of problems arising in digital signal processing and control engineering is well known. This is, for example, illustrated by considering optimal/adaptive filtering from the former area and optimal/adaptive control from the latter which are well established, both theoretically and in applications, and are based on recursive least-squares techniques which are structurally very similar. Further, infinite impulse response (IIR) filtering is an example of techniques whose various forms have found wide application across both areas. Note also that, from an implementation standpoint, a large number of widely applicable digital feedback control schemes can, in effect, be regarded as a real-time digital signal processor embedded within a global feedback loop through the plant to be controlled. Computing systems are also, of course, an indispensable part of off-line operation functions such as simulation, controller design and knowledge based control systems. A more comprehensive treatment of the interconnections between these two areas can, for example, be found in the book by Willsky (1979). Appropriate exploitation of this link is clearly a feasible way forward with a high possibility of reasonably rapid progress.

1.2.2 Related Work

Conventionally, digital control/compensation systems are implemented on a single general-purpose computer which is based on the von Neumann model. As discussed in Section 1.2.1, however, the increase in speed of this class of architecture is bounded by the bus structure, etc., and hence the processor–memory communication bottleneck, even if the logic circuit speed could be improved to its maximum realizable value. In order to exploit concurrency fully and resolve the problem of the memory–processor communication bottleneck, totally novel architectures must be used or, alternatively, multiple memory–processor input/output (I/O) links must be constructed. Realizations of multiple I/O links, however, are difficult in practice, since the associated control circuitry can be very complex and require a much larger silicon area. Alternatively, distributed or local memory can be introduced into processor array architectures to increase the I/O bandwidth, and a task partitioned into many subtasks which can be executed by assigned PEs with the result that data access times are significantly reduced.

The rapid development of VLSI technology has, since the early 1980s, attracted the attention of control engineers seeking effective alternatives to the traditional methods of designing and implementing digital control systems. Chen (1982) published a control systems design using a special-purpose VLSI DSP chip (Intel 2920), which was programmed with its own instruction set. Similar contributions have also been reported, with the general aim of providing (effective) high performance within a compact framework.

In the same period, work aimed at developing control-dedicated monolithic 'digital control processors' (DCPs) began (see, for example, Lang, 1984). These special-purpose systems overcame some of the inefficiency resulting from using general-purpose processors such as the programming overhead and the need for multibyte operations, to provide the combined requirement of accuracy and dynamic range (Farrar and Eidens, 1980; Lang, 1984). Note, however, that all such special-purpose systems are still based on uniprocessor architectures and hence the increase in speed is limited by the maximum speed of a single processor.

In order to provide higher performance, early efforts concentrated on partitioning a large task into several subtasks to distribute the complexity of the computing load (Anderson and Linnemann, 1987). Some parallel algorithms for optimal estimation, Kalman filtering, identification and simulation have been developed based on decentralized or distributed architectures – see, for example, Datta (1986). Others have used multiprocessor systems in an attempt to exploit parallelism in control systems implementation – see, for example, Kalyaev and Kalyaev (1986). More systematically, Larsen and Evans (1988) employed graph theory to develop a structural design method for certain classes of control problems which uses Boolean matrix operations to supplement the numerically based methods. Various applications of distributed/decentralized architectures have been reported (see, for example, Kalyaev and Kalyaev, 1986). Further, massively parallel architectures, and the associated 'dual look-ahead' computation algorithm, have been developed for the design of a generic class of multivariable real-time control systems (Rogers and Li, 1988). Generally, spreading of arithmetic operations has resulted in a significant increase in computational speed.

Work has also been directed towards VLSI realization of these architectures for control engineering oriented applications. For example, pipelined IIR filter architectures have been used to implement feedback compensation schemes and problems related to the associated system latencies addressed (Moroney, 1983). Jones (1988) (and in subsequent publications such as Xu *et al.*, 1992) has also reported the development of a VLSI-oriented 'programmable adaptive computing engine', which is specifically aimed at control and robotics applications.

Traditional multiprocessor approaches are, however, based on several processors accessing a global memory via a data bus. Consequently, the processor–memory communication bandwidth is still a bottleneck, especially for compute-bound problems, to which most control computations belong. These architectures also lack regularity and modularity which are highly desirable in terms of VLSI implementation. All of these factors, therefore, impose a very significant limitation on the performance improvement possible.

System-specific systolic/wavefront array architectures provide very fine partitioning of the algorithm to be implemented and map this onto corresponding hardware structures. Algorithms which do not require recursive computation can easily be implemented on an architecture which is pipelined at an arbitrarily low level and this provides an arbitrarily high throughput rate (Li, 1990). In the case of Figure 1.1, however, the algorithm is recursive.

Work on applying systolic architectures to control problems has already been reported – see, for example, papers in Irwin and Fleming (1990) and Rogers (1991). One option here is to attempt to exploit work already reported for structurally similar DSP problems. For example, suppose, as from Section 1.2.4 onwards here, that the controller computation in Figure 1.1 has the structure of an IIR filter. Then, in principle, the architectures reported by Kung (1984), Lin (1986) and Kwan (1987) could be used. The basic difficulty with these architectures is that the system latency, or the processing delay, is always a crucial factor in the case of control systems implementation. In particular, both Lin's and Kwan's architectures have a system latency of $2n$ cycles, which is not only 'too long' for control applications but is also dependent on the filter order, n, and hence they would require modification for use in implementing closed-loop systems. If, however, the modified controller architecture results in an increase in the controller order, the design becomes inconsistent and hence these architectures are not applicable to feedback control problems (the details of this point will be discussed in Section 1.3.1).

The systolic architecture for auto-regressive moving average (ARMA), or IIR, filtering developed by Kung (see Kung, 1984, Figure 4), can be directly applied to feedback control system implementation. In their paper, Wang and Lin (1986) directly applied this architecture to single-input single-output (SISO) control systems. Note, however, that this architecture in effect alters the controller transfer function by altering its denominator and hence the required recursive computation (see Section 1.4.1 where a solution is given). Further, this implementation introduces a computational delay of 2.5 sampling periods, i.e. a system latency of 2.5 cycles. Based on this design, however, extensions to multivariable systems by possible utilization of the emerging 3D VLSI technology are feasible.

Work on applying other systolic and concurrent architectures to control problems in general can be found in other chapters of this text and in special issues of learned journals, such as Irwin and Fleming (1990) and Rogers (1991). This chapter will concentrate on the application of systolic/wavefront architectures to feedback control/compensation problems and, in particular, the subclass defined in Section 1.2.4. Other relevant work in this particular case includes that reported by Li (1990), Li *et al.* (1991), Li and Rogers (1989, 1990) and Rogers and Li (1988, 1989, 1990, 1991). As essential background, the next section summarizes the basic properties of this class of architecture.

1.2.3 Systolic/Wavefront Arrays

The systolic/wavefront array architecture (Kung, 1988) remains a very active research topic. A major factor here is the fact that this class of array belongs to the generation of very high-speed architectures and is suitable for VLSI/WSI (wafer scale intergration) implementation. In such an array, the processing elements or cells have their own memory and are locally connected. The following is the accepted formal definition of a systolic array.

Definition 1.1

A systolic array is a computing network possessing the features of (a) synchrony of data flow and computation; (b) modularity and regularity; (c) spatial and temporal locality; and (d) linear-rate pipelinability.

Such an array differs from a single-instruction multiple-data (SIMD) array in the fact that data input to a systolic array is pipelined through boundary PEs whilst in a SIMD array the data are preloaded from the data bus to every PE. Note that, if N PEs are used, then the linear pipelinability of an implementation provides an $O(N)$ speedup of processing rate.

The *wavefront* array is very similar to the systolic array, except that the control of data flow is self-timed and data-driven, rather than by a global clock. Consequently, temporal locality is no longer needed, since the data operations are simply controlled by the arrival of data from neighbouring elements. Further, the most distinctive difference between a wavefront array and a multiple-instruction multiple-data (MIMD) array is the same as that for systolic arrays and SIMD arrays, namely, a different data input mode.

A systolic/wavefront array permits multiple-processing for each memory access and thus speeds up execution of *compute-bound* computations (Kung, 1988) without increasing the communication requirements. The term 'compute-bound' refers to computation problems where the number of computations required is greater than the number of data inputs. This also serves to highlight the appeal of such architectures to the work reported here, since most control algorithms belong to this category. A typical example is the transfer function of a SISO controller, where there is actually only one piece of input data, say $e(k)$, which must be successively delayed a finite number of times, say n, during operation to form $e(k-1), e(k-2), \ldots, e(k-n)$. The counterpart of the compute-bound class is those with the so-called *I/O bound*, i.e. more input data than operations, as in an addition of two matrices. Further, high-level algorithms can be directly mapped onto cell-level hardware through systolic/wavefront arrays, which obviously saves time in the design procedure. In principle, therefore, a control task can be mapped onto a small processing element, such as a multiplication-accumulation (MAC) unit, and the massive number of PEs in the array will then produce a very high speed throughput.

In comparison to other special-purpose designs, systolic/wavefront array architectures benefit from those characteristics described in their definitions and the following:

1. High degree of parallelism and pipelining.
2. Single access for multiple use of data per sample.
3. Very high speed.
4. Simple timing circuitry.
5. Ability to map high level algorithms onto VLSI architectures.
6. Cost-effectiveness.

In common with conventional computing hardware, a control algorithm is naturally written in sequential code, since there are dependence relationships between the computation steps. Further, the order in which data are to be processed must be specified in the algorithm. If there is no data dependence in the operations for a particular special case, *parallelism* is possible. In general, however, this is impossible and a minimum amount of serialism is required. Potential concurrency in these algorithms does, however, often exist and appropriate reorganization of them could lead to an implementation with maximal concurrency, which can then be realized by *pipelining* (Kogge, 1981). This is also one of the major reasons why systolic array architectures have received, and continue to receive, much attention in real-time processing. Parallelism and pipelining operations constitute *concurrent processing*.

As noted previously, systolic/wavefront array architectures with very high concurrency have been extensively researched, with particular attention to problems such as recursive IIR filtering, correlation and convolution, fast Fourier transform (FFT) and other transformations, LU decomposition, orthogonal triangularization, least-squares estimation, Kalman filtering and adaptive beamforming, etc. (McCanny and White, 1987; Kung, 1988). One measure of the maturity of this area is the fact that some of the architectures developed have reached the stage of commercial availability and, see Section 1.2.2, work on applying them to real-time control-related problems has already been reported. Note, however, that no work on using wavefront arrays in this general area has yet been reported.

1.2.4 A Typical Control Computation

In this section, a broad-based example of the computations which typically arise in control engineering applications and its computation graph will be discussed, first to provide a basis for architectural development. Further, it is expected that solving the problems associated with recursive filtering will lead to control systems implementations with better performance and vice versa.

Consider, therefore, a SISO plant described by the state-space model

$$\dot{\mathbf{x}}_p(t) = A_p\mathbf{x}_p(t) + B_pu(t) + \mathbf{w}_1(t) \tag{1.1a}$$

$$y(t) = C_p\mathbf{x}_p(t) + \mathbf{w}_2(t) \tag{1.1b}$$

Here $\mathbf{x}_p(t) \in \mathscr{R}^n$ is the plant state vector, $u(t)$ is the plant input, $y(t)$ is the plant output, and $\mathbf{w}_1(t)$ and $\mathbf{w}_2(t)$ are Gaussian white noise disturbances. The matrices A_p, B_p and C_p are of appropriate dimensions with elements which (for the remainder of this chapter) are assumed to be either known or computed off-line. Suppose also that this plant is to be controlled on-line using the (typically used) scheme of Figure 1.1. Then the following state-space description of the controller is assumed here since it includes a large number of well-known choices as special cases:

$$\mathbf{x}(k+1) = A\mathbf{x}(k) + Be(k) \tag{1.2a}$$

$$u(k+1) = C\mathbf{x}(k+1) \tag{1.2b}$$

where

$$e(k) = r(k) - y(k) \tag{1.2c}$$

Here $\mathbf{x}(k) \in \mathscr{R}^n$ is the controller state vector and $r(k)$ is the command or reference signal. In the case of a (commonly encountered) proportional plus integral (PI) controller, n is equal to 2; and in adaptive/self-tuning control, the coefficient matrices of the controller are time-varying. Further, it is easily seen that computing (1.2a)–(1.2c) requires $n(n+2)$ multiplications and $n(n+1)$ additions, which includes the calculation time for (1.2c).

Consider now the operation of Figure 1.1 in real-time. Then in order to ensure synchronized sampling, the computation of (1.2a)–(1.2c) must be completed in one sampling period. Hence the minimum computational time required is the lower bound on the sampling period. Suppose therefore, for example, that the *word-level* multiplication and accumulation times required for a single processor execution are T_m and T_a seconds respectively. In which case the minimum sampling period possible is bounded above by $n(n+2)T_{\mathrm{m}} + n(n+1)T_{\mathrm{a}}$ seconds if *serialism*, i.e. a single processor, is used in the implementation.

The number of multiplications and additions necessary in this computation can be immediately reduced by application of the z-transform to yield

$$\frac{U(z)}{E(z)} = H(z^{-1}) = \frac{\sum_{i=1}^{n} a_i z^{-i}}{1 - \sum_{i=1}^{n} b_i z^{-i}} \tag{1.3a}$$

or, in difference equation terms,

$$\begin{aligned} u(k+1) = {} & a_1 e(k) + a_2 e(k-1) + \ldots + a_n e(k-n+1) \\ & + b_1 u/k) + b_2 u(k-1) + \ldots + b_n u(k-n+1) \end{aligned} \tag{1.3b}$$

where z^{-1} is used to represent either the argument of the z-transform or the shift operator of a one-cycle delay, as appropriate. This computation is in fact a real-time recursive filtering problem which corresponds to the IIR or ARMA digital filter (Oppenheim and Schafer, 1975), except for the fact that in this case the filter is embedded within the feedback loop of the overall control system.

Serial implementation of such a system requires $2n$ multiplications and $(2n-1)$ additions. Further, it is easily seen that a totally serial implementation of (1.3) could require the processor to fetch and decode the data and instructions for every multiplication or addition and store the intermediate results. Hence it is liable to suffer program overheads and inefficient software management so, because of the processor/memory communication bandwidth, heavily constraining the processing speed. For example, suppose that the bandwidth is 30 megabytes per second, which is rather high for present technology, and at least two bytes must be read or written for every operation in (1.3). Then the minimum sampling period (in μs) possible is

$$T_{\mathrm{w}} = (4n-1)/15 + 2nT_{\mathrm{m}} + (2n-1)T_{\mathrm{a}} \tag{1.4}$$

where T_{m} and T_{a} are measured in microseconds. Compared with concurrent approaches, serialism requires the minimum hardware modules but takes the maximum amount of computational time to complete the same task, and hence offers the minimum throughput rate. It is clear that increasing the hardware by a factor of $2n$ would result in a $2n$ times improvement in throughput for computation of (1.3). Further, by partitioning this computation into subtasks suitable for a basic MAC function, the throughput rate would ultimately approach the theoretical maximum of $(T_{\mathrm{a}} + T_{\mathrm{m}} + T_{\mathrm{t}})^{-1}$ for word-level pipelining, where T_{t} represents the time required for data accessing and transmission.

In effect, *parallelism* achieves high speed by replicating the hardware structure of some basic functions many times and providing each replication with one piece of the input data. Alternatively, pipelining partitions the task into many interconnected subfunctions and allocates separate hardware to each piece (Kogge, 1981). It is not always possible to execute a task in a totally parallel manner, since there is a minimum amount of serialism required according to the data dependence of the algorithm, and hence concurrency can be achieved by pipelining if appropriate decoupling can be obtained. Both techniques, however, have the same origin and are often hard to separate in practice where, typically, a mixture of both approaches is used in a so-called overlapped design. A common example in this case is array processors (Kung, 1988) which attempt to improve performance by increasing the number of hardware modules, with all units in the system run concurrently. Parallelism is more apparent in SIMD and semi-systolic arrays, which are discussed in detail in Section 1.3.2. Pipelining is more apparent in systolic/wavefront arrays, which are considered in detail in Section 1.4.

1.3 ARCHITECTURAL CONSIDERATIONS

1.3.1 State Augmentation and Predictive Control

In a feedback control system, the processed data for (1.2) or (1.3) must be fed back to the plant at the beginning of the next sample, i.e. the system latency must be one cycle. As discussed in Section 1.2.1, a processing delay of several cycles is of no major consequence in digital filtering or in open-loop real-time applications, but is of critical importance in feedback control schemes. This is mainly because the unplanned pure time delay introduced could lead to a severe degradation in implemented system performance.

The computation of (1.3b) has a canonical direct IIR filtering structure and an associated computation graph (see Kung, 1988, for a comprehensive treatment of this technique) shown in Figure 1.2 with the particular feature of 'long' adders. The unit delay of z^{-1} in (1.3) can be implemented as a clocked or buffered register or a latch, represented by D in the graph. It is then clear that simply applying a processor array with $2n$ multiplication elements of buffered inputs will directly yield

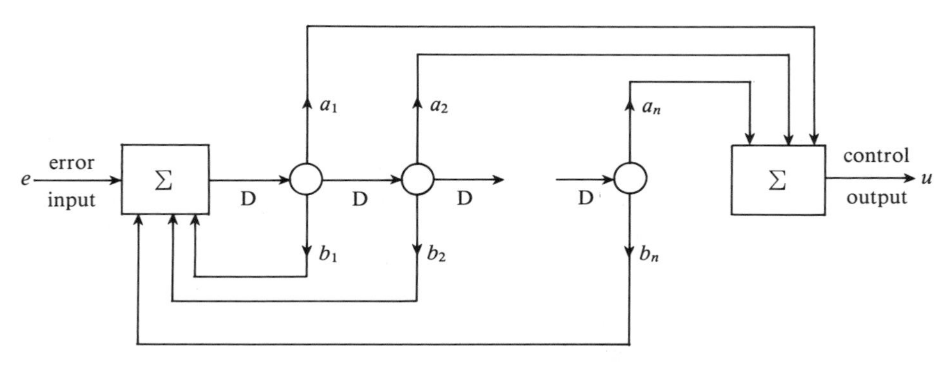

Figure 1.2 Computation graph of the IIR filter of (1.3)

'high' concurrency. The additions, however, have to be carried out by two lumped adders as shown in the figure. In his work, Moroney (1983) directly pipelined this structure in order to implement a high-speed optimal compensator for an F8 fighter model of sixth order. The system latency in this directly pipelined implementation is, however, two cycles, since the operations on the coefficient a_1 at node 1 occur only after its state value is available, and this takes one cycle to complete after the sampled data and command values have been compared.

An obvious way to avoid this undesirable feature is to double the sampling period and incorporate the additional delay into this time. This, however, is clearly not the best choice in terms of speed to yield high throughput. One possibility is to compensate this delay using a phase lead network but such an approach would (typically) require a high gain and a 'broad' bandwidth which, in turn, degrades robustness. If such an approach is to be adopted, however, Åström and Wittenmark (1984) have proposed the use of sensitivity studies based on the root locus to model the effects of the resulting computational delay.

An effective alternative to these approaches is to incorporate this delay into the plant model at the initial design stage (Åström and Wittenmark, 1984) by use of the well-known *state augmentation* technique (Moroney, 1983). This delay is regarded as a particular transport delay of the system studied here, since placing these delay elements serially anywhere within the loop will not alter the overall transfer function. The rational approximation of this delay in analog systems can be modelled by *Padé approximates* (Middleton and Goodwin, 1990) and its only contribution is an increase in the system order in digital systems (i.e. in z-transform terms). The additional orders resulting from the augmented states will be compensated for by the implemented controller which has additional delay units. For example, if these delays require a total of n^1 states then the augmented system vector has dimension $n + n^1$. Applying this technique to Moroney's implementation changes (1.2b) to

$$u(k+2) = C\mathbf{x}(k+1) \tag{1.5}$$

and (1.3a) becomes

$$\frac{U(z)}{E(z)} = H'(z^{-1}) = \frac{\sum_{i=1}^{n} a_i z^{-i-1}}{1 - \sum_{i=1}^{n} b_i z^{-i}} \tag{1.6}$$

after augmenting the plant model and redesigning the controller.

An alternative is to employ the so-called '*predictive control*' technique (Åström and Wittenmark, 1984). This is very similar to the 'state augmentation' approach, in that the processing delay is also incorporated into the plant model. The differences between these two approaches are that the delay unit is more explicit in the latter and that this delay is compensated for more explicitly by the design of a predictor, or a predictive controller.

In the cases when the sampling rate is 'relatively high' compared with the A/D conversion rate (normally, D/A is much faster), the conversion period is no longer negligible. This time delay can also be considered as a number of augmented states of the plant. Controller design is then performed for this augmented model which will obviously have a higher order. The disadvantage of this method, however, is that the hardware requirements will be increased. Consequently, a basic means of providing improved performance is to develop architectures which both offer high speed and only introduce the shortest possible system latency (in cycles). This development begins in the following section with the 'semi-systolic' architectures derived from, in effect, reversing the computation graph – termed graph reversal.

1.3.2 Graph Reversal and the Semi-systolic Array

To begin, the *reversal* of a signal-flow (computational) graph (Mason and Zimmermann, 1960) and its properties are reviewed here since they are central to what follows. The reversal has the same underlying graph as the corresponding original but the direction of every edge is reversed and the input and output vertex nominations are interchanged. It is known that the reversal of a linear SISO graph is also computable and, more importantly, that its transfer function is invariant under this operation.

Consider again Figure 1.2, whose graph reversal, termed the reversed computation graph (RCG), is shown in Figure 1.3. In this case, the reversal operation distributes the lumped adders at the I/O nodes to the other n nodes. A semi-systolic array architecture now directly arises from this graph by short-circuit connection (or *broadcasting* in implementation terms) of the input and output node values. The resulting array and cell configurations are shown in Figure 1.4. This array is termed a Type 1 semi-systolic array and this broadcasting arrangement

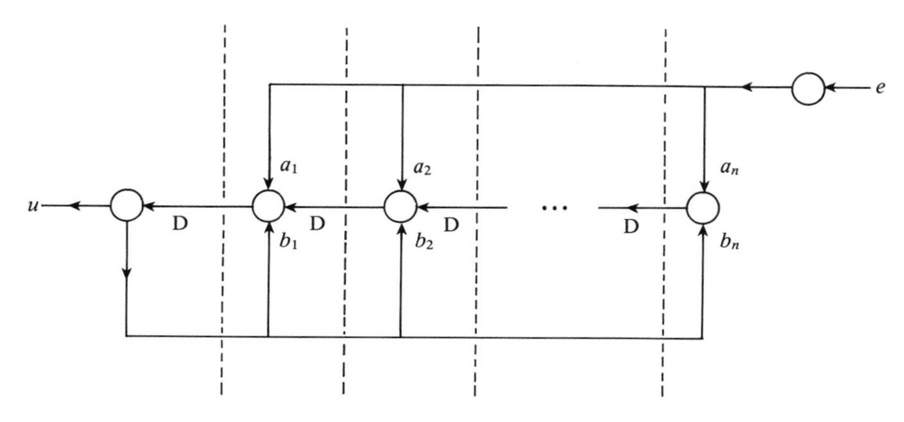

Figure 1.3 An RCG of a canonical IIR filter (with distributed adders)

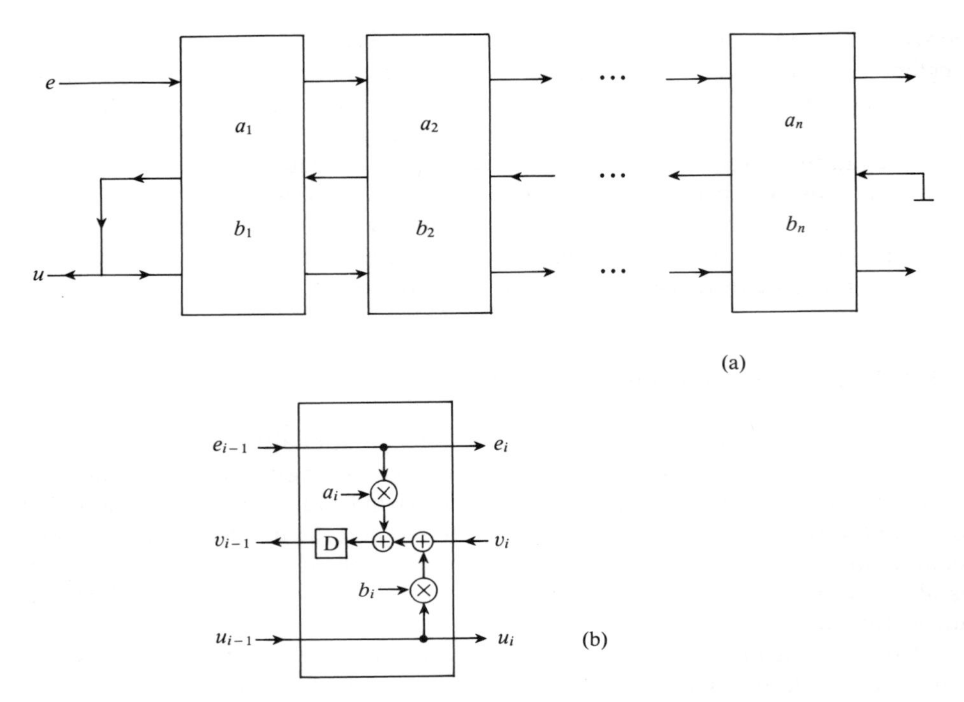

Figure 1.4 A Type 1 semi-systolic array based on the reversed computation graph of Figure 1.3: (a) array configuration and (b) cell configuration

guarantees that the input data will be processed (or computed) and output after one cycle has elapsed.

Compared with the directly pipelined alternative developed by Moroney (1983), this architecture provides a modular array structure, which is very attractive in terms of implementation by VLSI technology or transputer networks, and the schedule control is simpler. Hence by modular pipelining the word-level iteration period in this case has been reduced to

$$T_{\mathrm{w}} = T_{\mathrm{m}} + 2T_{\mathrm{a}} + T_{\mathrm{b}} \tag{1.7}$$

where T_{b} denotes the broadcasting time which is always negligible in comparison to T_{m} or T_{a}. Further, by arranging direct broadcasting lines the system latency has been reduced to one cycle (iteration period), which is the minimum possible.

Combining the above approach with the so-called *Delay-transfer rule* for systolization (Kung, 1988, pp. 208–10), provides a much more powerful technique which can be used to develop alternative architectures for a given case. Note also that this rule was intuitively employed in Section 1.3.1 to develop the state augmentation based implementation where it was stated in the form that 'a delay unit could be placed anywhere around the feedback loop'. This rule also implies that the system latency remains the same if the input and output are located on the same side of the systolization cut-set.

Application of the RCG approach to the processing element of Figure 1.4(a) immediately yields the computation graph shown in Figure 1.5(a). Further, applying the delay-transfer rule to this element yields the candidate PE architecture of Figure 1.5(b). Note, however, that these operations alter the structure of Figure 1.4(a) to that of Figure 1.5(c) but the system latency remains one cycle. This array is termed the Type 2 semi-systolic array.

In this new array, the error input is summed with the state output of the first PE and then fed back into the array by broadcasting. The two semi-systolic architectures allow data to be broadcast to every PE in parallel and avoid creating a memory/processor communication bottleneck, by arranging hardware interconnections through the broadcasting 'channels' and using intercell communication. Broadcasting forces the PEs to sample the data simultaneously on receiving a 'data ready' signal from the host. In many practical control applications, the controller order is often not very high (for example, proportional plus integral) and hence broadcasting lines would be acceptable in an implementation. If,

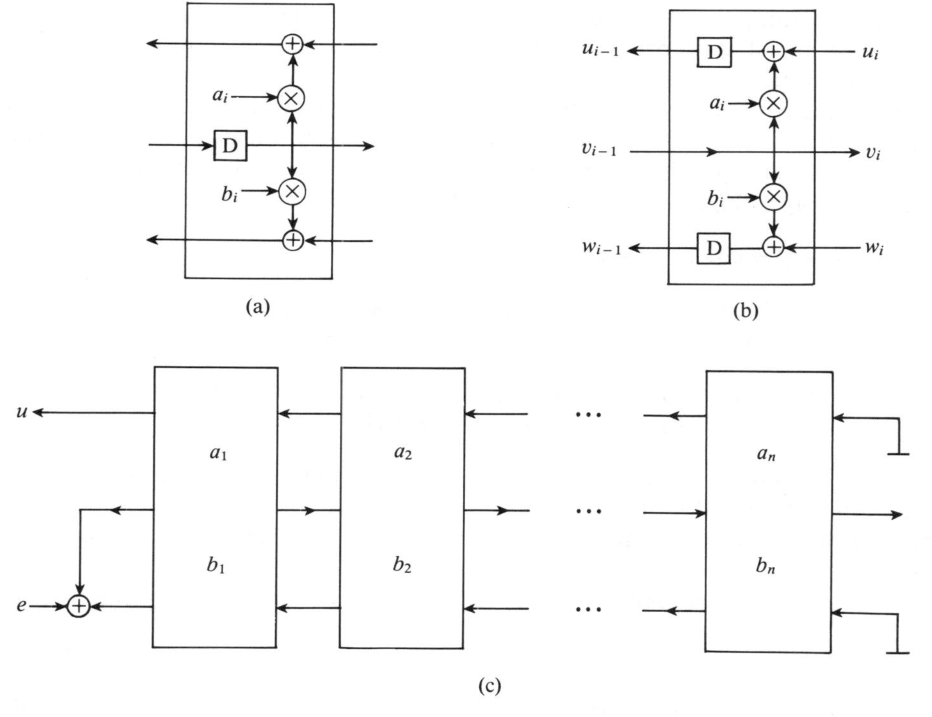

Figure 1.5 (a) RCG of the cell of Figure 1.4; (b) its PE configuration; and (c) Type 2 semi-systolic architecture

however, broadcasting must be avoided, fully systolic arrays can be used and these will be developed in Section 1.4. This work will use these semi-systolic architectures as basic building blocks.

1.3.3 Transfer-function Verification

Obviously, the controller transfer function must be invariant under architectural manipulations and reconnection of cells, i.e. the new architecture must have the same transfer function as the controller. To verify this, Lin (1986) and Kwan (1987) have developed the so-called 'snap-shot' method to check the behaviour of the implementation either after every cycle or after several cycles have elapsed. In his book, Kung (1988) states that 'this is the most natural tool a designer can adopt to check or verify a new array algorithm'. This method is, however, difficult to use in practice since the designer has to check many time cycles and deduce from this the general behaviour of the array. Further, when using the snap-shot approach, it is impossible to identify whether the array is pipelinable (or expandable). In this section, alternative methods for verifying an implemented function in the time-domain will be developed.

The method proposed here is, in effect, similar to the snap-shot approach. To begin, the signal flow property of the cell is described in terms of difference equation(s) which can be obtained by inspection. Note also that each cell in the array has the same equation(s) and these iterate n times, where n denotes the number of cells. Then if there is convergence to (1.3b), the design yields a correct transfer function. To illustrate this approach, consider the time-domain behaviour of a cell from the array of Figure 1.4 (Type 1), which is described by the difference equation:

$$v_{i-1}(k+1) = v_i(k) + a_i e(k) + b_i u(k) \tag{1.8}$$

Note that (1.8) holds for all the cells of the array and iterate this n times, and hence push v_i to v_n. Then

$$\begin{aligned} v_0(k+1) = v_n(k-n) + a_1 e(k) + a_2 e(k-1) + \ldots + a_n e(k-n+1) \\ + b_1 u(k) + b_2 u(k-1) + \ldots + b_n u(k-n+1) \end{aligned} \tag{1.9}$$

which is the same as (1.3b) since $v_0 = u$ and $v_n = 0$. Further, this iteration method also explicitly shows that the array is expandable since, if $n+1$ cells are cascaded, the time-domain description also takes the structure of (1.3b) with n replaced by $n+1$.

A similar analysis applied to the cell of Figure 1.5(b) (Type 2) yields

$$u_{i-1}(k+1) = u_i(k) + a_i v(k) \tag{1.10a}$$

$$w_{i-1}(k+1) = w_i(k) + b_i v(k) \tag{1.10b}$$

which describes each cell in the Type 2 semi-systolic array. Iteration of these equations now yields

$$u_0(k+1) = u_n(k-n) + a_1v(k) + a_2v(k-1) + \cdots + a_nv(k-n+1) \qquad (1.11a)$$

and

$$\begin{aligned} v_0(k+1) - e(k+1) &= w_0(k+1) \\ &= w_n(k-n) + b_1v(k) + b_2v(-1) + \cdots + b_n(k-n+1) \end{aligned} \qquad (1.11b)$$

respectively. Eliminating the variable v, by combining (1.11a) and (1.11b), gives (1.3b) or (1.3a) in z-transform terms since $u_0 = u$, $u_n = 0$ and $w_n = 0$. This iteration method will also be used to verify the transfer functions of the systolic/wavefront arrays developed in the following section.

1.4 SYSTOLIC AND WAVEFRONT ARCHITECTURES

VLSI/WSI based systolic/wavefront array architectures offer very high performance for digital signal/image processing applications. Given the structural similarities with large classes of control algorithms, this technology can therefore be expected to play a similar role in mapping such algorithms onto real-time control processing architectures. Note again, however, that in this application both 'short' processing time and 'fast' system response are required.

This section begins with a review of existing word-level systolic architectures for recursive IIR filters and discusses the possibility of applying them to feedback control problems. Following this, the analysis of Section 1.3 will be used as a starting point to develop word-level systolic/wavefront architectures (in effect, one-dimensional IIR filters), suitable for real-time control use.

1.4.1 Use of Existing Architectures

As noted in Section 1.2.2, work in the DSP area has led to systolic arrays for realizing IIR filters (Kung, 1984; Lin, 1986; Kwan, 1987). In Kwan's realization, however, certain transformations of the transfer function are required and a number of restrictions on the filter coefficients must be invoked, such as certain of them must be nonzero. Lin's architecture is not constrained by these factors, but requires different types of cells to construct the array. More critically, both architectures have a system latency of $2n$ cycles, which is not only 'too long' for control applications but is also dependent on the filter order n. Further, this delay cannot be compensated for using the predictive control or state augmentation techniques, since use of these would result in an inconsistent design (Moroney, 1983). If this unplanned delay is ignored in controller implementation, it will degrade closed-loop system performance and, in the worst case, could result in an unstable implementation (Shin and Cui, 1988).

The architecture proposed by Kung (1984), which is redrawn and shown here in Figure 1.6, could, however, be applied to feedback compensation/control problems, as this has a fixed latency of 2.5 sampling cycles. Note, however, that the throughput rate is reduced by a factor of two after the systolization procedure. Wang and Lin (1986) directly applied this architecture to control problems but this use transforms the original transfer function to

$$H'(z^{-1}) = \frac{z^{-2}\sum_{i=1}^{n} a_i z^{-i}}{1 - z\sum_{i=1}^{n} b_i z^{-i}} \tag{1.12}$$

owing to the inclusion of the outside adder in a cell structure (refer to Figure 1.6). One possible solution to this problem is to use the technique of 'M-expanded' pipelining (Rogers and Li, 1988; Li, 1990). This basically involves adding an additional cycle delay unit into the array, which has a cycle time of half that of the

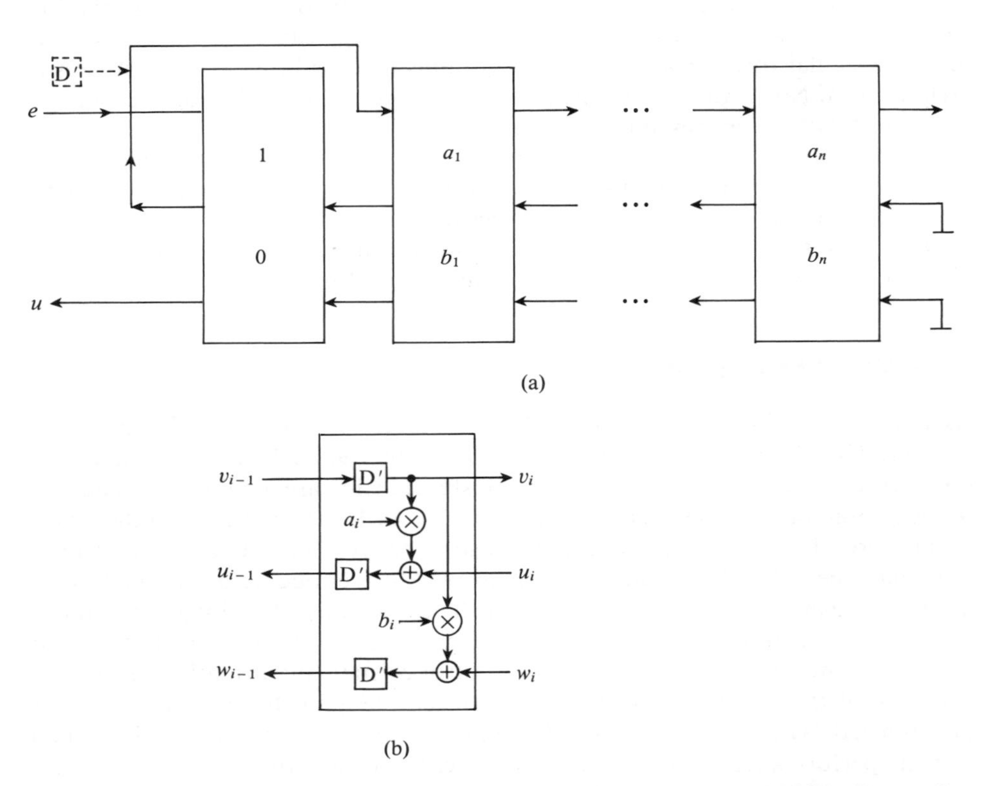

Figure 1.6 Kung's ARMA filter architecture (redrawn from Kung, 1984, Figure 4)

new array, after first transforming the ARMA transfer function to

$$H(z^{-1}) = \frac{\sum_{i=1}^{n+M} e_i z^{-i}}{1 - z^{-M} \sum_{i=1}^{n} f_i z^{-i}} \tag{1.13}$$

with $M = 1$ in this case. Note, however, that this modified realization still requires second-order state augmentation compensation and the array still has irregular interconnections.

1.4.2 Type 1 Architecture

The preliminary semi-systolic arrays discussed in Section 1.3.2 make use of some 'long' broadcasting lines. This broadcasting process has to be in series, propagating gradually from the first cell to the last (the use of Occam and transputer based architectures to simulate this case are detailed in Li, 1990). For VLSI/WSI implementation, however, it is much more appropriate to avoid these broadcasting lines. In this section, therefore, the objectives are to avoid the use of unbuffered data transmission in these semi-systolic arrays and to develop them into 'fully systolic' arrays.

Consider again the Type 1 semi-systolic cell shown in Figure 1.4(b). Here, unbuffered transmission, and thus broadcasting, occurs in the upper and lower lines whilst, see Figure 1.5(b), exactly the opposite is the case in the Type 2 cell but here the data transmission directions are also reversed. This suggests that combining these two cells to produce fully systolic arrays should be possible, provided of

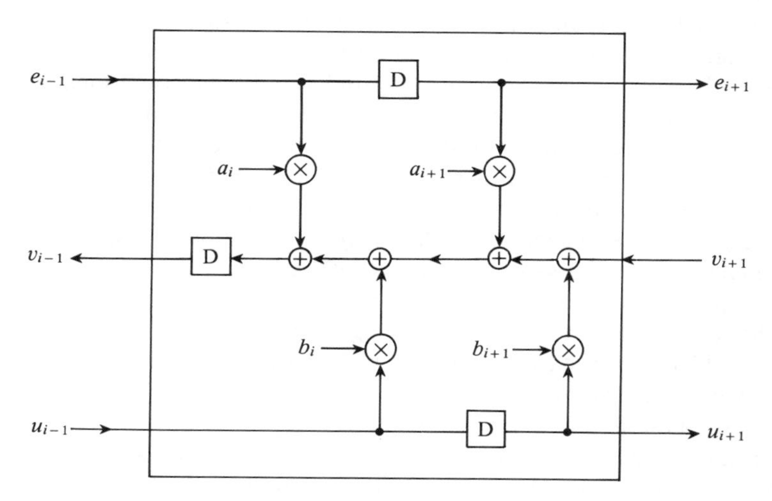

Figure 1.7 Architecture of Type 1 fully systolic cell

course that the implemented transfer function is invariant. In Section 1.3.2, it has been shown that a reversal procedure does not change the transfer function (matrix) or the cell latency. Hence, placing the reversed Type 2 cell to the right-hand side of Type 1 cell gives the candidate cell configuration shown in Figure 1.7 and this is termed a Type 1 fully systolic cell. This forms a second-order filter section and the array structure also takes the form of Type 1 semi-systolic array as shown in Figure 1.4(a). An alternative would be to place the reversed Type 2 cell on the right-hand side but this would not give buffered outputs and is, of course, not of interest here. Mirroring rather than reversing a Type 2 cell and then combining as above will transform the transfer function and is again of no interest here.

In effect, a reversed Type 2 semi-systolic cell also acts as a first-order section similar to a Type 1 cell and consequently it is to be expected that the result should then yield an invariant transfer function. This follows immediately on use of the iteration method for the cell difference equations and the details are hence omitted.

To establish transfer function invariance for the Type 1 case, consider again Figure 1.7. Then it follows immediately that

$$\begin{aligned} v_{i-1}(k+1) &= v_{i+1}(k) + a_i e_{i-1}(k) + a_{i+1} e_{i+1}(k) + b_i u_{i-1}(k) + b_{i+1} u_{i+1}(k) \\ &= v_{i+1}(k) + a_i e_{i-1}(k) + a_{i+1} e_{i-1}(k-1) \\ &\quad + b_i u_{i-1}(k) + b_{i+1} u_{i-1}(k-1) \end{aligned} \tag{1.14}$$

since $e_{i+1}(k) = e_{i-1}(k-1)$. Further, (1.14) holds for all cells in the array and iterating n times yields

$$\begin{aligned} v_0(k+1) = v_n(k) &+ a_1 e_0(k) + a_2 e_0(k-1) + \dots + a_n e_0(k-n+1) \\ &+ b_1 u_0(k) + b_2 u_0(k-1) + \dots + b_n u_0(k-n+1) \end{aligned} \tag{1.15}$$

which is equal to (1.3b) since $v_0(k+1) = u_0(k+1)$ and $v_n(k) = 0$. this verifies that the design does indeed leave the transfer function invariant.

A proposed hierarchical arithmetic/logic level design with timing control is shown in Figure 1.8. As suggested in Figure 1.8(a), Kung's proposed semi-custom-design (Kung, 1988) is used for adder, multiplier and logic component details. The adder and multiplier units shown in Figure 1.8(b) form a dual multiply-accumulation (DMAC) element as shown in Figure 1.8(c) and two of these form a Type 1 fully systolic cell which is shown in Figure 1.8(d).

The operation of this array is similar to that of Type 1 semi-systolic array. At rest all the state values (which are all latched) and the input are set to zero and, on the rising edge of the first sampling clock ('GO' signal in the figure), $e_0(0)$ is input to the first cell of the array. As demonstrated in Figure 1.8 and Figure 1.9, which shows the timing order, this signal triggers the start of four multiplication operations which run in parallel. On completion of these, the '*m*!' lines are raised to 'high', which signals that the system is ready to begin the consecutive additions. Inspection of these figures shows that only the addition on the result of the b_{i+1} multiplication can be enabled to operate, since the others are still waiting for their respective enabling signal. Once this addition is completed, it triggers the addition on the result of a_{i+1}. The operations in a cell continue in this manner until the fourth

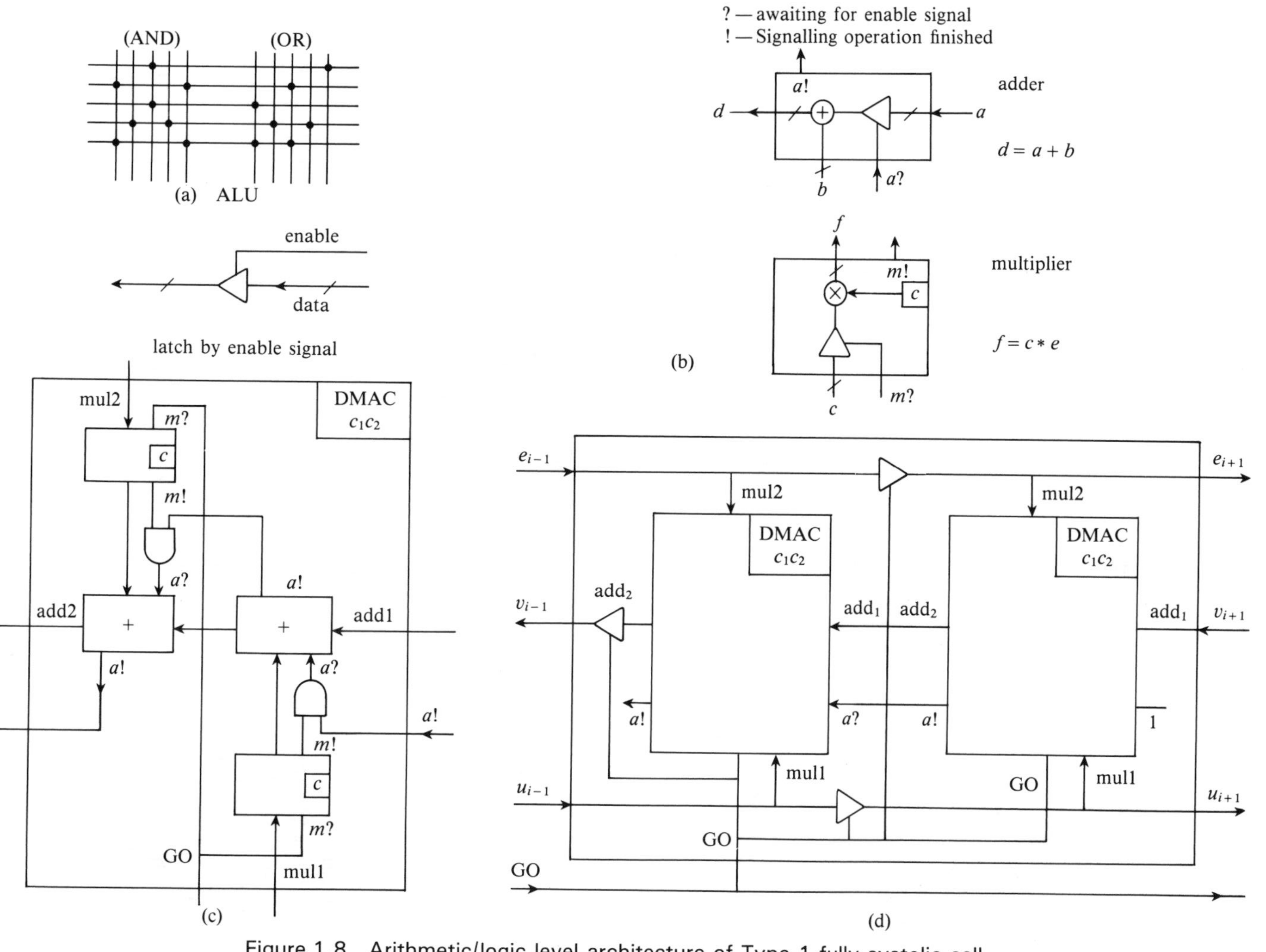

Figure 1.8 Arithmetic/logic level architecture of Type 1 fully systolic cell

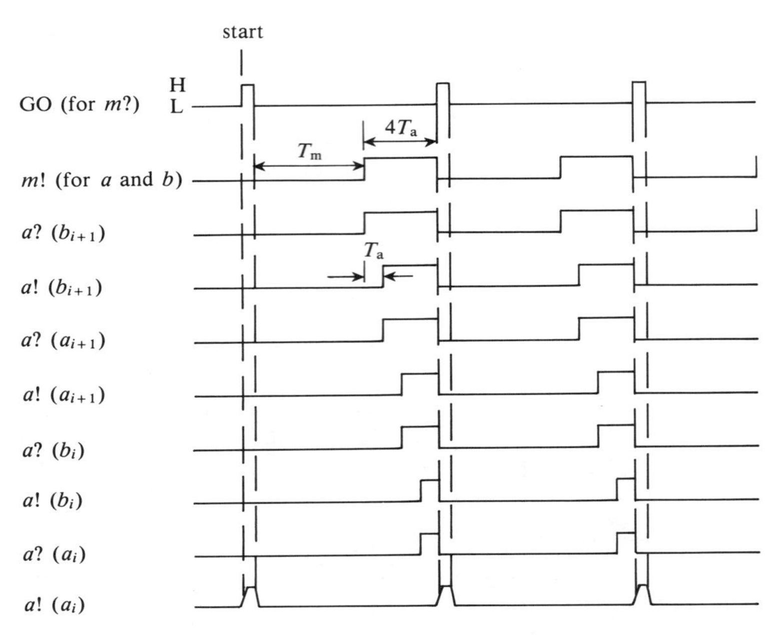

Figure 1.9 Timing characteristics of Type 1 fully systolic cell

addition (i.e. the processing of a_i) is completed. The last 'a!' signal of addition-completion is actually not required in this array and the terminal (on the right-hand side) requiring this is put to 'high' (floating). This output signal, however, rises to 'high' when the addition is finished and the next clock has to be arranged to follow after this. The clock period here must not be less than

$$T_w = T_m + 4T_a \tag{1.16}$$

in order to complete the four parallel multiplication and four serial addition operations.

Once the next rising edge of the 'GO' cycle arrives, $e_{i-1}(k)$, $u_{i-1}(k)$ and $v_{i-1}(k)$ are enabled and passed to the neighbouring cells and e_{i-1}, u_{i-1}, v_{i+1} are updated. These operations are then followed by a series of new computations on a newly sampled input. The complete array operates as a two-directional pipeline, with each cell, both pulling and pushing data at the rising edge of every 'GO' cycle at a high rate. It is easy to see that the computed result on newly sampled data is available from the array in one cycle and hence the system latency is one cycle.

It has been shown, therefore, that the Type 1 systolic array architecture is highly regular and modular. Data flow and computation in it are synchronous, interconnection between cells is local, and data transmission is latched (hence no unbuffered or global data transmission). This arrangement hence not only saves broadcasting time but also reduces transmission interferences. Overall, the array fetches the error data items once per sample but stores them and operates on them many times. This avoids a potential communication bottleneck and yields a high throughput. Since the cells are identical, it follows that the timing control circuitry is very simple. Note, however, that this signal needs to be broadcast to every cell and will result in a certain time skew at some 'remote' cells when the sampling rate is 'very high'. One possible solution in this case is to use a special arrangement such as tree structures. Another option is to employ equivalent wavefront arrays, which are data-driven, rather than globally timed, and this will be discussed in Section 1.4.4.

As shown earlier in this section, combining one reversed Type 2 semi-systolic cell to the right of a Type 1 semi-systolic cell leaves the transfer function invariant. Similarly, it can be proved that combining N such reversed cells simply expands the size of the cell and (1.14), which still yields (1.15) and hence an invariant transfer function. The expanded cell size, however, will cause the order of the controller to be increased to

$$\left\lceil \frac{n}{N+1} \right\rceil \cdot (N+1)$$

For example, if three such cells are combined with one Type 1 cell to implement a fifth order filter/compensator, this number must then be increased to eight by inserting zeros for the coefficients a_i and b_i corresponding to the sixth, seventh and eighth orders.

1.4.3 Type 2 Architecture

By following a similar development to that of Type 1 array, a fully systolic version of the Type 2 array can also be obtained. In particular, consider again the Type 2 semisystolic cell shown in Figure 1.5(b). Here, unbuffered transmission, and thus broadcasting, only occurs in the middle line. In the Type 1 semi-systolic cell, whose RCG is shown in Figure 1.5(a), exactly the opposite happens and exactly the same data transmission directions are present as in the Type 2 cell. Placing this to the right-hand side of Type 2 cell should give the same transfer function and system latency. This gives the candidate fully systolic cell as shown in Figure 1.10, which is termed a Type 2 fully systolic cell and the resulting array also takes the form of Type 2 semi-systolic array as shown in Figure 1.5(c). Verification of the transfer function of this array can be found in Li (1990), pp. 100–2. Note, however, that placing a reversed Type 1 cell at the right-hand side will not give buffered outputs and this combination is of no interest here. Further, mirroring rather than reversing a Type 1 cell and then combining as above will alter the transfer function.

Using the same DMAC element as shown in Figure 1.8(c), the timing control of this cell can be configured as shown in Figure 1.11. Again, the DMAC element is modular and two are needed to configure a Type 2 cell.

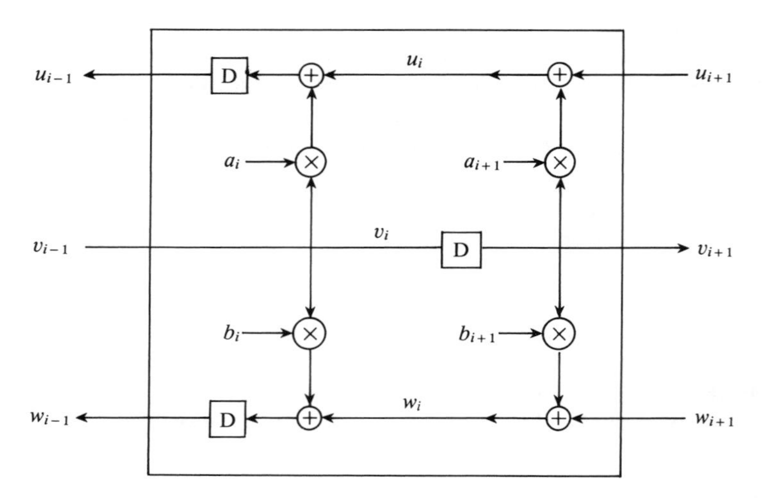

Figure 1.10 Architecture of Type 2 fully systolic cell

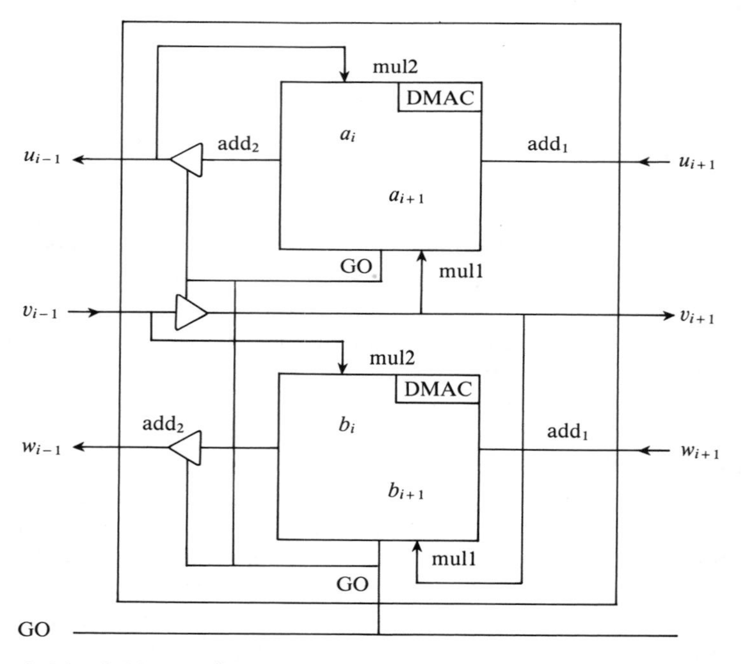

Figure 1.11 Arithmetic/logic level architecture of Type 2 fully systolic cell

The operation of this array is similar to that of a Type 2 semi-systolic array. At rest all the state values (buffered/latched) are set to zero. After the error is added with the state value $w_0(0)$ in the first cell, operations on this result begin. In each cell, see also Figure 1.11, multiplications of v_{i-1} on a_i and b_i and those of v_i on a_{i+1} and b_{i+1} all take place in parallel. The additions on feedforward or feedback results, however, must be serial. Hence the complete cycle time is $T_m + 3T_a$, including the addition time of $r - y$. On the trigger edge of next cycle, the results of u_{i-1} and w_{i-1} are passed to the left-hand side neighbouring cell and those of v_{i+1} to the right. Inside the cell v_{i+1} is also updated to v_i. Consequently, the latency of this system is one cycle and it is also easy to see that the Type 2 array architecture has similar advantages to Type 1 array. The throughput is a little higher, but the advantages of this are compromised by the use of an irregular outside adder.

In a similar manner to the multiple combinations used for the Type 1 array, the controller transfer function and the system latency (processing delay) is invariant under this arrangement of the Type 2 array. The proof of this follows in an almost identical manner to the Type 1 case and hence the details are omitted. Finally, note that the Types 1 and 2 architectures developed here share certain structural similarities with those of Luk and Jones (1988) and Rogers and Li (1988) and that these arrays can also be pipelined to sub-word level to provide even higher speed by use of so-called M-expanded pipelining algorithms (Rogers and Li, 1988).

1.4.4 Wavefront Architectures

In digital signal processing, wavefront arrays are also amenable to VLSI implementation for compute-bound problems (Kung, 1984, 1988) and they share advantages with systolic arrays such as modularity, regularity and local interconnections. Further, they have asynchronous or data-driven, rather than global timing, control and hence they have the potential to overcome disadvantages of the systolic array, such as clock skew and peak power drawn (at the rising time of every clock) which are caused by global timing control (Kung, 1984).

The purpose of this section is to show by an example how a systolic array can be transformed into a wavefront array for the implementation of control schemes. In particular, only the Type 1 array is considered since the required analysis for all other cases (fully or semi-systolic arrays) is virtually identical.

Consider again Figure 1.8. Inside a DMAC element, the two adders become active when their handshaking terminal, denoted by 'a?', receives an enable signal from the prior processing units. The multiplications and hence the whole cell are, however, triggered by the arrival of the 'm?' signals, i.e. the rising edge of the synchronous 'GO' clock. In a wavefront array implementation, this can be arranged in a similar way to the operations of the adders. Hence the multiplication operations are data-, rather than clock-, driven. Consequently, a clock-controlled latch (representing a delay unit in computation graphs) is replaced by a 'separator

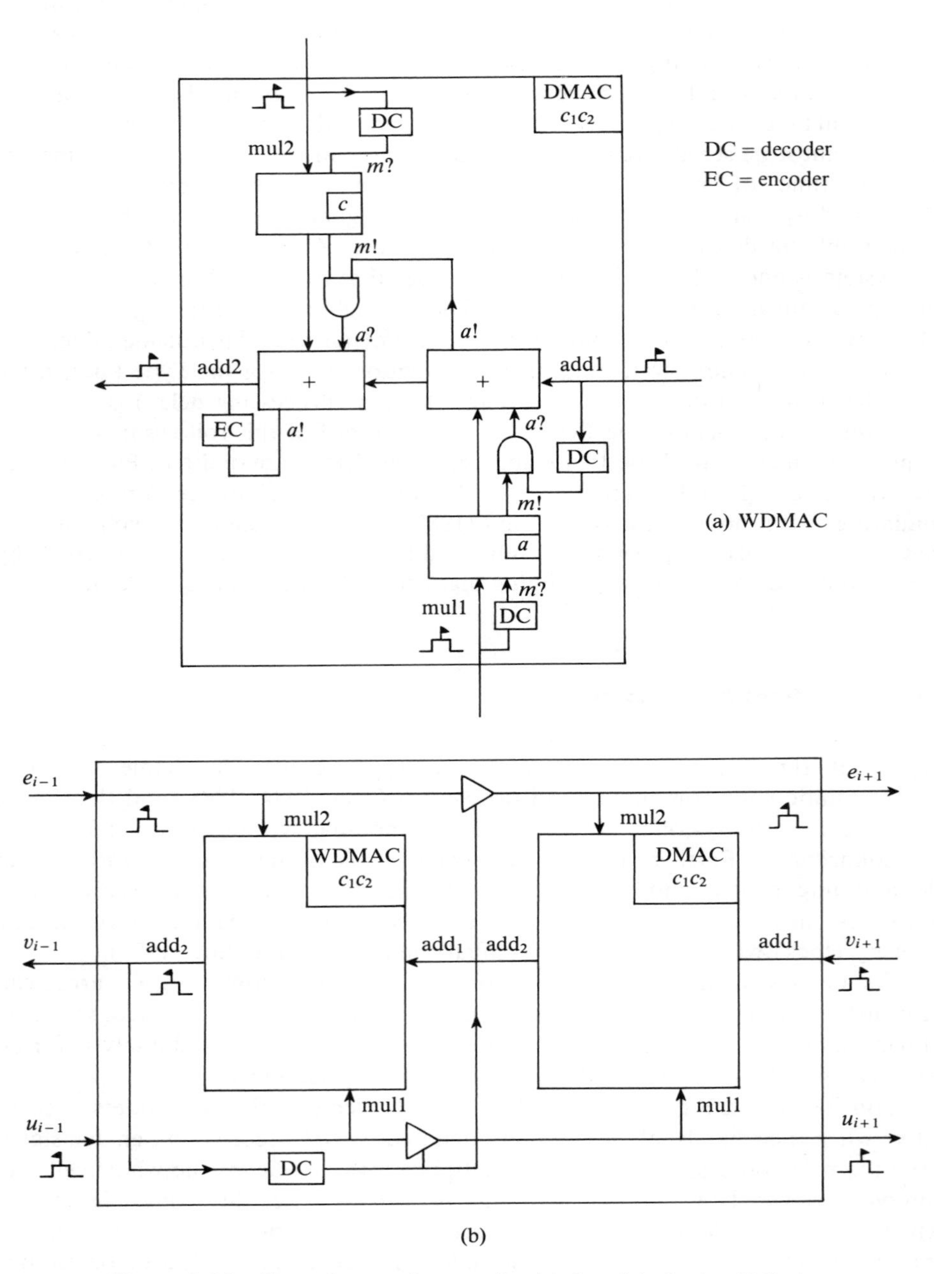

Figure 1.12 Type 1 wavefront cell with handshaked self-timing

register' (Kung, 1984), which is locally controlled by a handshaking flag rather than a clock, to signal the arrival of the relevant operand.

The resulting wavefront cell is termed a Type 1 wavefront cell and is shown in Figure 1.12(b), with the modified DMAC elements for wavefront operation (termed 'WDMAC') shown in Figure 1.12(a). Further, it follows immediately from the discussion above that the operation of this array is similar to that of the Type 1 systolic array. Here, however, all data are encoded with an additional bit for handshaking and the input is encoded by the host, which also supervises A/D and D/A conversions. If, however, encoding is not recommended for certain implementations, handshaking is then announced by a separate bit with a separate hard wire.

On starting computation, encoding occurs first and includes the reset zero states. As shown in Figure 1.12, the instant the data are decoded (or received if separate handshaking wires are used), four multiplications take place and run concurrently. Once these are completed, the 'm!' lines are raised 'high' and the additions then follow, where these again are completed serially. The v_{i-1} data output to a neighbouring cell is encoded with the 'a!' handshaking signal of the last adder and this allows e_{i-1} and u_{i-1} to be passed to a neighbouring cell.

The minimum cycle time in this case is therefore also bounded by (1.16) and the array has a system latency of one such 'cycle'. All such 'cycles' may not, however, be identical in the absence of appropriate hardware or software schedules between the cells. In practical implementations of these control systems, the A/D and D/A conversions must be synchronized eventually, otherwise the compensator output signal fed into the plant will be slightly and unpredictably shifted due to the asynchronized data arriving instant. These may be ignored, however, in the presence of appropriately defined accuracy ranges.

Summarizing, therefore, this section has briefly considered the development of wavefront arrays for real-time control purposes. In particular, the possibilities for converting systolic arrays to wavefront arrays for this general purpose have been briefly discussed and this leads to the conclusion that every systolic architecture can, in effect, be converted into a wavefront version. Both the systolic and wavefront arrays developed can also be modified to accommodate adaptive/self-tuning algorithms, by providing an additional bank of coefficient registers as the update coefficient register (UCR) of the A100 component which will be used in Chapter 9.

1.5 CONCLUSIONS AND FURTHER WORK

The major output of the work reported here has been the production of candidate concurrent processing architectures, and associated implementation techniques, for a generic class of real-time feedback compensation and control schemes, where a short system latency (in processing cycles) is the most important performance characteristic. This has yielded architectures in the form of systolic and wavefront arrays with well-defined general performance characteristics. In particular, their throughput or speed (at the corresponding pipelined level) is close to the maximum possible and they also have the shortest possible system latency, or processing delay, in cycles.

In essence, the background to all the work reported here lies in the well-documented structural links between certain problems in digital signal processing and control engineering. This general topic has been discussed in Section 1.2.1 by way of motivation as have the limitations associated with conventional implementations. A critical overview of 'state of the art' in concurrent architectures, particularly in systolic/wavefront arrays, has been given with particular emphasis on those aspects of performance, such as processing and response speed, which are crucial to the work reported here.

To provide essential background for the other work reported here, Section 1.3 has considered methods for realizing concurrent architectures using the transfer function and computation graph description of a controller which is representative of a general class of SISO control schemes. Further, the state augmentation and predictive control techniques, which can be used to compensate for processing delay, have also been discussed. Following this, two simple methods for verifying the transfer functions of an array architecture or the transfer function matrices of a particular cell have been presented. These have then been used to develop two preliminary semi-systolic array architectures.

Based on this analysis, Section 1.4 has developed fully systolic and wavefront array architectures, compatible with VLSI realization, which belong to the fine-grain category and are more system-specific. Problems associated with applying existing signal processing oriented systolic arrays to control computations were considered and then the preliminary semi-systolic architectures were used to develop fully systolic arrays which avoid data broadcasting lines. Subsequently the techniques used to develop systolic arrays were employed again to develop wavefront arrays. Note also that these two families of architectures are very similar, except that one of them is globally timed and the other is data-driven.

These systolic/wavefront architectures should offer 'sufficiently high' speed for most control engineering applications. Consider, however, the case where even higher speed is required. This problem has been addressed here by suitable transformation of the controller transfer function to reduce the iteration bound to a (theoretically) arbitrarily low limit by use of a particularly simple calculation law for the coefficients in the transformed transfer function given in Rogers and Li (1988) and Li (1990).

Section 1.4 has provided a range of alternative architectures for both SISO and, see below, multi-input multi-output or multivariable (MIMO) compensation/control system implementations. Trade-offs exist between all the architectures developed here in terms of speed, response or hardware limitations for a given application. Consider, for example, the case when a 'short' processing delay is required and/or desirable, or compensation for a 'long' processing delay cannot be achieved due to hardware limitations. Then the obvious, most acceptable, choice is one of the architectures with the shortest latency possible, i.e. one sampling period. Further, suppose that broadcasting lines are allowed in the realization. Then a 'semi-systolic' array can be used which has a simpler circuit configuration than a fully systolic array implementation.

In order to assist with the selection of an appropriate architecture for a given application, Table 1.1 gives a comparative summary of the (expected) performance of the architectures developed. This provides a cell and array library and also includes a comparison with the existing DSP architectures discussed in Section 1.4.1. The following are the major points arising from a detailed study of this table:

1. Most of the architectures developed in this chapter have the shortest latency at word-level pipelining.
2. The cycle times of all developed architectures are 'close' to the theoretically minimum possible value of $T_m + 2T_a$.
3. Type 1 arrays require only one cell type and Type 2 require at most a particular modification of one cell. This indicates a high degree of modularity which is particularly appealing for expansion and for VLSI implementation.
4. The hardware requirements in terms of multipliers and accumulators are 'close' to the minimum of $2n$.
5. Compared with existing DSP-oriented systolic architectures, the developed architectures clearly provide the minimum response time and retain high throughput.

Suppose also that applications arise which require even higher speed. Then this case can be dealt with by lower level pipelining using the so-called 'M-expanded' computation method detailed in Rogers and Li (1988) and Li (1990). Further, it is

Table 1.1 Performance comparison between existing and proposed architectures

Architectures		*Latency (cycles)*	*Cycle time*[a]	*Cell types*	*Multipliers/ adders needed*	
Type 1	Semi-systolic	1	$T_m + 3T_a + T_b$	1	$2n$,	$2n$
	Fully systolic	1	$T_m + 4T_a$	1	$\leqslant 2n+2$,	$\leqslant 2n+2$
	Wavefront	1	$T_m + 4T_a$	1	$\leqslant 2n+2$,	$\leqslant 2n+2$
Type 2	Semi-systolic	1	$T_m + 3T_a + T_b$	1[b]	$2n$,	$2n+1$
	Fully systolic	1	$T_m + 3T_a$	1[b]	$\leqslant 2n+2$,	$\leqslant 2n+3$
Lin (1986)		$2n$	$T_m + 3T_a$	2	$4n$,	$4n$
Kwan (1987)		$2n$	$T_m + 3T_a$	1	$4n$,	$4n$
Kung (1980)		n	$2(T_m + 3T_a)$	1	$4n$,	$4n$
Kung (1984)		5/2	$2(T_m + 3T_a)$	1	$2n+2$,	$2n+2$

[a] Cycle time includes the time taken for calculation of (1.2c), and T_m and T_a are counted for word-level operations.

[b] Indicates that an outside adder is additional to the arrays.

expected that the fabrication of such hierarchical architectures will mature rapidly with the development of VLSI techniques. In practical implementations of these architectures, the speed of the data streams is 'very high' and consequently the effect of this high-rate clocking must be taken into consideration. For example, most A/D converter circuits are based on a comparison with a digital counter and would be deemed 'slow' compared with the clock speed and hence would produce a 'long' response time. Consider also cases which, for example, require 'very good' stability margins or a 'highly' accurate feedback or return time for the plant input signal. Then this conversion time must be compensated for using a state augmentation technique. Alternatively, predictive control methods (Li, 1990) could be used but it should be noted that this area is still very much in the development stage. Other potentially performance degrading factors resulting from the high clock rate include items such as high frequency interference and peak power drawn.

Consider now the direct extension of these linear architectures to MIMO feedback control schemes, i.e. the need to complete maxtrix/vector operations. Then this would result in a long and (crucially) system order dependent latency and hence they are only applicable to cases where, for example, the processing delay can be ignored. Suppose, however, that the particular application under consideration is such that (a) the plant is multiple-input and single-output, or (b) the plant can be adequately controlled by a diagonal (decoupled) controller. Then in these cases the developed architectures could be directly employed to form the individual processing units of a multiple-instruction multiple-data (MIMD) system.

Architectures for the general MIMO case can, however, be developed by using parallel vector reduction processing modules to restrict the processing delay to the minimum possible. This, in turn, has led to the development of an SIMD architecture for MIMO systems which directly process matrix-vector based computations and can also be pipelined to a level below word-level if even higher speed is required. As expected, however, this increase in speed must be 'traded-off' against the system latency. The undesired pure time delay can, however, be reduced by a factor of two using the 'dual look-ahead' computation technique.

Use of this technique still results in a single-instruction multiple-decision architecture which is not 'optimized' for VLSI implementation and therefore would not necessarily yield 'high' performance. To overcome this, so-called macro-systolic arrays have been developed and the SIMD architecture mapped onto them. Complete details of all this work for the MIMO case can, for example, be found in Li (1990) and Rogers and Li (1989, 1991).

In addition to the increase in speed and error tolerance resulting from the provision of an increased sampling rate, use of concurrent architectures for control engineering processing also provides many other benefits. For example, loosely coupled parallel systems offer a potential fault-tolerance capability in the implemented control system. As one processing unit fails, the others still support some degree of operation rather than a complete shut-down of the system. Fault-tolerant parallel systems for control applications are discussed in Chapter 10.

Indeed, parallel systems can be specially arranged to accommodate this function when reliability of the control system is very important.

Concurrent architectures obviously offer a better mapping between algorithms expressed in terms of block diagrams and multiprocessor hardware. Consequently, VLSI technology offers the prospect of 'low-cost' concurrent array processing and this can be exploited with these architectures to offer an increased degree of versatility in a given control scheme specification. For example, suppose that controller hardware with more than the minimum required number of processing components is available. Then different software can be loaded and mapped onto the hardware to meet different requirements whilst retaining high performance although, in some cases and/or time periods, some of the processing units could be redundant. Note that, for example, the systolic/wavefront arrays can also be extended easily to implement adaptive/self-tuning control schemes. In this case, only two-bank coefficient register structures are required. In the implementation, estimates of the plant parameters and updating of the controller coefficients do not require a high throughput but the processing of control signal does (Åström and Wittenmark, 1989). Hence these major subtasks can also be mapped onto different elements of the processing system and executed at a different speed.

The following areas are obvious candidates for short to medium term future research/development:

1. Further detailed architectural development design studies are required. Particular aspects which should be addressed include programmability, reconfigurability and fault tolerance of the chip as required by a given application.
2. The VLSI oriented architectures are at the so-called architectural level and hence the obvious next stage is to consider VLSI realization aspects. These include design, simulation, verification and testing at logic/register, circuit and geometric (routeing and layout) levels and, finally, microelectronic fabrication.
3. Further work to produce a fine degree of systolization for MIMO architectures. This should yield a better structure for realization of the 'short' processing delay architectures using current VLSI technology and should mature with the development of three-dimensional IC (integrated circuit) technology.
4. In depth theoretical investigation of the consequences of not compensating for the processing delay, which may appear in different forms if the same architecture is used to implement various control schemes, such as those, for example, relating to the convergence properties of adaptive/self-tuning algorithms. One approach to this has been proposed by Åström and Wittenmark (1984), using the root locus technique. Work on developing more effective methods of compensating for this delay should also be undertaken.
5. Appropriate theoretical developments should lead to work on extending the

developed architectures to other control related topics such as non-linear and/or self-learning control systems. These architectures are also expected to apply to general digital signal processing problems, particularly in the cases where 'short' system latency is required.

6. In a general setting, work should be directed towards the use of other architectures such as neural networks (for comparative purposes at least).

REFERENCES

ANDERSON, B. D. O. and LINNEMANN, A. (1987) 'Control of decentralised systems with distributed controller complexity', *IEEE Trans. Auto. Contr.*, **AC-32**, 625–9.

ÅSTRÖM, K. J. and WITTENMARK, B. (1984) *Computer Controlled Systems: Theory and Design*, Englewood Cliffs, NJ: Prentice Hall.

ÅSTRÖM, K. J. and WITTENMARK, B. (1989). *Adaptive Control*, Reading MA: Addison-Wesley.

CHEN, C. S. (1982) 'Application of one-chip signal processor in digital controller implementation' *IEEE Control Systems Mag.*, **2** (9), 16–22.

DATTA, B. N. (1986) 'Efficient parallel algorithms for controllability and eigenvalue assignment problems', in *Proceedings of the 25th IEEE International Conference on Decision and Control, Athens*, Vol. 3, pp. 1611–16.

FARRAR, F. A. and EIDENS, R. S. (1980) 'Microprocessor requirements for implementing modern control logic.', *IEEE Trans. Auto. Contr.*, **AC-25**, 461–8.

IRWIN, G. W. and FLEMING, P. J. (eds) (1990) *IEE Proc., Pt.D: Parallel Processing for Real-Time Control*, special issue.

JONES, S. (1988) 'PACE: A VLSI architecture', in *Parallel Processing for Control: The transputer and other architectures* (ed. P. Fleming), London: Peter Peregrinus, pp. 233–40.

KALYAEV, I. A. and KALYAEV, A. V. (1986). 'The homogeneous structures for optimal control problem solving', in *Proceedings of the 3rd International Workshop on Parallel Processing by Cellular Automata and Arrays, Berlin*, pp. 218–25.

KOGGE, P. M. (1981) *The Architecture of Pipelined Computers*, New York: McGraw Hill.

KUNG, S. Y. (1984) 'On supercomputing with systolic/wavefront processors', *Proc. IEEE*, **72**, 867–84.

KUNG, S. Y. (1988) *VLSI Array Processors*, Englewood Cliffs, NJ: Prentice Hall.

KWAN, H. K. (1987) 'New systolic array for realising second-order recursive digital filters', *Electronics Lett.* **23**, 442–3.

LANG, J. H. (1984) 'On the design of special-purpose digital control processors', *IEEE Trans. Auto. Contr.*, **AC-29**, 195–201.

LARSEN, P. M. and EVANS, F. J. (1988) 'Structural design of decentralised control systems', in *Advanced Computing Concepts and Techniques in Control Engineering* (eds M. J. Denham and A. J. Laub), Berlin: Springer-Verlag pp. 257–86.

LI, Y. (1990) 'Concurrent architectures for real-time control', PhD Thesis, University of Srathclyde, Glasgow, UK.

LI, Y., NIESSEN, K. and ROGERS, E. (1991) 'Mapping systolic structures onto transputer/A100 based parallel processors for adaptive/self-tuning control', *Int. J. Contr.*, **55** (6), 1399–411.

LI, Y. and ROGERS, E. (1989) 'High throughput and shortest latency systolic IIR filters and their applications to control engineering' in *Systolic Array Processors* (eds J. V. McCanny, J. McWhirter and E. Swartzlander), Hemel Hempstead: Prentice Hall, pp. 96–104.

LI, Y. and ROGERS, E. (1990). 'A systolic array information processing strategy for real-time automatic control, in *Advanced Information Processing in Automatic Control* (ed. R. Husson), Oxford: Pergamon Press, pp. 68–74.

LIN, H. (1986) 'New VLSI systolic array design for real-time digital signal processing', *IEEE Trans. Circuits and Syst.*, **CAS-33**, 673–6.

LUK, W. and JONES, G. (1988) 'Systolic recursive filters', *IEEE Trans. Circuits and Syst.*, **CAS-35**, 1067–8.

McCANNY, J. V. and WHITE, J. C. (1987) *VLSI Technology and Design*, London: Academic Press.

MASON, S. J. and ZIMMERMANN, H. J. (1960) *Electronic Circuits, Signals, and Systems*, New York: John Wiley.

MIDDLETON, R. H. and GOODWIN, G. C. (1990) *Digital Control and Estimation: A Unified Treatment*, Englewood Cliffs, NJ: Prentice Hall.

MORONEY, P. (1983) *Issues in the Implementation of Digital Feedback Compensators*, Cambridge, MA: MIT Press.

OPPENHEIM, A. V. and SCHAFER, R. W. (1975) *Digital Signal Processing*, Englewood Cliffs, NJ: Prentice Hall.

ROGERS, E. (ed) (1991) *Int. J. Contr.: Parallel Processing Techniques in Control*, **54**(6), special issue.

ROGERS, E. and LI, Y. (1988) 'Systolic array based concurrent processing for real-time high performance control', in *Proceedings of the 27th IEEE International Conference on Decision and Control, Austin*, pp. 2236–7.

ROGERS, E. and LI, Y. (1989) 'Concurrent array processing for linear multivariable control systems', in *Proceedings of the IFAC International Symposium on Adaptive Systems in Control and Signal Processing, Glasgow*, April, Vol. 2, pp. 559–63.

ROGERS, E. and LI, Y. (1990) 'Real-time control systems design using a high-speed rapid response systolic array, in *Proceedings of the 29th IEEE International Conference on Decision and Control, Hawaii*, pp. 947–9.

ROGERS, E. and LI, Y. (1991) 'Systolic array based processing for linear multivariable feedback control schemes', in *Proceedings of the 30th IEEE International Conference on Decision and Control, Brighton*, pp. 2366–7.

SHIN, K. C. and CUI, X. (1988) 'Effects of computing time delay on real-time control systems', in *Proceedings of the American Control Conference*, pp. 1071–6.

WANG, S. S. and LIN, T. P. (1986) 'The systolic architectures of digital control processors', in *Proc. ISMM Symposium on Software and Hardware Applications of Microcomputers* (eds M. H. Hanza and G. K. F. Lee), pp. 54–8.

WILLSKY, A. S. (1979) *Digital Signal Processing and Control and Estimation Theory: Points of tangency, areas of interaction, and parallel directions*, Cambridge, MA: MIT Press.

XU, Y., JONES, S. and SPRAY, A. (1992) 'The application of PACE within control' in *Algorithms and Architectures for Real-time Control* (eds P. J. Fleming and D. I. Jones), Oxford: Pergamon, pp. 93–8.

2

Systolic Architectures for Adaptive Control

L. Chisci and G. Zappa

2.1 INTRODUCTION

Adaptive control is an example of an area which has been supported by the revolutionary advances in microelectronics. This area is characterized by a wealth of approaches and algorithms, often of an 'ad hoc' nature, and hence it appears impossible to embrace within a common framework all the control systems referred to as 'adaptive'. Hence it is necessary to consider specific control schemes from the outset. In this chapter an adaptive controller is taken to be a nonlinear control system, shown schematically in Figure 2.1 (taken from Åström and Wittenmark, 1989, p. 429) consisting of the following three control loops:

1. *The regulation loop*: this applies digital linear control to the plant by exploiting both feedback and feedforward action. It includes analog to digital (A/D) and digital to analog (D/A) converters, possibly analog anti-aliasing and post-sampling filters (G_{aa} and G_{ps}), and a *regulator* with 'tunable' parameters.
2. *The adaptation loop*: this is responsible for the adaptation of the regulator parameters. It consists of a recursive parameter *estimator* which estimates on-line the plant model parameters from input/output data, and of a controller *design* block which computes the regulator parameters from the estimated plant parameters according to some design criterion.
3. *The supervision loop*: this monitors data from the other blocks and appropriately tunes some 'knobs' in the estimation and controller design blocks. It is also responsible for data exchange with external devices and for human operator interfacing.

These loops are hierarchically structured in terms of the following:

1. *Signal bandwidth*: a prerequisite for successful application of adaptive control is the 'recommended spectrum' for the time-varying quantities involved, shown in Figure 2.2 (taken from Johnson, 1990, p. 147). Hence

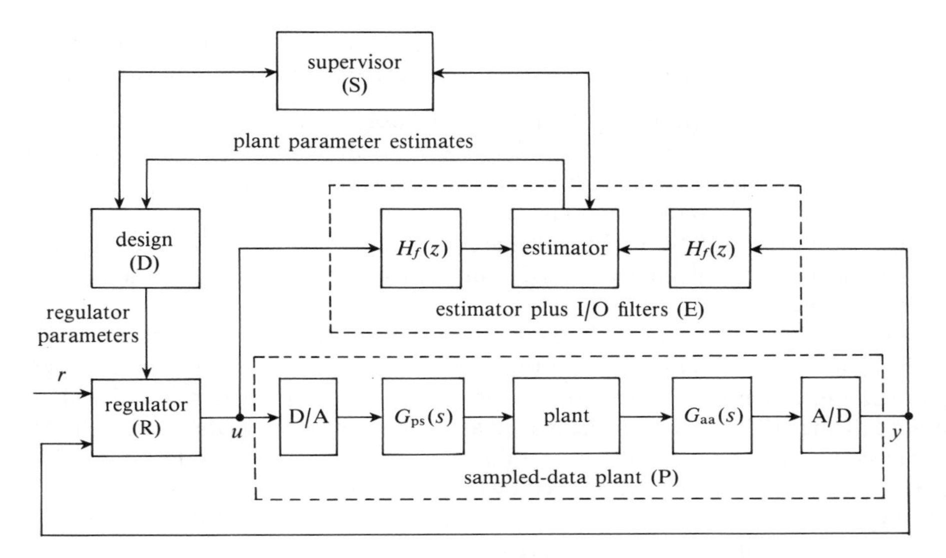

Figure 2.1 An adaptive control system

the adaptive system bandwidth, i.e. the convergence rate of the plant parameter estimates and of the regulator parameters, should fall well below the desired closed-loop bandwidth. In turn, adjustments of the 'tuning knobs' should be made occasionally to follow plant parameter variations.

2. *Computational complexity and regularity*: the regulator block simply performs digital linear filtering of the reference and output signals. Conversely, the estimator and control-design blocks involve somewhat intensive computations (e.g. Riccati iterations, spectral factorizations, matrix factorizations). Finally the supervisor performs complex

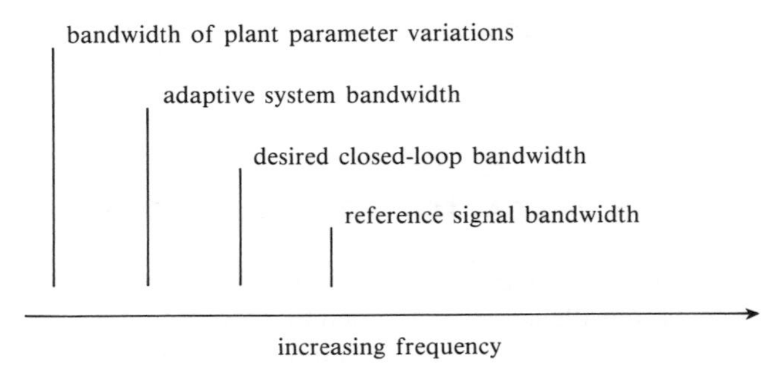

Figure 2.2 Recommended spectrum for robust adaptive control

mathematical computations (e.g. statistical tests on input/output data, matrix rank determination) and a variety of data communication tasks.

The above description yields useful indicators as to the needs, benefits and difficulties of parallel processing in adaptive control. The fine-grain parallelization of digital linear filters has already been extensively investigated in the *digital signal processing* (DSP) literature, and has also been considered in a control environment (Rogers and Li, 1989). Several parallel architectures have been developed, pipelined up to the bit-level,which can satisfy any speed requirement demanded by the regulation loop.

Alternatively, the complexity and irregularity of the supervision tasks, coupled with the low speed requirements of the supervision loop, advise against a fine-grain parallel implementation of the supervisor. Conversely, the regularity and compute-bound (Kung, 1988) nature of the matrix algorithms involved in the adaptation tasks suggests the exploitation of fine-grain parallelism to achieve high speed and performance in this phase. Motivated by these considerations, this chapter will focus on the parallelization of the adaptation loop by using systolic architectures for the estimator and controller design blocks.

A considerable volume of work (see, for example, McWhirter, 1983; Chisci and Zappa, 1988; Varvitsiotis *et al.*, 1989; Moonen, 1990; Ward *et al.*, 1986) has been undertaken on systolic arrays for *recursive least-squares* (RLS) parameter estimation; while, to the best of the authors' knowledge, a unique contribution (Chisci and Zappa, 1991) has addressed systolic arrays for controller design. In all these references, however, little attention has been devoted to the integration of such arrays in the overall adaptive control system which is the main subject of this chapter. In Section 2.2 some performance figures characterizing a parallel implementation are introduced and discussed. Some theoretical limitations are also detailed. In section 2.3 an algorithm is developed for solving constrained least-squares problems, amenable to systolic implementation, which, suitably specialized, can perform both estimation and controller design tasks. In Section 2.4 a systolic array for RLS estimation is developed with emphasis on adaptive control requirements. In Section 2.5 a general algorithm for designing the parameters of predictive control laws is developed, and in Section 2.6 some systolic implementations of this algorithm are proposed. Finally, Section 2.7 summarizes the chapter.

2.2 PARALLELIZATION OF ADAPTIVE CONTROL SYSTEMS

The aim of this section is to decide on the maximum possible benefit which can be achieved by applying parallelizing and pipelining computations to an adaptive control system. The objectives here can be summarized as follows:

1. A sampling rate $1/T_s$ as fast as required.

2. The shortest possible computational delay, Δ_s, between sampling and generation of the control signal.
3. An updating rate of the regulator parameters, $1/T_u$, which is as fast as possible.
4. The shortest possible computational delay, Δ_u, between sampling and updating of the regulator parameters.

The first two objectives relate to the ability of the adaptive controller to follow fast plant dynamics, whilst the latter two relate to the ability to track plant changes. Recall also that for successful use of adaptive control, the dynamics of the plant parameter variations must be much slower than the plant dynamics.

For the subsequent analysis, it is convenient to represent the adaptive control system as the directed graph shown in Figure 2.3. The nodes represent the *plant P* and the processing blocks R (*regulator*), E (*estimator*) and D(*design*), whilst the arcs represent unidirectional data links $L_j (j = 1, 2, \ldots, 6)$ divided into *signal* links $(j = 1, 2, 3, 4)$ and *parameter* links $(j = 5, 6)$. It is assumed, for simplicity, that the system is synchronous, i.e. there exists a global clock, whose rate equals the sampling rate $1/T_s$, which times data transfers between blocks. Since the computations are recursive, the data along each link represent different time instances of the same variable, each of which is indexed by an integer k.

Each block is characterized by a triple $(\delta_i, \delta_o, \tau)$ where δ_i denotes the *input iteration interval* (i.e. the time that must be allowed between the input of two successive instances), δ_o the *output iteration interval* and τ the *latency* (i.e. the computational delay from input to output). It is assumed these quantities are negligible for the plant, which is subsequently regarded as a digital processing block. Further, each link L_j is characterized by a tuple of integers (r_j, Δ_j) where: $1/r_j$ denotes the data-transfer rate along the link, measured in terms of the sampling rate $1/T_s$, i.e. $r_j T_s$ is the time interval between two successive instances; Δ_j is the index shift between the exiting and entering instances. Clearly the tuples (r, Δ) characterize the algorithm, and the triples $(\delta_i, \delta_o, \tau)$ characterize the hardware of each block.

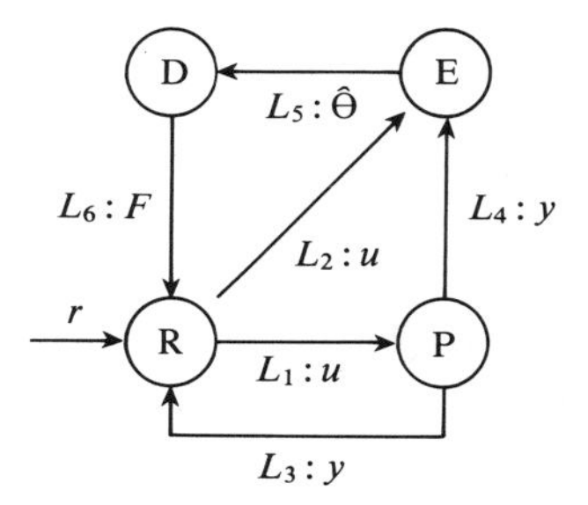

Figure 2.3 Graph representation of an adaptive control system

An algorithm is consistent with the hardware limitations if the following constraints are satisfied:

1. *For each block*:

$$\begin{cases} \delta_i < T_s r_j \text{ for each link } L_j \text{ entering the block} \\ \delta_o < T_s r_j \text{ for each link } L_j \text{ exiting the block} \end{cases} \tag{2.1}$$

2. *For each loop*:

$$T_s \sum_j r_j \Delta_j > \sum_i \tau_i \tag{2.2}$$

where Σ_j and Σ_i extend over the links and the blocks of the loops respectively.

To investigate the influence of the above constraints on the performance figures T_s, Δ_s, T_u and Δ_u it is first assumed that

$$r_1 = r_2 = r_3 = r_4 = 1 \tag{2.3}$$

i.e. the input/output signals are transferred at a rate equal to the sampling rate. Further, it is also assumed that

$$\Delta_1 = \Delta_2 = 1, \quad \Delta_3 = \Delta_4 = 0 \tag{2.4}$$

i.e. there is a single delay in the regulation loop and at least one delay in the adaptation loop. Note also that for a conventional sequential implementation, $\Delta_5 = \Delta_6 = 0$: i.e. at each sampling interval the following tasks are performed in sequence:

1. Estimate updating.
2. Regulator parameter computation.
3. Control signal generation.

Given these assumptions, consider first a *single-rate implementation*, i.e. new plant parameter estimates and regulator parameters are computed and transferred at each sampling period. This implies that the following condition holds:

$$r_5 = r_6 = 1 \tag{2.5}$$

Using (2.1) and (2.2) now yields the following lower bounds:

$$T_s = T_u > \max\left\{\delta(\mathrm{R}), \delta(\mathrm{E}), \delta(\mathrm{D}), \tau(\mathrm{R}), \frac{\tau(\mathrm{E}) + \tau(\mathrm{D}) + \tau(\mathrm{R})}{1 + \Delta_5 + \Delta_6}\right\} \tag{2.6}$$

$$\Delta_s > \tau(\mathrm{R})$$

$$\Delta_u > \tau(\mathrm{E}) + \tau(\mathrm{D})$$

where $\delta(\bullet) := \max\{\delta_i(\bullet), \delta_o(\bullet)\}$. (The notation $O(\bullet)$ stands for 'of the order of $\bullet$'. In particular $O(1)$ means independent of the size of the problem.) Further, (2.6) yields conditions for the pipelinability of the overall system, i.e. for independence of the sampling rate from the computational complexity of the adaptive controller. These are stated formally as follows.

Proposition 2.1

The adaptive control system of Figure 2.1 is pipelined if, and only if,

1. $\delta(\mathrm{R}), \delta(\mathrm{D}), \delta(\mathrm{E})$ are $O(1)$
2. $\tau(\mathrm{R})$ is $O(1)$
3. $\Delta_{\mathrm{al}} = \Delta_5 + \Delta_6$ is $O\left(\frac{\tau(\mathrm{E}) + \tau(\mathrm{D})}{T_{\mathrm{s}}}\right)$

At this stage it is clear how the presence of recursive loops hinders pipelining of the conventional adaptive control scheme. In particular, a necessary condition for pipelinability is that the estimator be implemented with complexity-independent latency and iteration interval. To the best of the authors' knowledge, this requirement cannot be met by any of the known (parallel) implementations of the estimator. Conversely, suppose that an appropriate index-shift, Δ_{al}, is introduced into the adaptation loop so as to satisfy condition 3 of Proposition 2.1. Then the inequality of (2.6) becomes

$$T_{\mathrm{s}} > \max \{\delta(\mathrm{E}), \delta(\mathrm{D}), \delta(\mathrm{R}), \tau(\mathrm{R})\} \tag{2.7a}$$

provided

$$\Delta_{\mathrm{al}} > \frac{\tau(\mathrm{E}) + \tau(\mathrm{D}) + \tau(\mathrm{R})}{T_{\mathrm{s}}} - 1 \tag{2.7b}$$

and hence T_{s} is independent of the computational delays $\tau(\mathrm{E})$ and $\tau(\mathrm{D})$ of the estimator and controller design blocks respectively.

On the assumption that the plant dynamics change slowly with respect to the sampling period, the delay Δ_{u} will not cause any appreciable degradation in self-tuning performance. Hence, the introduction of an index-shift Δ_{al} represents a viable solution in order to increase the sampling rate. Conversely, any additional index-shift within the regulation loop ($\Delta_2 > 1$) would severely degrade control performance, reducing, in particular, the stability margin. Thus the regulation loop remains the real bottleneck of the overall scheme, and for this reason the regulator, in addition to being a pipelined digital filter, should also have the shortest possible latency. Usually the regulator is a linear infinite impulse response (IIR) filter with variable coefficients for which there exist parallel implementations based on *transversal* filter realizations which exhibit both $O(1)$ latency and iteration interval (see, for example, Kung, 1985; Woods *et al.*, 1988). Conversely, cascade realizations are not applicable to pipelined control systems because of their complexity-dependent latency.

The maximum achievable sampling rate is also limited by the iteration interval, $\delta(\mathrm{D})$, of the design block. This is a severe limitation whenever the design of the control-law is based on complex iterative procedures, such as in predictive control. In such cases, even exploiting maximum parallelism, $\delta(\mathrm{D})$ may in fact be considerably larger than the other entries in (2.7a), thus drastically reducing the sampling rate. To overcome this limitation, a *multirate* implementation can be used with slower updating rates for both the plant parameter estimates and the regulator parameters, i.e. $r_5 > 1$ and $r_6 > 1$. This means that the plant parameter estimates and the regulator parameters are computed once every r_5 and r_6 sampling periods respectively. Note also that, since $r_2 = r_4 = 1$, all the available input/output data are still used by the estimator, i.e. no information is lost.

In this case, use of (2.1) and (2.2) gives the following lower bounds:

$$
\begin{aligned}
&T_s > \max \{\delta(\mathrm{R}), \tau(\mathrm{R}), \delta_i(\mathrm{E})\} \\
&\Delta_s > \tau(\mathrm{R}) \\
&T_u > \max \{\delta_o(\mathrm{E}), \delta(\mathrm{D})\} \\
&\Delta_u > \tau(\mathrm{E}) + \tau(\mathrm{D})
\end{aligned}
\tag{2.8}
$$

Hence the computational speed of the design block does not directly influence the maximum achievable sampling rate, which is only a function of the computational speed (both the rate and the delay) of the regulator and of the input computational rate of the estimator. Conversely, the computational speeds of the estimator and controller design blocks are the primary factors governing the 'alertness' of the adaptation.

In the case of the regulator, it suffices to note that by suitable systolic implementations it is possible to make $\delta(\mathrm{R})$ and $\tau(\mathrm{R})$ equal to the computing time of a few multiply-and-add operations. In particular

$$
\delta(\mathrm{R}) = 2(T_+ + T_\times), \quad \tau(\mathrm{R}) = T_+ + 2T_\times \tag{2.9}
$$

where $T*$ denotes the computing time of the (scalar) operation '$*$' ($* = +, \times, /, \surd$). Hence the remainder of this chapter will develop implementations of the estimator and controller design blocks which speed up computations, i.e. reduce $\delta(\mathrm{E})$, $\tau(\mathrm{E})$, $\delta_i(\mathrm{D})$, $\delta_o(\mathrm{D})$ and $\tau(\mathrm{D})$. By choosing appropriate algorithms for recursive parameter estimation and controller design, it will be shown how to design parallel computing architectures for these blocks which drastically reduce the above figures with respect to a sequential implementation.

2.3 BASIC ALGORITHM AND SYSTOLIC ARCHITECTURE

2.3.1 The Basic Algorithm

It will be shown in what follows that parameter estimation and controller design amount, in many approaches, to the solution of the following *constrained least-squares* (CLS) optimization problem: *minimize with respect to* x_1, *as a function of* x_2, *the quadratic cost*

$$
J(\Omega, R, x_1) := \left\| \Omega R \begin{bmatrix} x_1 \\ x_2 \end{bmatrix} \right\|^2 \tag{2.10}
$$

subject to the linear constraints

$$(I-\Omega)R\begin{bmatrix}x_1\\x_2\end{bmatrix}=0 \tag{2.11}$$

In (2.10) and (2.11): $x_1 \in \mathscr{R}^{N_1}$, $x_2 \in \mathscr{R}^{N_2}$, $\Omega \in \mathscr{R}^{MxM}$ and $R \in \mathscr{R}^{Mx(N_1+N_2)}$, with $M \geqslant N_1$. Further, set $x=[x_1^{\mathrm{T}}, x_2^{\mathrm{T}}]^{\mathrm{T}}$ and partition Ω and R as

$$\Omega=\begin{bmatrix}\Omega_1 & 0\\0 & \Omega_2\end{bmatrix},\quad R=\begin{bmatrix}R_{11} & R_{12}\\R_{21} & R_{22}\end{bmatrix} \tag{2.12}$$

where Ω_1 and R_{11} are $N_1 \times N_1$ matrices, and introduce the following assumptions:

1. R_{11} is upper-triangular.
2. $\Omega = \mathrm{diag}\{\omega_1, \omega_2, \ldots, \omega_M\}, \omega_i \in \{0, 1\}$.
3. For all $x_2 \in \mathscr{R}^{N_2}$, the CLS problem (2.10)–(2.11) has a solution, i.e. the linear manifold $\mathscr{M}(\Omega, R, x_2) := \{x_1 : (I-\Omega)R[x_1^T, x_2^T]^T = 0\}$ is non-empty.

Remark 2.1

Characterization of costs and constraints: by assumption 2, each row r_i of the matrix R, $i=1, 2, \ldots, M$, corresponds either to the *quadratic cost* $(r_i x)^2$ in (2.10) if $\omega_i=1$, or to the *linear constraint* $r_i x=0$ in (2.11) if $\omega_i=0$. Hence, on reordering the rows of R, the CLS problem (2.10)–(2.11) can be rewritten as

$$\min_{x_1}\left\| R_1 \begin{bmatrix}x_1\\x_2\end{bmatrix}\right\|^2 \text{ subject to } R_o \begin{bmatrix}x_1\\x_2\end{bmatrix}=0 \tag{2.13}$$

where

$$\begin{bmatrix}R_1\\R_o\end{bmatrix}=PR \tag{2.14}$$

and P is the permutation matrix such that

$$P\Omega P^{\mathrm{T}}=\begin{bmatrix}I_{\mathrm{rank}\,\Omega} & 0\\0 & 0\end{bmatrix} \tag{2.15}$$

The fact that the same CLS problem can be represented by many different tuples (Ω, R) motivates the following:

Definition 2.1

Equivalence of representations: (Ω, R) and $(\bar{\Omega}, \bar{R})$ are said to be *equivalent* if, for all $x_2 \in \mathscr{R}^{N_2}$ the following two conditions hold:

1. $\mathscr{M}(\Omega, R, x_2)=\mathscr{M}(\bar{\Omega}, \bar{R}, x_2)$
2. $J(\Omega, R, x_1)=J(\Omega, R, x_1), \quad \forall x_1 \in \mathscr{M}(\Omega, R, x_2)=\mathscr{M}(\bar{\Omega}, \bar{R}, x_2)$

Definition 2.2

Canonical representation: the tuple $(\bar{\Omega}, \bar{R})$ is said to be a *canonical* representation of the CLS problem (2.10)–(2.11), if $\bar{R}$ is of the form

$$\bar{R} = \begin{bmatrix} \bar{R}_{11} & \bar{R}_{12} \\ 0 & \bar{R}_{22} \end{bmatrix} \tag{2.16}$$

Remark 2.2

Solution of the CLS problem: any solution to the CLS problem represented by the canonical tuple $(\bar{\Omega}, \bar{R})$ must satisfy the relation

$$\bar{R}_{11} x_1 + \bar{R}_{12} x_2 = 0 \tag{2.17}$$

On the assumption that $\bar{R}_{11}$ is nonsingular, there exists a unique solution $\hat{x}_1$ given by

$$\hat{x}_1 = \hat{L} x_2 \tag{2.18}$$

where

$$\hat{L} := -\bar{R}_{11}^{-1} \bar{R}_{12} \tag{2.19}$$

In what follows, $\hat{L}$, instead of $\hat{x}_1$ will be referred to as the solution of the CLS problem (2.10)–(2.11). The existence assumption 3 (after (2.12)) implies that, for $j = N_1 + 1, \ldots, M, \bar{r}_j = 0$ whenever $\bar{\omega}_j = 0$.

It is now possible to prove that a CLS optimization problem can always be represented by an equivalent canonical tuple $(\bar{\Omega}, \bar{R})$. This requires the following elementary row transformations on the tuple (Ω, R) defined in terms of the rows r_i, r_j of R and the corresponding diagonal entries, ω_i, ω_j of Ω.

Givens Rotation G_{ij}

$$\begin{bmatrix} \bar{r}_i \\ \bar{r}_j \end{bmatrix} = G_{ij} \begin{bmatrix} r_i \\ r_j \end{bmatrix}, \quad \bar{\omega}_i = \omega_i, \bar{\omega}_j = \omega_j$$

where

$$G_{ij} = \begin{bmatrix} c & s \\ -s & c \end{bmatrix}, \quad c^2 + s^2 = 1$$

and is admissible whenever $\omega_i = \omega_j$.

Gauss transformation L_{ij}

$$\begin{bmatrix} \bar{r}_i \\ \bar{r}_j \end{bmatrix} = L_{ij} \begin{bmatrix} r_i \\ r_j \end{bmatrix}, \quad \bar{\omega}_i = \omega_i, \bar{\omega}_j = \omega_j$$

where

$$L_{ij} = \begin{bmatrix} \gamma & 0 \\ \beta & 1 \end{bmatrix}, \ \gamma \neq 0$$

and is admissible whenever $\omega_i = 0$.

Interchange J_{ij}

$$\begin{bmatrix} \bar{r}_i \\ \bar{r}_j \end{bmatrix} = J_{ij} \begin{bmatrix} r_i \\ r_j \end{bmatrix}, \ \bar{\omega}_i = \omega_i, \bar{\omega}_j = \omega_j$$

where

$$J_{ij} = \begin{bmatrix} 0 & 1 \\ 1 & 0 \end{bmatrix}$$

and is always admissible.

The importance of the above admissible elementary transformations is that they do not change the underlying CLS problem, as stated by the following:

Theorem 2.1

Let $(\bar{\Omega}, \bar{R})$ be obtained from (Ω, R) via a sequence of admissible elementary transformations. Then (Ω, R) and $(\bar{\Omega}, \bar{R})$ are equivalent.

Proof

First note that elementary transformations affect only two rows of R and possibly interchange two diagonal entries of Ω. Admissible Givens rotations either replace the linear constraints $r_i x = 0$ and $r_j x = 0$ by the equivalent ones $\bar{r}_i x = 0$ and $\bar{r}_j x = 0$, or re-express the quadratic cost-term $(r_i x)^2 + (r_j x)^2$ as $(\bar{r}_i x)^2 + (\bar{r}_j x)^2$. Admissible Gauss transformations replace the constraint $r_i x = 0$ by the equivalent one $(\gamma r_i)x = 0$, and the cost-term $(r_j x)^2$ by $((\beta r_i + r_j)x)^2$, the two terms being equal on the manifold $\mathcal{M}(\Omega, \mathcal{R}, x_2)$. Hence, admissible Givens rotations, Gauss transformations and, obviously, interchanges do not modify the associated CLS problem.

Next it is shown that, by using admissible elementary transformations, it is possible to transform any given CLS representation into an equivalent canonical one.

Theorem 2.2

Any tuple (Ω, R) can be transformed into an equivalent canonical tuple $(\bar{\Omega}, \bar{R})$ by a suitable sequence of admissible elementary transformations.

Proof

It suffices to zero out the elements of the block R_{21} by the same annihilation strategy using QR decomposition and selecting at each step the appropriate admissible elementary transformations (with the appropriate coefficients) according to the values of the Ω-entries.

Remark 2.3

Generalized triangularization: the transformation

$$(\Omega, R) = \left(\begin{bmatrix} \Omega_1 & 0 \\ 0 & \Omega_2 \end{bmatrix}, \begin{bmatrix} R_{11} & R_{12} \\ R_{21} & R_{22} \end{bmatrix} \right) \rightarrow (\bar{\Omega}, \bar{R}) = \left(\begin{bmatrix} \bar{\Omega}_1 & 0 \\ 0 & \bar{\Omega}_2 \end{bmatrix}, \begin{bmatrix} \bar{R}_{11} & \bar{R}_{12} \\ 0 & \bar{R}_{22} \end{bmatrix} \right) \quad (2.20)$$

which transforms (Ω, R) into the equivalent canonical representation $(\bar{\Omega}, \bar{R})$ according to Theorem 2.2, will be referred to as *generalized triangularization* in the sequel. If Ω_1 and Ω_2 are both identity matrices it reduces to an orthogonal triangularization; while if Ω_1 and Ω_2 are null and identity matrices respectively, it reduces to a Gaussian elimination. In the general case, annihilation of R_{21} requires a combination of orthogonal (Givens) and non-orthogonal (Gauss) elementary transformations. This generalized triangularization can be used to solve 'constrained' least squares (LS) problems in the same way as orthogonal triangularization is used to solve 'ordinary' LS problems.

Remark 2.4

Normalized algorithm: this work uses an unnormalized algorithm, i.e. it is based on an unnormalized upper-triangular matrix R_{11}. A normalized version could equally well be used by ensuring that R_{11} has *unit* diagonal elements throughout (Gentleman, 1973) by using 'fast' rather than 'standard' Givens rotations.

2.3.2 The Basic Systolic Array

A systolic trapezoidal array, abbreviated *trapezarray*, for solving the CLS problem (2.10)–(2.11) is shown in Figure 2.4. This array constitutes the core of the systolic implementations of both the estimator and the controller design blocks which will be presented in Sections 2.4 and 2.6 respectively.

This is essentially the array of Gentleman and Kung (1981) with the functions of the processing elements extended in order to perform generalized triangularization (cf. Remark 2.3). The array consists of two types of processing elements whose functions are both illustrated in Figure 2.5. Diagonal processors

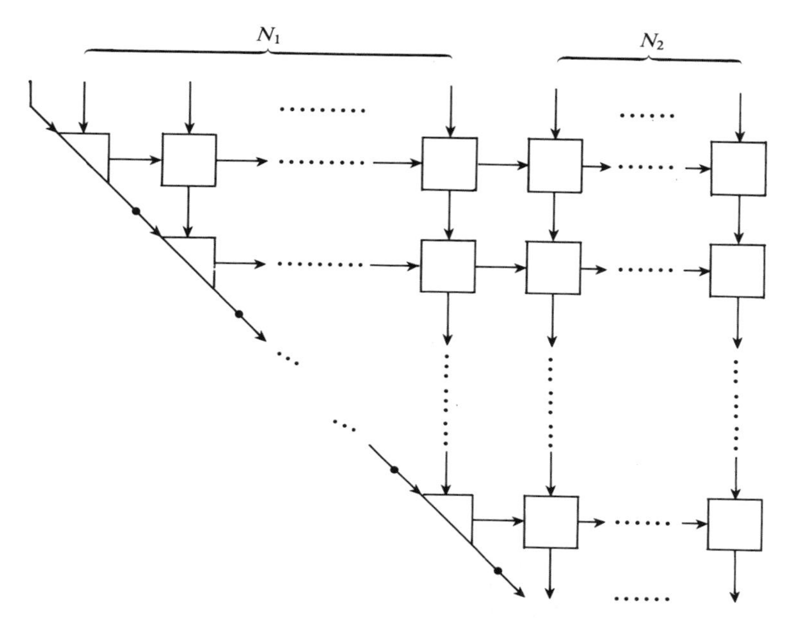

Figure 2.4 Basic trapezarray (N_1, N_2) for constrained least-squares computations. It consists of an orthogonally connected triangular array of $N_1(N_1 + 1)/2$ processors plus an orthogonally connected rectangular array of $N_1 N_2$ processors. Triangles represent diagonal processors, squares represent off-diagonal processors, while dots represent delay elements

(cf. Figure 2.5(a)) select an admissible elementary transformation to annihilate a component of the input vector, compute the relative parameters, and transmit them rightwards. Off-diagonal processors (cf. Figure 2.5(b)) apply the transformations received from the left, transmit the transformed input vector component downwards, and the transformation parameters rightwards.

With the above processor functions, the CLS problem (2.10)–(2.11) can be implemented on the trapezarray as follows. Assume that at the outset the matrices $[R_{11} \ \ R_{12}]$ and Ω_1 are stored in the array (one R-entry for each processor and one Ω-entry for each diagonal processor) and that the matrix $[R_{21} \ \ R_{22}]$ is input in skewed fashion from the top while Ω_2 is fed-in diagonally (Figure 2.6(a)). Then, after a latency of $2N_1$ clock cycles, the matrices $\bar{R}_{22}$ and $\bar{\Omega}_2$ will flow out from the bottom. Further, the matrices $[\bar{R}_{11}, \bar{R}_{12}]$ and $\bar{\Omega}_1$ will be stored in place of $[R_{11} \ \ R_{12}]$ and Ω_1 (Figure 2.6(b)).

Note that the array operates in a fully pipelined way in that (skewed) input vectors are fed in and the corresponding output vectors are taken out at the rate of one every clock cycle. Hence the array is 100% efficient, i.e. any processor is active on any clock cycle.

It now remains to show how to compute the CLS solution $\hat{L}$ in (2.19) which uses

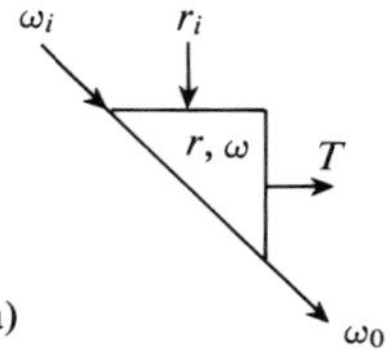

(a) $(r, \omega, \omega_0, T) := \text{GENERATE}(r, r_i, \omega, \omega_i)$

(b) $(r, r_0) := \text{APPLY}(r, r_i, T)$

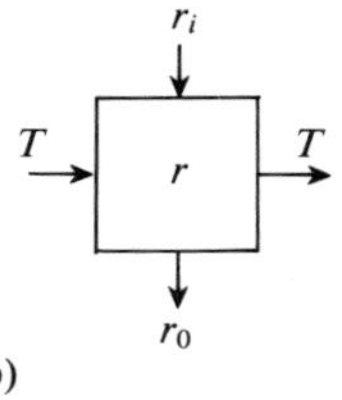

$(\bar{r}_1, \bar{r}_2) :=$ function APPLY(r_1, r_2, T);
begin

$$\begin{bmatrix} \bar{r}_1 \\ \bar{r}_2 \end{bmatrix} := T \begin{bmatrix} r_1 \\ r_2 \end{bmatrix}$$

end.

$(\bar{r}_1, \bar{\omega}_1, \bar{\omega}_2, T) :=$ function GENERATE$(r_1, r_2, \omega_1, \omega_2)$;
{Compute a 2×2 admissible elementary transformation T such that $[\bar{r}_1 \ 0] = [r_1 \ r_2]\, T^{\mathrm{T}}$}
begin

if $|r_2| < \varepsilon$ then begin {Identity}

$$\bar{r}_1 := r_1;\ \bar{\omega}_1 := \omega_1;\ \bar{\omega}_2 := \omega_2;\ T := \begin{bmatrix} 1 & 0 \\ 0 & 1 \end{bmatrix};\ \text{exit}$$

end{Identity}
if $|r_1| < \varepsilon$ then begin{Interchange}

$$\bar{r}_1 := r_2;\ \bar{\omega}_1 := \omega_2;\ \bar{\omega}_2 := \omega_1;\ T := \begin{bmatrix} 0 & 1 \\ 1 & 0 \end{bmatrix};\ \text{exit}$$

end {Interchange}
if $\omega_1 = \omega_2$ then begin {Givens}
$(\bar{r}_1, c, s) := \text{GIVENS}(r_1, r_2)$; $\bar{\omega}_1 := \omega_1$; $\bar{\omega}_2 := \omega_2$;

$$T := \begin{bmatrix} c & s \\ -s & c \end{bmatrix};\ \text{exit}$$

end{Givens}
if $\omega_1 = 0$ then begin{Gauss}

$$\bar{r}_1 := r_1;\ \bar{\omega}_1 := 0;\ \bar{\omega}_2 := 1;\ T := \begin{bmatrix} 1 & 0 \\ -r_2/r_1 & 1 \end{bmatrix};\ \text{exit}$$

end{Gauss}
{Interchange + Gauss}

$$\bar{r}_1 := r_2;\ \bar{\omega}_1 := 0;\ \bar{\omega}_2 := 1;\ T := \begin{bmatrix} 0 & 1 \\ 1 & -r_1/r_2 \end{bmatrix}$$

end.

$(\bar{r}_1, c, s) :=$ function GIVENS(r_1, r_2);
begin
$\bar{r}_1 := \sqrt{(r_1^2 + r_2^2)}$;
$c := r_1/\bar{r}_1$; $s := r_2/\bar{r}_1$
end.

Figure 2.5 Functions of the processing elements of the CLS trapezarray: (a) diagonal processor; (b) off-diagonal processor

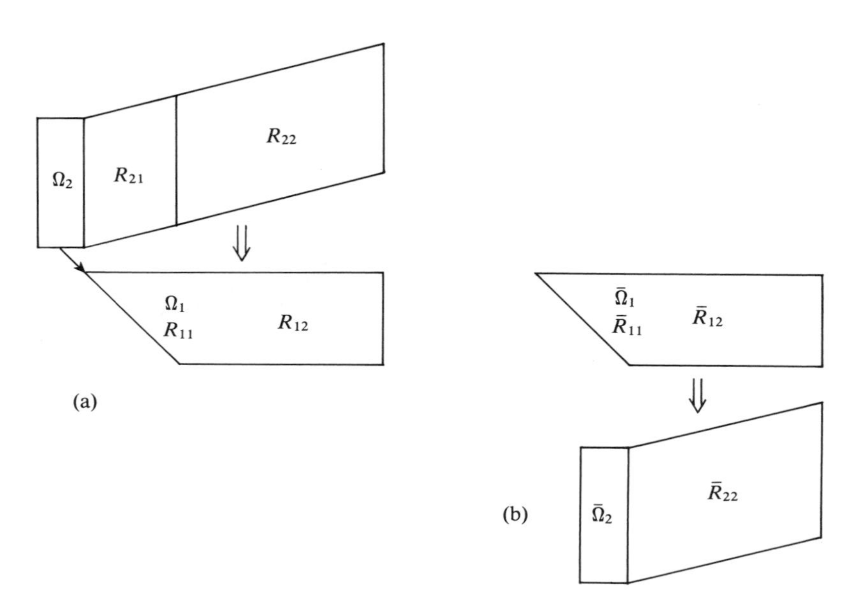

Figure 2.6 The data arrangements for the generalized triangularization procedure: (a) data input (the trapezarray has initially resident matrices Ω_1 and $[R_{11}R_{12}]$); (b) data output (the trapezarray has finally resident matrices $\bar{\Omega}_1$ and $[\bar{R}_{11}\bar{R}_{12}]$)

the well-known Faddeeva algorithm (Quinton and Robert, 1989). First note that the generalized triangularization

$$\left(\begin{bmatrix} \bar{\Omega}_1 & 0 \\ 0 & I_{N_1} \end{bmatrix}, \begin{bmatrix} \bar{R}_{11} & \bar{R}_{12} \\ I_{N_1} & 0 \end{bmatrix}\right) \rightarrow \left(\begin{bmatrix} \tilde{\Omega}_1 & 0 \\ 0 & \tilde{\Omega}_2 \end{bmatrix}, \begin{bmatrix} \tilde{R}_{11} & \tilde{R}_{12} \\ 0 & \tilde{R}_{22} \end{bmatrix}\right) \tag{2.21}$$

yields

$$\tilde{R}_{22} = -W\bar{R}_{11}^{-1}\bar{R}_{12} = \begin{bmatrix} \omega_1 & & & \\ & \omega_2 & & \\ & & \ddots & \\ & & & \omega_{N_1} \end{bmatrix} \hat{L} \tag{2.22}$$

where the diagonal scaling factors ω_i, $i = 1, 2, \ldots, N_1$, are the products of the cosines of the Givens rotations used to annihilate the ith row vector of the identity matrix. Hence the CLS solution can be extracted from the trapezarray as follows. The matrix $[I_{N_1}\ 0_{N_1 x N_2}]$ can be fed into the array from the top immediately after $[R_{11}\ R_{12}]$ while N_1 unit elements are being fed in diagonally (Figure 2.7(a)). Thus the matrix $\tilde{R}_{22} = W\hat{L}$ will flow out of the array immediately after R_{22}. The components, w_i, of W, which can be easily computed across the diagonal processors, are then fed into a linear scaling array, together with the matrix $\tilde{R}_{22}$ to product the desired matrix $\hat{L} = W^{-1}\tilde{R}_{22}$ (Figure 2.7(b)).

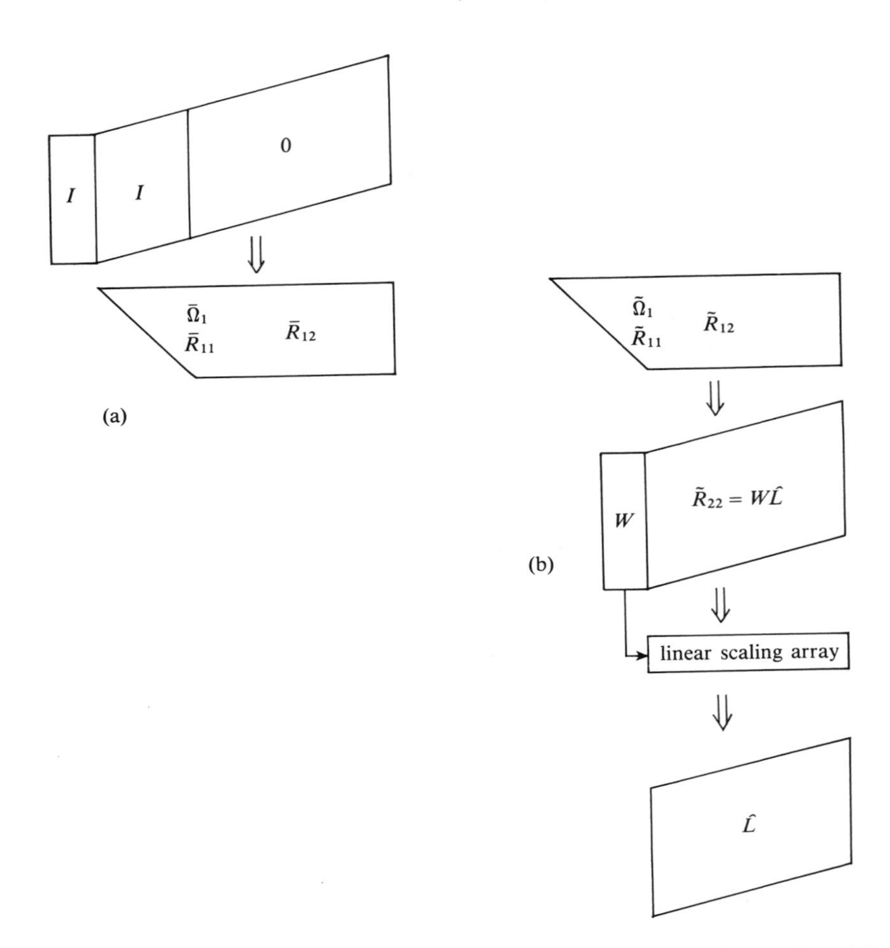

Figure 2.7 The data arrangements for the computation of the CLS solution: (a) data input; (b) data output. If $\bar{\Omega}_1 = 0$, a Gaussian elimination is performed so that $\tilde{R}_{11} = \bar{R}_{11}$, $\tilde{R}_{12} = \bar{R}_{12}$ and $\tilde{R}_{22} = \hat{L}$ ($W = I$), i.e. the scaling is not needed

Note that the transformation (2.21) also modifies the matrices $\bar{R}_{11}$ and $\bar{R}_{12}$ stored in the array. This can be a problem whenever the CLS problem is to be solved in a continuously recursive fashion as, for example, in recursive parameter estimation. To avoid this, a Gaussian elimination can be used instead of (2.21), i.e.

$$\left(\begin{bmatrix} 0 & 0 \\ 0 & I_{N_1} \end{bmatrix}, \begin{bmatrix} \bar{R}_{11} & \bar{R}_{12} \\ I_{N_1} & 0 \end{bmatrix}\right) \rightarrow \left(\begin{bmatrix} \tilde{\Omega}_1 & 0 \\ 0 & \tilde{\Omega}_2 \end{bmatrix}, \begin{bmatrix} \bar{R}_{11} & \bar{R}_{12} \\ 0 & \hat{L} \end{bmatrix}\right) \tag{2.23}$$

which leaves $\bar{R}_{11}$ and $\bar{R}_{12}$ unchanged, and also directly yields the CLS solution $\hat{L}$ without rescaling.

If $\bar{R}_{11}$ is singular, then (2.17) admits infinite solutions. In such a case, both methods of extraction (2.21)–(2.22) and (2.23) provide the unique minimum-norm solution $\hat{L}$ such that

$$\hat{x}_1 = \hat{L}x_2 \text{ minimizes } \|x_1\|^2 \text{ subject to } \bar{R}_{11}x_1 + \bar{R}_{12}x_2 = 0$$

with $\hat{L} = \bar{R}_{11}^* \bar{R}_{12}$, R_{11}^* denoting the unique Moore–Penrose pseudo-inverse of $\bar{R}_{11}$.

2.4 THE RECURSIVE PARAMETER ESTIMATOR

Recursive least-squares (RLS) is the parameter estimation algorithm most commonly used in adaptive control. In fact, it exhibits faster convergence rates compared to gradient-based algorithms such as *least-mean squares* (LMS) and *normalized* LMS. In addition, it does not require complicated stability checking procedures, unlike other more general algorithms such as *extended least-squares* (ELS) and *recursive maximum-likelihood* (RML) (see Ljung, 1987).

In RLS estimation the possibly multi-input multi-output (MIMO) plant is modelled by the discrete-time input/output representation

$$A(d)y(t) = B(d)u(t) + e(t) \tag{2.24}$$

where: $y(t) \in \mathscr{R}^p$ is the output; $u(t) \in \mathscr{R}^m$ is the input; $e(t) \in \mathscr{R}^p$ is a noise term; and $A(\bullet), B(\bullet)$ are polynomial matrices of compatible dimensions in the *unit delay operator* $d(\mathrm{d}x(t) := x(t-1))$, viz.

$$A(d) = I + \sum_{i=1}^{n_a} A_i \, d^i$$

$$B(d) = \sum_{i=1}^{n_b} B_i \, d^i \tag{2.25}$$

Note that $u(t)$ and $y(t)$ may represent either the actual or band-pass filtered input/output data.

The model (2.24) can be rewritten as the linear regression:

$$y(t) = \Theta\nu(t) + e(t) \tag{2.26}$$

where the regressor $\nu(t)$ consists of past input/output samples

$$\nu(t) := [y^{\mathrm{T}}(t-n)u^{\mathrm{T}}(t-n)\ldots y^{\mathrm{T}}(t-1)u^{\mathrm{T}}(t-1)]^{\mathrm{T}} \in \mathscr{R}^{n_p}$$

$$n_p = (p+m)n, n = \max\{n_a, n_b\} \tag{2.27}$$

and the parameter matrix Θ is defined by

$$\Theta = [-A_n B_n \ldots - A_1 B_1] \tag{2.28}$$

Least-squares estimation amounts to computing

$$\Theta(t) = \arg\min_{\Theta} J(t,\Theta),\ J(t,\Theta) := \sum_{k=1}^{t} \| y(k) - \Theta\nu(k) \|^2 \tag{2.29}$$

It can be easily shown that the LS estimate $\hat{\Theta}(t)$ coincides with the solution of the CLS optimization problem (2.10)–(2.11) obtained by setting:

$$\begin{cases} \Omega = I_{n_p+t} \\ [R_{11} R_{12}] = 0_{n_p \mathrm{x}(n_p+p)} \\ [R_{21} R_{22}] = \begin{bmatrix} \nu^{\mathrm{T}}(1) & -y^{\mathrm{T}}(1) \\ \nu^{\mathrm{T}}(2) & -y^{\mathrm{T}}(2) \\ \vdots & \vdots \\ \nu^{\mathrm{T}}(t) & -y^{\mathrm{T}}(t) \end{bmatrix} \end{cases} \tag{2.30}$$

Note that the above problem is actually *unconstrained*, since Ω is an identity matrix. Hence (cf. Remark 2.3) it can be solved by orthogonal triangularization using only Givens rotations.

The problem (2.30) is amenable to recursive solution. It suffices to perform, at each sampling interval k, the following orthogonal triangularization:

$$\underbrace{\begin{bmatrix} R_{11}(k-1) & R_{12}(k-1) \\ \nu^{\mathrm{T}}(k) & -y^{\mathrm{T}}(k) \end{bmatrix}}_{\text{pre-array}} \rightarrow \underbrace{\begin{bmatrix} R_{11}(k) & R_{12}(k) \\ 0 & R_{22}(k) \end{bmatrix}}_{\text{post-array}} \tag{2.31}$$

having initialized the matrices R_{11} (upper-triangular) and R_{12} as

$$[R_{11}(0) R_{12}(0)] = 0_{n_p \mathrm{x}(n_p+p)} \tag{2.32}$$

The LS estimate is then given by

$$\hat{\Theta}'(k) = -R_{11}^{-1}(k) R_{12}(k) \tag{2.33}$$

as soon as $R_{11}(k)$ becomes non-singular. The matrix $R_{11}(k)$ updated through (2.31) is the Cholesky square-root factor of the information (inverse covariance) matrix, i.e.

$$P^{-1}(k) := \sum_{i=1}^{k} \nu(i)\nu^{\mathrm{T}}(i) = R_{11}^{\mathrm{T}}(k) R_{11}(k) \tag{2.34}$$

Further, McWhirter (1983) has noted that the row vector $R_{22}(k)$ in (2.31) is a scaled version of both the predicted and the filtered residuals, i.e.

$$R_{22}(k) = -w(k)[y(k) - \hat{\Theta}(k-1)\nu(k)] = -w^{-1}(k)[y(k) - \hat{\Theta}(k)\nu(k)] \tag{2.35}$$

where the scaling factor $w(k)$ is the product of the cosines of all the Givens rotations used to perform (2.31).

Given that it is equivalent to a special case of CLS problem, RLS estimation can also be implemented on the trapezarray of Figure 2.4. The R- and Ω-entries in the array must be initialized with zeros and unit entries respectively, while the input vectors $[\nu^T(k) - y^T(k)]$, together with unit Ω-entries, are continuously supplied to the array. Hence, $R_{11}(k)$ and $R_{12}(k)$ are updated in place and the vectors $R_{22}(k)$ emerge from the array bottom. The parameter estimates $\hat{\Theta}(k)$ in (2.33) can be extracted from the bottom of the array column-by-column using the Faddeeva scheme as explained in Section 2.3. In the case when the extraction is performed by Givens rotations, care must be taken to prevent the auxiliary vectors from modifying the matrices R_{11} and R_{12} stored in the array.

In the case when the array implementation detailed in Figures 2.4 and 2.5 is used, the estimator requires some specific features as detailed below.

Remark 2.5

Simultaneous updating and parameter extraction: to enable continuous adaptive operation, parameter extraction should be performed in parallel with the processing of new input vectors. This implies that the data vectors $[\nu^T(k) - y^T(k)]$, $k = 1, 2, \ldots,$ and the auxiliary vectors e_i^T must be processed simultaneously.

Remark 2.6

Tracking of time-varying parameters: in order to track possibly time-varying parameters, it is necessary to introduce a mechanism for discounting old data. This can be easily accomplished by exponential forgetting of past data in the LS performance index, i.e. modifying $J(t, \Theta)$ in (2.29) to

$$J(t, \Theta) = \| y(t) - \Theta \nu(t) \|^2 + \lambda(t) J(t-1, \Theta), \quad 0 < \lambda(t) \leqslant 1 \tag{2.36}$$

This, in turn, modifies the basic recursion (2.31) to

$$\underbrace{\begin{bmatrix} \lambda^{1/2}(k) R_{11}(k-1) & \lambda^{1/2}(k) R_{12}(k-1) \\ \nu^T(k) & -y^T(k) \end{bmatrix}}_{\text{pre-array}} \rightarrow \underbrace{\begin{bmatrix} R_{11}(k) & R_{12}(k) \\ 0 & R_{22}(k) \end{bmatrix}}_{\text{post-array}} \tag{2.37}$$

Hence, exponential forgetting can be easily incorporated into the systolic array implementation. This simply requires that the matrices R_{11} and R_{12} stored in the array are rescaled by $\lambda^{1/2}(k)$ as each recursion.

The forgetting factor $\lambda(k)$ can be continuously supplied to the array and hence it can be changed on-line. If, for instance, infrequent abrupt changes of parameters occur then λ should be kept 'small' immediately after the change and then increased. Consequently the LS residual can be extracted from the array, monitored to detect parameter changes, and appropriate modification of λ made.

Remark 2.7

Robustness with respect to no-informative data: exponential forgetting together with non-informative input/output data may lead to pathological behaviour occurring in the estimator (covariance blow-up and bursts). These are due to 'small' eigenvalues of the information matrix $R^{\mathrm{T}}R$. Several modifications have been suggested (for example, periodic resetting, directional forgetting, constant trace). A simple remedy, within this parallel scheme, consists of imposing lower bounds on the values stored in each diagonal cell of the trapezarray.

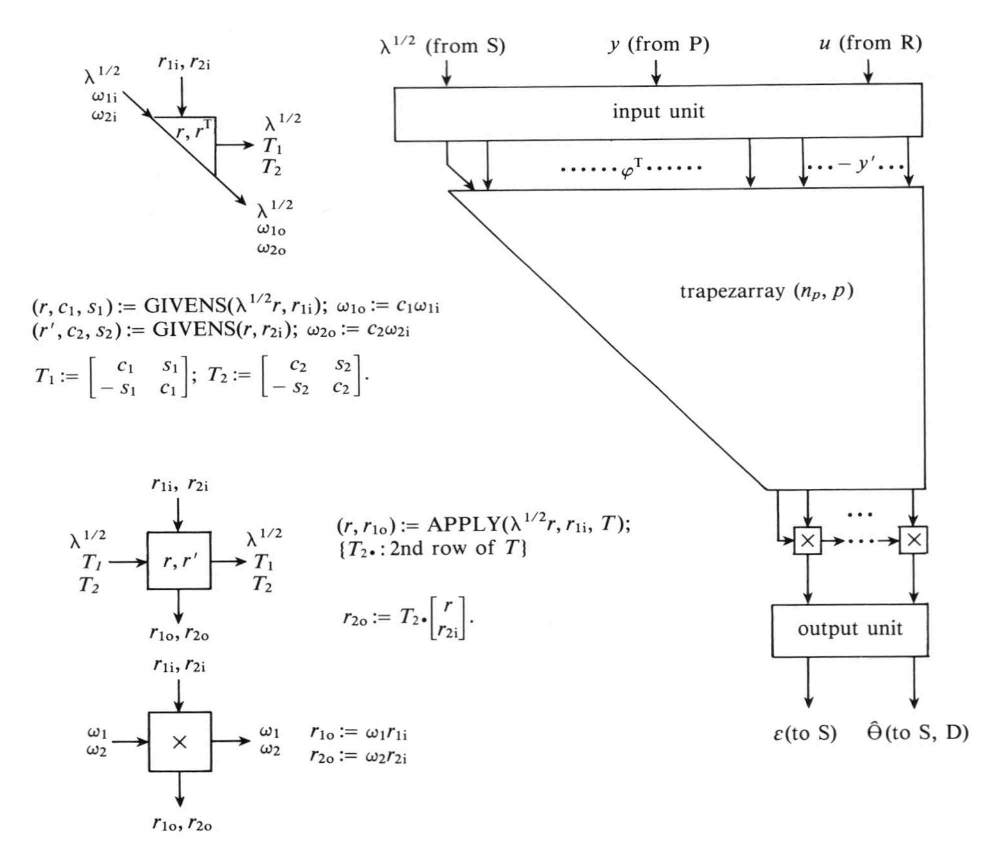

Figure 2.8 Systolic implementation of the recursive least-squares (RLS) estimator. The input unit provides data in the appropriate format for the systolic array operation, i.e. forms the data vectors $[\nu^{\mathrm{T}}(k) - y^{\mathrm{T}}(k)]$ from I/O data as well as the auxiliary vectors e_i^f $(i = 1, \ldots, n_p, 1, \ldots, n_p, \ldots)$ and inputs them in skewed fashion. The array continuously outputs (in skewed fashion) both the LS residuals $\varepsilon(k)$ and columns $\hat{\Theta}_{\bullet i}$ of the estimated parameter matrix $(i = 1, \ldots, n_p, 1, \ldots, n_p, \ldots)$. The functions of the diagonal, off-diagonal and scaling processors are detailed

A complete scheme for the estimator block is shown in Figure 2.8, where the processor functions required to incorporate all the above features are also detailed. The figures $\delta_i(E), \delta_o(E)$ and $\tau(E)$ characterizing the performance of the estimator implementation in accordance with the analysis of section 2.2, can easily be derived for this systolic implementation. Let T_E denote the clock period of the array, then

$$\delta_i(E) = T_E, \delta_o(E) = n_p T_E, \tau(E) = 2n_p T_E \qquad (2.38)$$

The clock period T_E is clearly determined by the computing time of the (slower) diagonal processor. Hence, for the specific unnormalized implementation of Figure 2.8, which used Givens rotations for both updating and parameter extraction,

$$T_E = 7T_x + 4T_/ + 2T_{\sqrt{}} + 2T_+ \qquad (2.39)$$

Note, however, that T_E can be drastically reduced by using a normalized implementation (based on fast Givens rotations) and performing parameter extraction using Gaussian elimination. In this case, the above computing time reduces to

$$T_E = 6T_x + T_/ + 2T_+ \qquad (2.40)$$

2.5 CONTROLLER DESIGN

In advanced control algorithms, the design of the control law is mostly commonly based on any one of the following three design approaches:

1. *Pole-zero placement* (Åström and Wittenmark, 1989).
2. *Linear quadratic (LQ) control* (Grimble, 1984).
3. *Predictive control* (Peterka, 1984; Clarke *et al.*, 1987).

The first approach shapes the closed-loop transfer function between the reference and the output. Conversely in LQ and predictive control, the design objective is the minimization of a quadratic cost-function, involving input and tracking error variances, extended over an infinite (LQ control) or finite (predictive control) horizon. In pole-zero placement and LQ control, computations are performed in the frequency domain by solving diophantine and/or spectral factorization polynomial equations. Alternatively, predictive control laws are computed in the time domain by solving a sequence of quadratic optimization problems (in particular, Riccati iterations).

Over the last few years, predictive control has become increasingly popular (Bitmead *et al.*, 1990) for several reasons. First, it has a simple control objective with only a few 'tuning knobs'. Secondly, it is flexible in that it can easily cope with time-varying situations, input–output constraints and knowledge of future reference behaviour. Further, it includes, to a certain extent, the other design approaches. Consequently, what follows considers the implementation of predictive control, by, in effect interpreting this design approach as a CLS optimization problem.

A (typical) predictive control problem is characterized by the following plant model, reference signal, cost function and terminal constraints.

Plant model

The plant is usually modelled by the discrete-time input/output representation (2.24), i.e.

$$A(d)y(t) = B(d)u(t) + e(t) \tag{2.41}$$

In the adaptive case, the coefficients of the polynomial matrices $A(\bullet)$ and $B(\bullet)$ are provided by the estimator.

Reference signal

It is assumed that the reference signal $r(t)$ can be modelled by an auto-regressive (AR) equation, i.e.

$$C(d)r(t) = w(t), \quad C(d) = I + \sum_{i=1}^{n_c} C_i d^i \tag{2.42}$$

where $w(t)$ is a white-noise sequence. This model allows for prediction of future samples of the reference. If no information is available on $r(t)$, one option is to set $C(d) = (1 - d)I_p$ which amounts to assuming that the best prediction of the future values $r(t+i), i > 0$, coincides with the actual value $r(t)$.

Cost function

At each sampling instant, the control signal $u(t)$ is chosen to 'minimize' the expected value of the multistep quadratic cost function:

$$J = \sum_{k=1}^{N} [\| y(t+k) - r(t+k) \|_{Q_y}^2 + \| u(t+k-1) \|_{Q_u}^2] \tag{2.43}$$

Here N is the control horizon and Q_u and Q_y are 'suitably chosen' nonnegative-definite weighting matrices.

Terminal constraints

In several predictive control algorithms the cost function (2.43) is minimized subject to terminal constraints of the form

$$\left.\begin{aligned} u(t+k) &= 0, & N-n_u \leqslant k \leqslant N \\ y(t+k)-r(t+k) &= 0, & N-n_y < k \leqslant N \end{aligned}\right\} \quad (2.44)$$

Here n_u and n_y denote the number of terminal constraints on the input and output respectively.

The following points concerning the above formulation are now noted.

Remark 2.8

Reference signal: as an alternative to (2.42), it may be assumed that the future values of the reference are known (a 'tracking' problem). If this is the case, the complexity of the control law increases with the control horizon since all the reference samples included in the cost function must be used in order to generate the optimal control.

Remark 2.9

Cost function: in order to be meaningful, this should be referenced to a control system with integrators in the forward path of the regulation loop so that, asymptotically, a zero-offset in the reference tracking can be obtained with a zero input. If these integrators are not present in the plant, they must be introduced by the controller. In this case $u(t)$ in (2.41) denotes the *incremental input*, i.e. the difference of the input at times t and $t-1$. Further, in order to improve tracking properties, the cost function may involve filtered versions of the input and output signals, i.e.

$$J = \sum_{k=1}^{N} [\| y_f(t+k) - r(t+k) \|_{Q_y}^2 + \| u_f(t+k-1) \|_{Q_u}^2] \quad (2.45)$$

where $y_f(t) = W_y(d)y(t)$ and $u_f(t) = W_u(d)u(t)$. Assuming, for simplicity, that the shaping filters $W_y(d)$ and $W_u(d)$ are scalar, it is possible to replace (2.41) with

$$A_f(d)y_f(t) = B_f(d)u_f(t) + e_f(t) \quad (2.46)$$

where

$$A_f(d) = A(d)W_u(d), B_f(d) = B(d)W_y(d), e_f(t) = W_u(d)W_y(d)e(t) \quad (2.47)$$

Equations (2.45) and (2.46) constitute a standard predictive control problem in the filtered variables $y_f(t)$ and $u_f(t)$.

Remark 2.10

Terminal constraints: these have been introduced in order to reduce computational complexity and improve closed-loop behaviour (in particular stability) even for 'relatively small' control horizons. In particular generalized predictive control (GPC) (Clarke *et al.*, 1987) is characterized by

$$n_y = 0 \tag{2.48}$$

and n_u is a 'design knob' to be chosen to trade-off performance against computational complexity. Conversely, the work of Clarke and Scattolini (1990) and Mosca *et al.* (1990) suggests using

$$n_y = n_u = \max\{n_a, n_b - 1\} \tag{2.49}$$

The extended horizon self-tuning control (EHSC) approach (Ydstie, 1984) simply sets

$$n_y = 1, n_u = 0 \tag{2.50}$$

Remark 2.11

Connections with other design criteria: as the control horizon N increases, the terminal constraints of (2.44) are superfluous, since they are automatically satisfied. Hence predictive control, in practice, encompasses LQ control and this relation becomes even tighter in the adaptive case. In fact, the solution of the predictive control problem is computed recursively with respect to the control horizon, proceeding backwards in time. Consequently, in the adaptive case, this recursive procedure can be performed indefinitely over the sampling intervals by employing new plant parameter estimates as soon as they become available. Then if the estimates eventually converge, the recursive optimization procedure yields the LQ control law. Additionally, by a proper choice of the dynamic weights $W_y(d)$ and $W_u(d)$, predictive control results in a combined pole-placement and LQ control design. In adaptive predictive control, the control signal is usually generated according to a *receding horizon* strategy, namely, at each sampling instant t, the input sequence $\{u(t), u(t+1), \dots, u(t+N-1)\}$ minimizing (2.43), possibly subject to (2.44), is computed but only the first input $u(t)$ is actually applied to the plant, the whole procedure being repeated at time $t+1$.

It can be shown that the control law which yields the optimal input $u(t)$ does not depend on the stochastic terms $e(t)$ in (2.41) and $w(t)$ in (2.42) provided they are assumed to be white. In this case only the deterministic counterpart of the problem needs to be considered.

In order to recast (2.41)–(2.44) as a CLS optimization problem, the following notation is introduced:

$$n := \max\{n_a, n_b - 1, n_c\}$$

$$l := 2p + m$$

$$z(t) = [y^{\mathrm{T}}(t)r^{\mathrm{T}}(t)u^{\mathrm{T}}(t-1)]^{\mathrm{T}} \in \mathscr{R}^l$$

$$s(t) = [z^{\mathrm{T}}(t)z^{\mathrm{T}}(t-1)\ldots z^{\mathrm{T}}(t-n+1)]^{\mathrm{T}} \in \mathscr{R}^{nl}$$

$$\Gamma_k := \begin{bmatrix} A_k & 0 & -B_{k+1} \\ 0 & C_k & 0 \end{bmatrix} \in \mathscr{R}^{(l-m)xl}, \quad k = 0, 1, \ldots, n, \; A_0 = C_0 = I_p$$

$$\Gamma := [\Gamma_0\Gamma_1\ldots\Gamma_n] \in \mathscr{R}^{(l-m)\mathrm{x}(n+1)l} \tag{2.51}$$

$$Q_y = R_y^{\mathrm{T}}R_y, Q_u = R_u^{\mathrm{T}}R_u\text{: } R_u \textit{ and } R_y \text{ upper} - \text{triangular}$$

$$R_z := \begin{bmatrix} R_y & -R_y & 0 \\ & 0 & 0 \\ & & R_u \end{bmatrix} \in \mathscr{R}^{lxl}$$

$$\Omega_z := \begin{bmatrix} I_p & & \\ & I_p & \\ & & I_m \end{bmatrix}, \Omega_u := \begin{bmatrix} I_p & & \\ & I_p & \\ & & 0_{mxm} \end{bmatrix}, \Omega_y := \begin{bmatrix} 0_{pxp} & & \\ & I_p & \\ & & I_m \end{bmatrix}, \Omega_{uy} := \Omega_u\Omega_y$$

Using this formulation, the plant model (2.41) and with the reference model (2.42) can be interpreted as the set of linear constraints on the samples of the joint process $z(t)$ of the form

$$\Gamma\begin{bmatrix} (t+k+1) \\ s(t+k) \end{bmatrix} = 0 \quad k \geqslant 0 \tag{2.52}$$

Hence, the general predictive control problem stated in this section is a special case of the CLS optimization problem (2.10)–(2.12) obtained by setting:

$$N_1 = Nl$$

$$N_2 = nl$$

$$M = N(2l - m)$$

$$x_1 = [z^{\mathrm{T}}(t+N)\ldots z^{\mathrm{T}}(t+2)z^{\mathrm{T}}(t+1)]^{\mathrm{T}}, \quad x_2 = s(t)$$

$$\Omega_1 = \begin{cases} \text{block} - \text{diag}\left\{\underbrace{\Omega_{uy}, \ldots, \Omega_{uy}}_{n_y \text{ times}}, \; \underbrace{\Omega_u, \ldots, \Omega_u}_{n_u - n_y \text{ times}}, \; \underbrace{\Omega_z, \ldots, \Omega_z}_{N - n_u \text{ times}}\right\}, & \text{if } n_u \geqslant n_y \\ \text{block} - \text{diag}\left\{\underbrace{\Omega_{uy}, \ldots, \Omega_{uy}}_{n_u \text{ times}}, \; \underbrace{\Omega_y, \ldots, \Omega_y}_{n_y - n_u \text{ times}}, \; \underbrace{\Omega_z, \ldots, \Omega_z}_{N - n_y \text{ times}}\right\}, & \text{if } n_u \leqslant n_y \end{cases}$$

$$\Omega_2 = 0 \tag{2.53}$$

$$R_{11} = \text{block} - \text{diag}\left\{\underbrace{R_z, \ldots, R_z}_{N \text{ times}}\right\}, \quad R_{12} = 0$$

$$[R_{21}R_{22}] = \begin{bmatrix} & & & \boxed{\Gamma} \\ & & \ddots & \\ & \boxed{\Gamma} & & \\ \boxed{\Gamma} & & & \end{bmatrix} \quad (l \leftrightarrow)$$

Note that the dynamic constraints (2.52) are expressed by the matrix $[R_{21} R_{22}]$, but the terminal constraints are expressed by the rows of $[R_{11} R_{12}]$ corresponding to the null diagonal entries of Ω_1 (i.e. diagonal zeros of Ω_u, Ω_y and Ω_{uy}).

Remark 2.12

The CLS representation (2.53) refers to the general formulation (2.41)–(2.44) and can be drastically simplified for special cases. For example, in regulation problems, i.e. when $r(t) = 0$ for all t, it is possible to set

$$l = p + m, \quad x(t) = [y^{\mathrm{T}}(t) u^{\mathrm{T}}(t-1)]^{\mathrm{T}}, \quad R_z = \text{block} - \text{diag}\{R_y, R_u\} \tag{2.54}$$

and modify $\Omega_z, \Omega_u, \Omega_y$ and Ω_{uy} accordingly.

The matrix

$$R = \begin{bmatrix} R_{11} & R_{12} \\ R_{21} & R_{22} \end{bmatrix}$$

in (2.53) possesses a special structure which permits major simplifications in the (generalized) triangularization algorithm. In fact, this algorithm can be partitioned into N steps, each step being devoted to the annihilation of a Γ-block in R_{21}. Starting this process at the bottom, the algorithm then reduces to a sequence of generalized transformations:

$$\underbrace{\begin{bmatrix} & \Gamma \\ U_{k+1} & 0 \\ & R_z \end{bmatrix}}_{\text{pre-array}} \rightarrow \underbrace{\begin{bmatrix} & 0 \\ \tilde{U}_k & \bar{U}_k \\ & U_k \end{bmatrix}}_{\text{post-array}}, \quad k = N-1, \ldots, 1, 0 \tag{2.55}$$

initialized by

$$U_N = \text{block} - \text{diag}\{\underbrace{R_z, \ldots, R_z}_{n \text{ times}}\} \tag{2.56}$$

The Ω matrices corresponding to (2.55), which have not been detailed here for simplicity, can be deduced from (2.53). Notice that *dynamic programming* is exploited here since the transformation (2.55) amounts essentially to a square-root version of the Riccati iteration. At the end of the annihilation procedure.

$$[\bar{R}_{11} \ \bar{R}_{12}] = \begin{bmatrix} \tilde{U}_{N-1} & \boxed{\bar{U}_{N-1}} & & & & \\ & & \cdots & & & \\ & & \vdots & & & \\ & & & \tilde{U}_1 & \boxed{\bar{U}_1} & \\ & & & & \tilde{U}_0 & \boxed{\bar{U}_0} \end{bmatrix} \tag{2.57}$$

Using Remark 2.2 the relation

$$\tilde{U}_k z(t+k+1) + \bar{U}_k s(t+k) = 0 \quad k = 0, 1, \ldots, N-1$$

defines the solution to the CLS problem (2.53). Further, by assumption, the matrices $\tilde{U}_k$ are nonsingular, and hence the optimal control law

$$u(t+k) = -F_k s(t+k) \quad k = 0, 1, \ldots, N-1 \tag{2.58}$$

where the control gains F_k are given by

$$\tilde{U}_k^{-1}\bar{U}_k = \begin{bmatrix} /// \\ F_k \end{bmatrix} \}m \tag{2.59}$$

and /// denotes entries which are of no interest.

Summarizing, therefore, the solution to the predictive control problem is provided by a sequence of transformations defined by (2.55). Further, the optimal control gains F_k can be obtained from the matrix $[\bar{R}_{11}\bar{R}_{12}]$ by the Faddeeva algorithm.

2.6 SYSTOLIC IMPLEMENTATION OF PREDICTIVE CONTROLLER DESIGN

This section addresses the systolic implementation of predictive controller design. The crucial point here is the mapping of the controller design algorithm onto an architecture of fixed size with respect to the control horizon N. The latter is, in fact, a design parameter in predictive control to be chosen, possibly on-line, to meet specific features of the controlled plant and the required control performance. Two different implementations will be developed.

The first implementation, referred to as *implementation I* below, uses the *orthogonally connected trapezarray* of Figure 2.4 as the basic building block. Here the partitioning with respect to N is achieved by passing data many times over the array, which requires that the array output is cylindrically connected to the input.

The second implementation, referred to as *implementation II*, uses a *hexagonally connected triarray* (triangular array). Here the partitioning is resolved by propagating data wavefronts in opposite directions. It will be shown that in order to improve the efficiency of this implementation, a suitable merging of contiguous processors and a coarser processor granularity must be adopted.

2.6.1 Implementation I: Orthogonally Connected Trapezarray

In the previous section it was shown that predictive controller design amounts to solving a suitable CLS optimization problem. Hence, in principle, it is possible to implement the overall design procedure on a *trapezarray* (Nl, nl) as explained in

Section 2.3. This solution would, however, be highly inefficient as it does not exploit in any way the specific structure of the matrix R. In addition, it would be highly impractical owing to the dependence of the array size on N.

In fact, it is possible to partition the algorithm to operate on a much smaller trapezarray (l, ln) resulting in a much higher utilization rate of the processors. This is schematically illustrated in Figure 2.9 where a Γ-block is completely annihilated by $n+1$ passes over the array. A requirement for this scheme to work is that the array must be equipped with a multiplexing device at the input to allow either feed-in from the host or feedback from the array itself, plus a feed-in/feed-out (FIFO) buffer of suitable size at the output to feed the output of the array back to the input.

To trade-off efficiency against throughput rate, it is feasible to link multiple, say q, *trapezarrays* (l, ln). In this case q consecutive Γ-blocks can be fed in from the host.

The architecture shown in Figure 2.10 can implement the overall predictive controller design procedure for an arbitrary horizon N as described in the following steps:

1. Initialize the trapezarray with the first q blocks of R_{11} and of Ω_1.
2. For cycle $i = 1, 2, \ldots, [N/q] - 1$ do:
   (a) feed in q Γ-blocks (spaced apart by l vertical positions);
   (b) re-initialize the array with the next q blocks of R_{11} and of Ω_1;
   (c) feedback $q(L + 2l)$ output vectors.

   Next cycle:

   (a) feed in the remaining ($\leqslant q$) Γ-blocks;
   (b) extract the regulator gain F_0 from the bottom sub-trapezarray (m, ln).

The array operation is therefore cyclic modulo $T_q := q(3l - m + L)$ and $T_q/q = 3l - m + L$ is the pipelining time (the inverse of the throughput rate), i.e. the

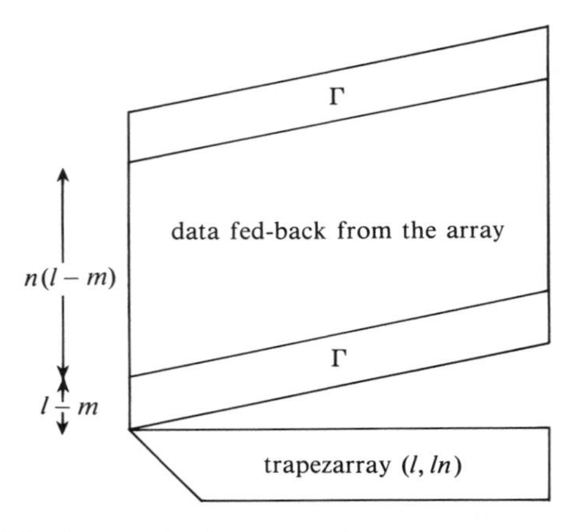

Figure 2.9 Partitioning predictive controller design on a trapezarray (l, ln)

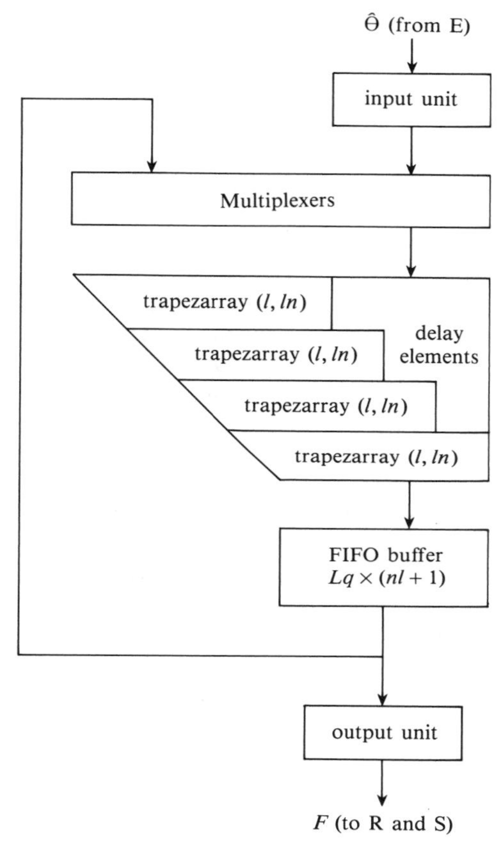

Figure 2.10 Overall architecture for partitioning predictive controller design on $q \geqslant 1$ ($q = 4$ in the figure) trapezarrays (l, ln) (implementation I)

time required to perform one iteration of the algorithm. Note that here time is measured in terms of clock periods.

To yield a correct partitioning, the FIFO buffer at the array bottom should provide enough storage capability to prevent overrun of data during the feed-in stage. In particular, given l, m, n and q, the parameter L (Lq is the FIFO length) should not be less than $\max\{0, [n/q](l-m) - 2l\}$. Hence the minimum pipelining time for given l, m, n and q is:

$$T_p = \max\{3l - m, (l-m)([n/q] + 1)\} \tag{2.60}$$

and the pipelining time (2.60) achieves its minimum value of $3l - m$ provided that q is 'suitably large'. The actual requirement is $q \geqslant q^*$ where:

$$q^* = \min_q \left\{ q \text{ integer: } [n/q] < \frac{2l}{l-m} \right\} \tag{2.61}$$

Efficiency is a function of l, m, n and q and, in particular,

$$\varepsilon = \frac{\text{pipelining time on a single processor}}{T_p \times \text{number of processors}}$$

$$= \frac{(l-m)(n+1)(ln+l+1)}{q(2ln+l+1)\max\{3l-m, (l-m)([n/q]+1)\}} \tag{2.62}$$

Examination of (2.60) and (2.62) shows that, for fixed (l, m, n), there is no reason for choosing q larger than q^* in (2.61). In fact, for $q \geqslant q^*$ the pipelining time remains constant ($T_p = 3l - m$) so that increasing q (i.e. the number of processors) beyond q^* produces no further improvement. Hence a good choice for the architectural design parameter q is $1 \leqslant q \leqslant q^*$; the closer q is to the lower bound, the higher the efficiency (the lower the throughput rate) and vice versa.

To give an indication of the performance achievable by implementation I, it is instructive to consider some specific cases. For SISO regulation ($l = 2$ and $m = 1$) the optimal throughput is $\frac{1}{5}$ of the iteration per clock period, $q^* = [n/4]$, and efficiency ranges over 40–60% depending on the values of n and q. For SISO tracking ($l = 3$ and $m = 1$) the optimal throughput drops to $\frac{1}{8}$, $q^* = [n/3]$, and the efficiency ranges approximately over the same interval. For two-input two-output tracking ($l = 6$ and $m = 2$) optimal throughput becomes $\frac{1}{16}$, $q^* = [n/3]$ and the efficiency again ranges approximately over 40–60%.

Hence it can be concluded that the efficiency of implementation I is not realistically affected by the parameters l and m. Note that the optimal throughput rate obtained for $q = q^*$, i.e., $1/(3l - m)$ equals the theoretical bound for the algorithm which results from the length of the critical path in its dependence graph. Hence, implementation I exhibits the best achievable throughput rate. In terms of efficiency, the utilization rate of the processors is about 50% on average, which is quite a satisfactory figure. Also notice that the efficiency is not significantly affected by l, m and n.

2.6.2 Implementation II: Hexagonally Connected Triarray

The hexagonally connected triarray of Figure 2.11 provides another basic module for performing predictive controller design. In contrast to implementation I, where the entries of the matrix $[R_{11} R_{12}]$ are stored in place, here they will be input from the left-hand side and propagated along the diagonals.

The required functions of the processors, both diagonal and off-diagonal, are virtually the same as in Figure 2.5 with a few modifications relating to the fact that, whilst one section of the data to be transformed is still received from the upper neighbour, the other piece is now received from the bottom-left neighbour instead of being stored in place. Care has also been exercised in programming the processor functions to ensure that data received at the diagonal input port of each processor

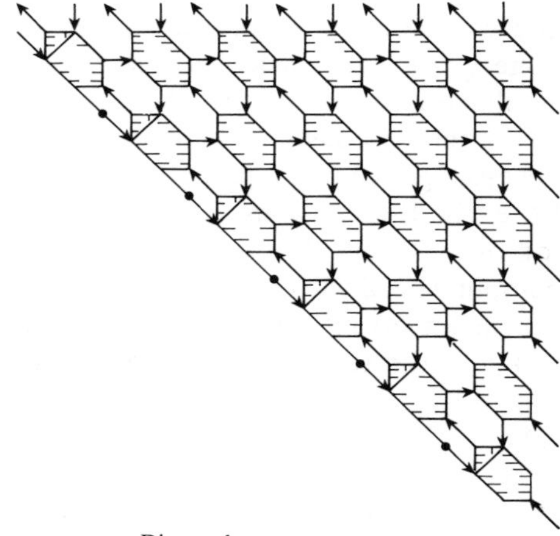

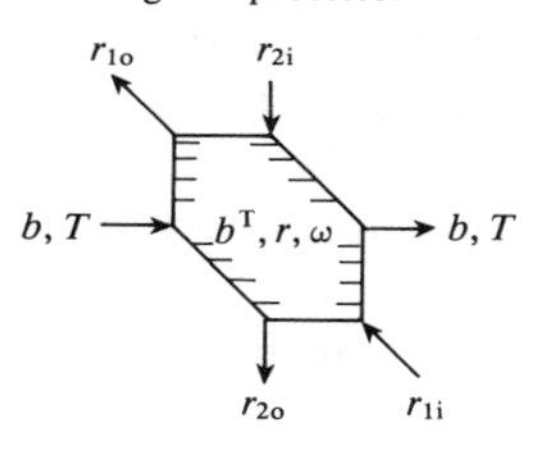


```
begin{off-diagonal processor}
case
  b = 0:  begin r1o := r1i; bT := b; exit end;
  b = 1:  begin
            if bT = 0 then (r, r2o) := APPLY(r1i, r2i, T)
                      else (r, r2o) := APPLY(r, r2i, T)
            bT := b;
            exit
          end;
  b = 2:  begin
            if bT = 0 then (r1o, r2o) := APPLY(r1i, r2i, T)
                      else (r1o, r2o) := APPLY(r, r2i, T)
            bT := b;
            exit
          end
endcase
end{off-diagonal processor}
```

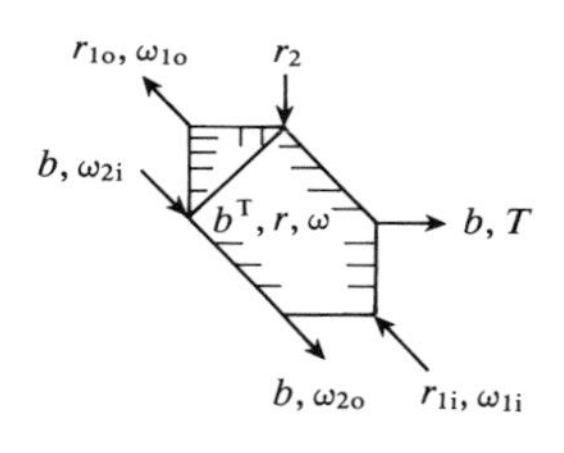


```
begin{diagonal processor}
{b: control
b = 0: current input item r2 absent;
b = 1: current input item r2 present, next present;
b = 2: current input item r2 present, next absent. }
case
  b = 0: begin r1o := r1i; ω1o := ω1i; bT := b? exit end;
  b = 1: begin
            if bT = 0  then (r, ω, ω2o, T) := GENERATE(r1i, r2, ω1i, ω2i)
                       else (r, ω, ω2o, T) := GENERATE(r, r2, ω, ω2i)
            bT := b;
            exit
         end;
  b = 2: begin
            if bT = 0  then (r1o, ω1o, ω2o, T) := GENERATE(r1i, r2, ω1i, ω2i)
                       else (r1o, ω1o, ω2o, T) := GENERATE(r, r2, ω, ω2i)
            bT := b;
            exit
         end
endcase
end{diagonal processor}.
```

Figure 2.11 Hexagonally connected triarray for predictive controller-design with the functions of the processing elements (implementation II)

are transformed in place with $l-m$ consecutive data pieces received at the top input port.

The predictive controller design algorithm can be mapped onto a triarray of $(n+1)l \times (n+1)l$ processors. Starting at time $t=0$, the entries of the matrix $[R_{11}R_{12}]$ are input from the left-hand side by diagonals, in a doubly skewed way, at the rate of one every three clock cycles, and are then propagated diagonally from right-bottom to left-up at the speed of one processor per clock cycle. Evolving from time $t=(n+1)l+1$, the entries of $[R_{21}R_{22}]$ (i.e. the Γ-blocks) are fed-in from the top, as in implementation I, at the rate of one Γ-block ($l-m$ vectors) every $4l-m-1$ clock cycles. In this way, the two computational wavefronts start meeting in the top row at time $t=(n+1)l-1$. Subsequent wavefronts coming from the left will meet the same wavefront at $t=(n+1)l+1$ in the second row, then at time $t=(n+1)l+3$ in the third row, and so forth.

Starting at time $t=(n+2)l-m-1$, the matrices $[\tilde{U}_k\bar{U}_k]$ will flow out diagonally from the top at the rate of one matrix every $4l-m-1$ clock cycles and, within each matrix, one (single skewed) row will be output every three clock cycles.

Note that each processor in the array is active for $l-m$ cycles over $4l-m-1$, hence the efficiency is

$$\varepsilon = \frac{l-m}{4l-m-1} \tag{2.63}$$

Further, the array can perform one iteration every $4l-m-1$ cycles, i.e.

$$T_p = 4l-m-1 \tag{2.64}$$

For SISO regulation it follows that $\varepsilon \approx 17\%$ and $T_p=6$; for SISO tracking, $\varepsilon = 20\%$ and $T_p=10$; while for two-input two-output tracking, $\varepsilon \approx 19\%$ and $T_p=21$. Hence, both efficiency and throughput rate turn out to be 'quite low'.

Efficiency can, however, be considerably increased by merging multiple processors into a smaller number. This is possible as all the processors remain inactive for most of the time. Consider, for example, the simple case of SISO regulation where each procesor works once every six clock cycles. It is then possible to merge three contiguous processors into a single one, as shown in Figure 2.12, and thus increase efficiency by a factor of three, i.e. $\varepsilon = 50\%$. Using Figure 2.12, it follows that each processor has six neighbours with which it must exchange data. Hence the resulting grid is hexagonally connected.

For SISO tracking, the situation is somewhat more complicated. To exploit the maximum achievable parallelism it is necessary to merge 3×3 triangles of contiguous processors into a cluster of three processors so that the efficiency is then 40%. The interconnection pattern between clusters is the same as the previous case. Further, to provide the correct interconnections within each cluster, all the three processors must be connected in both directions.

In the MISO (multiple-input single-output) case ($p=1$ and $m>1$), the above merging technique can be easily extended, where for MISO regulation ($l-m=1$) it is possible to merge an $l\times l$ triangle into a cluster of $[l/2]$ processors. Efficiency

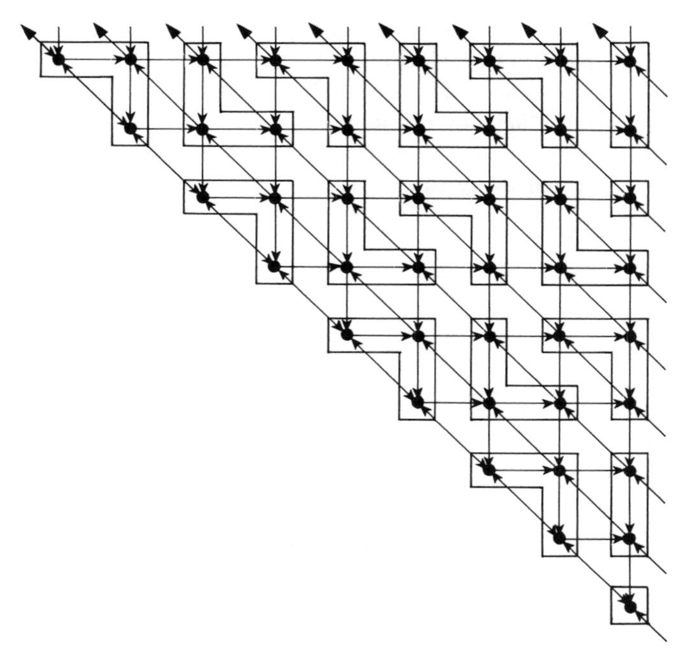

Figure 2.12 Merging three contiguous processors of the hex-connected triarray to increase efficiency

therefore increases by a factor l or $l+1$ depending on whether l is odd or even. Further, for MISO tracking $(l-m=2)$ it is possible to merge the same triangle into a cluster of l processors and hence increase efficiency by a factor of $(l+1)/2$.

As l is increased, it becomes more and more difficult to connect processors efficiently within each cluster so as to exploit parallelism locally, and, for multiple-output cases, the situation becomes even more intricate. To encompass the general MIMO case and to increase efficiency considerably, it is necessary to consider a coarser-grain implementation in which each processor sequentially performs block-processing of $l \times l$ blocks of data fed in from the diagonal input port with $(l-m) \times l$ blocks fed in from the top input port. Data must, of course, also be input in blocks of $l \times l$ from the left and in blocks of $(l-m) \times l$ from the top at the rate of one block every three clock cycles. It is easily seen that the resulting pipelining time is one iteration every three clock cycles where a clock period is now the time necessary to complete the annihilation of an $(l-m) \times l$ matrix with respect to an $l \times l$ matrix, i.e. measuring time in terms of elementary transformations

$$T_p = \frac{3}{2}(l-m)l(l+1) \tag{2.65}$$

Here each processor works once every three clock cycles, and therefore the efficiency is one-third, irrespective of the number of inputs and outputs. The same merging technique used in the SISO regulation case can be applied, i.e. three contiguous (block) processors can be merged into a single one so that, apart from the boundary processors, the overall efficiency is roughly 100%.

To conclude, it remains to be decided how to extract the regulator parameters F_k. This can be achieved using the matrix

$$[\tilde{U}_k \bar{U}_k] = [\tilde{U}_k \bar{U}_{k1} \bar{U}_{k2} \ldots \bar{U}_{kn}],\ \bar{U}_{ki} \in \mathscr{R}^{lxl},\ k = 0, 1, \ldots, N-1,\ i = 1, 2, \ldots, l$$

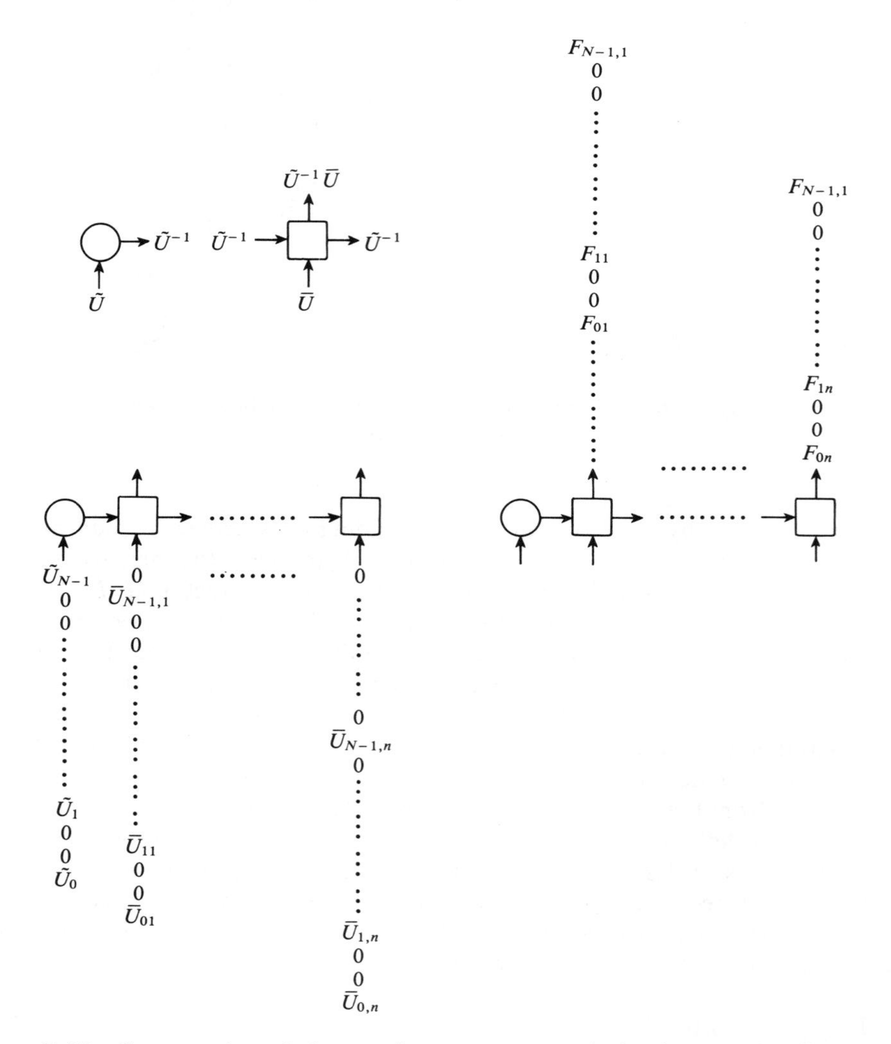

Figure 2.13 Computation of the regulator parameters in implementation II using an additional linear array (coarse-grain implementation II)

which flows out of the array top diagonally. In particular, suppose that this matrix is fed into a linear array as shown in Figure 2.13. Then the leftmost processor computes the inverse of $\tilde{U}_k$ and propagates it to the other processors on the right to be multiplied by $\bar{U}_k$ and thus produce F_k, which therefore emerges from the array at the top.

2.6.3 Performance Analysis

To conclude this section the performance of the two proposed implementations are summarized and compared. Implementation I achieves the optimal throughput rate at the expense of a quite complicated architecture involving multiplexers, FIFO buffers and cylindrical interconnections in addition to the array processor. The percentage utilization of the available computing power ranges over approximately 40–60% depending on the number of inputs, outputs and states (m, p and n), the architecture size (q), and on whether a regulation ($l = p + m$) or tracking ($l = 2p + m$) problem is considered. The efficiency of this implementation cannot, however, be improved by assigning coarser-grain tasks to the processors.

Alternatively, implementation II yields a lower (suboptimal) throughput rate by using a simpler architecture. The overall efficiency can be increased to roughly 100% by a suitable merging of contiguous processors and by a coarser granularity of the processor tasks.

For both implementations, the iteration interval of the controller design block is given by:

$$\delta(D) = \delta_i(D) = \delta_o(D) = T_p N T_D \tag{2.66}$$

and the latency $\tau(D)$ equals $\delta(D)$ plus a quantity which does not depend on N. In (2.66) the value of T_p is given by (2.60) for implementation I, and by (2.64) or (2.65) respectively for the finer- and coarser-grain versions of implementation II; the value of T_D, which is the computing time of the (slower) diagonal processor is approximately the same for both implementations. Hence $\delta(D)$ and $\tau(D)$ depend linearly on the design parameter N, and also on l and m. This is in contrast with the implementation of the regulator and of the estimator for which $\delta(R)$ and $\tau(R)$ (cf. (2.9)) plus $\delta_i(E)$ (cf. (2.37)) were constant. However, as noted in Section 2.2 whenever the application demands sampling rates faster than $1/\delta(D)$ a multirate implementation can be used in which the regulator parameters are not computed at each sampling interval.

2.7 CONCLUSIONS

This chapter has shown how systolic processing can be used to speed up adaptive control systems. As a starting point, the theoretical limits of parallelization have been investigated. These are due to the presence of computational loops, namely the regulation and the adaptation loops, which impose upper bounds on the sampling rate. Further, it is the more computationally intensive adaptation loop which provides the dominating bound. This analysis, however, has demonstrated how the adoption of a multirate implementation with different speeds in the regulation and in the adaptation loops, can make the sampling rate independent, to a certain extent, of the adaptation speed. Consequently this approach offers a viable solution whenever high sampling rates are demanded.

Using this analysis as a basis, the application of systolic processing to the adaptation loop has been studied. It has been shown how some algorithms commonly used for parameter estimation and controller design can be reduced to a common framework, i.e. sequential constrained least-squares minimization, for which a 'universal' systolic array has been proposed. This array can directly implement recursive least-squares estimation with 100% efficiency and full-rate pipelining of input data, producing parameter estimates at an $O(n_p^{-1})$ rate, where n_p is the regressor dimension.

Finally, the systolic implementation of predictive controller design procedures has been considered. It has been shown that a general class of predictive controller design methods can be regarded as a special constrained least-squares minimization with highly sparse matrices. This sparsity has been exploited in order to derive two novel systolic implementations both of which partition the design procedure with respect to the control horizon N, resulting in a high efficiency. By using $O(Nn^2)$ processors, both implementations reduce the processing time required from $O(Nn^2)$, for a sequential implementation, to $O(N)$, where n is the plant order.

REFERENCES

ÅSTRÖM, K. J. and WITTENMARK, B. (1989) *Adaptive control*, Reading, MA: Addison-Wesley.

BITMEAD, R., GEVERS, M. and WRTZ, V. (1990) *Adaptive optimal control*, Englewood Cliffs, NJ: Prentice Hall.

CHISCI, L. and ZAPPA, G. (1988) 'High-throughput parallel implementation of RLS', in *Proceedings of the 8th IFAC/IFORS Symposium on Identification and System Parameter Estimation, Beijing*, pp. 812–17.

CHISCI, L. and ZAPPA, G. (1991) 'A systolic architecture for iterative LQ optimisation', *Automatica*, **27**, 799–810.

CLARKE, D. W., MOHTADI, C. and TUFFS, P. C. (1987) 'Generalised predictive control:- Parts I and II. The basic algorithm', *Automatica*, **23**, 137–60.

CLARKE D. W. and SCATTOLINI, R. (1990) 'Constrained receding-horizon predictive control', *IEE Proc. Pt D*, **138**, no. 4, 347–54.

GENTLEMAN, W. M. (1973) 'Least-squares computations by Givens transformations without square roots', *J. Inst. Math. Appl.*, **12**, 329–36.

GENTLEMAN, W. M. and KUNG, H. T. (1981) 'Matrix triangularisation by systolic arrays', in *Proceedings of the SPIE Symposium on Real-Time Signal Processing IV*, Vol. 298. pp. 19–26.

GRIMBLE, M. J. (1984) 'Implicit and explicit LQG self-tuning controllers', *Automatica*, **20**, 661–9.

JOHNSON, C. R. JR (1990). *Lectures on adaptive parameter estimation*, Englewood Cliffs, NJ: Prentice Hall.

KUNG, S. Y. (1985) 'VLSI Signal Processing: from transversal filtering to concurrent array processing', in *VLSI and Modern Signal Processing* (eds S. Y. Kung, H. J. Whitehouse and T. Kailath), Englewood Cliffs, NJ: Prentice Hall, pp. 127–52.

KUNG, S. Y. (1988) *VLSI array processors*, Englewood Cliffs, NJ: Prentice Hall.

LJUNG, L. (1987) *System identification: theory for the user*, Englewood Cliffs, NJ: Prentice Hall.

McWHIRTER J. G. (1983) 'Recursive least-squares minimisation using a systolic array', in *Proceedings of the SPIE Symposium on Real-Time Signal Processing VI*, Vol. 430, pp. 105–12.

MOONEN, M. (1990) 'Jacobi-type updating algorithms', PhD Thesis, Katholieke Universiteit Leuven, Leuven, Belgium.

MOSCA, E., LEMOS, J. M. and ZHANG, J. (1990) 'Stabilizing I/O receding horizon control', in *Proceedings of the 29th IEEE International Conference on Decision and Control, Honolulu*, pp. 2518–23.

PETERKA, V. (1984) 'Predictor-based self-tuning control', *Automatica*, **20**, 39–50.

QUINTON, P. and ROBERT, Y. (1989) *Algorithmes et architectures sistoliques*, Paris: Masson.

ROGERS, E. and LI, Y. (1989) 'Concurrent array processing for linear multivariable feedback control systems', in *Proceedings of the IFAC Symposium on Adaptive Systems in Control and Signal Processing, Glasgow*, pp. 559–63.

VARVITSIOTIS A. P., THEODORIDIS, S. and MOUSTAKIDES, G. (1989) 'A novel structure for adaptive LS FIR filtering based on QR decomposition, in *Proceedings of the International Conference on Acoustics, Speech and Signal Processing-89*, Glasgow, pp. 904–7.

WARD, C. R., HARGRAVE, P. J. and McWHIRTER, J. G. (1986) 'A novel algorithm and architecture for adaptive digital beamforming, *IEEE Trans. on Acoustics, Speech and Signal Processing*, **34**, 338–46.

WOODS, R. F., KNOWLES, S. C., McCANNY J. V. and McWHIRTER, J. G. (1988) 'Systolic IIR filters with bit-level pipelining', in *Proceedings of the International Conference on Acoustics, Speech and Signal Processing*, New York, pp. 2072–5.

YDSTIE, B. E. (1984) 'Extended horizon self-tuning control', in *Proceedings of the 9th IFAC World Congress, Budapest*, Vol. 2, Hungary, pp. 915–19.

3

Parallel Processing for Kalman Filtering

R. W. Stewart

3.1 INTRODUCTION

The importance of the Kalman filter in control and signal processing applications can be measured by the very wide variety of applications available. Because the Kalman filter is a very computationally intensive algorithm, real-time implementation is often not possible with conventional microprocessor systems. Therefore parallel processing techniques must be used to make full use of the potential concurrency in the Kalman filter algorithm.

Using a set of states to represent a dynamic system, the Kalman filter is formulated using the state-space approach. With knowledge of the past and future values of the input, the future state and output of the system may be computed, given that the state contains all the necessary information about the system behaviour. The application of Kalman filter theory results in a set of difference equations which can be solved recursively. In fact each updated estimate can be computed from the previous estimate and the new input data, i.e. only the previous estimate must be stored. Hence there is no real need to store the entire past observed data. The Kalman filter algorithm is clearly more efficient than computing directly from the entire past observed data. Further, with appropriate modifications, it can be applied to both stationary and non-stationary environments.

Kalman filtering finds applications in target prediction and tracking, radar systems and many control applications. For most applications the computation is very demanding, and therefore parallel processing is the only way to achieve real-time performance. For example, satellite global positioning systems may require systems of order 10, and sampling rates of up to 100 kHz to 1 MHz. This requires a processing rate of gigaflops, which with today's technology can only be achieved with parallel processing.

In this chapter the Kalman filter is first presented, followed by a look at the potential concurrency in the Kalman update equations. When formulated as a least-squares computation the Kalman updates can be very efficiently implemented on a triangular array of processors as described in Section 3.5. This chapter also stresses the importance of numerical accuracy and stability of the Kalman filter, and therefore the matrix square root Kalman filter (SRKF) is considered with respect to its suitability for efficient parallel implementation. At the end of this chapter, *partitioning* of parallel arrays is discussed. Although algorithms can be successfully mapped onto parallel array designs, these designs are in the first instance only on paper, and the matrix dimension of the problem being solved is the same as the number of processors in the array. Therefore if 20×20 matrices are being manipulated, is it really necessary to have an array of $20 \times 20 = 400$ processors? Furthermore, is this practical? Partitioning large dimensioned problems onto smaller dimensioned parallel arrays is considered in more detail in Section 3.7.2.

3.2 THE KALMAN FILTER

The Kalman filter represents essentially a recursive solution of Gauss's original least-squares problem (Sorenson, 1970). The work of Kalman has been implemented in many wide and varied application domains, and a considerable number of theoretical studies and research programs have been inspired. The success of the Kalman filter can be put down to two main reasons: (a) Kalman filtering equations provide an extremely convenient procedure for digital computer implementation, using well-established numerical procedures; (b) Kalman stated and solved the estimation problem in a general framework that has an accordance with known results and problems. Kalman's classic paper 'A new approach to linear time filtering and prediction problems' was published in 1960 (Kalman, 1960).

Kalman filtering is widely applied in many signal processing and control applications. Limitations of some adaptive algorithms are that they do not make full use of all of the information available at the time of adaptation. Kalman (1960) provided stimulus and a formulation for the solution to a class of recursive minimum mean square estimation problems based on the entire past observed data (Papoulis 1984). In adaptive filtering, for example, Kalman filter theory is used to propose a new class of algorithms for obtaining rapid convergence of the weights of a transversal filter to their optimum settings (Haykin 1986). Kalman provides a much faster rate of convergence than, say, the least mean squares (LMS) algorithm at the expense of greatly increased computational complexity. The Kalman algorithm allows convergence in approximately $2M$ iterations as opposed to 20 M for the LMS (where M is the filter length). Quite simply, the Kalman algorithm converges faster because all of the available information from the start of the adaptive process up to the present time is exploited. LMS, on the other hand, relies only on currently available information in the tapped delay line. As the number of iterations approaches infinity the Kalman algorithm approaches the optimum Wiener solution. The LMS, however, is convergent in the mean as the iterations approach infinity, although then the ensemble average value and not the sample value of the tap weight vector estimate approaches the optimal Wiener value. The Kalman filter is inherently stable, whereas the LMS must have a step size chosen with certain bounds. Finally it should be noted that the Kalman filter pays the price for its performance, and is more complex in implementation than the LMS. Therefore parallel implementations of the Kalman filter are vitally important for many real-time applications.

3.2.1 Kalman Filtering Equations

The Kalman filter is a linear, discrete time, finite dimensional system that is an optimal minimum variance predictor (or estimator). The filter is formulated using a state-space approach, where an n-dimensional state vector, $\mathbf{x}(k+1)$, can be predicted recursively given each new m-dimensional measurement vector, $\mathbf{y}(k)$, in a discrete time-varying dynamic system:

$$\mathbf{x}(k+1) = F(k)\mathbf{x}(k) + \mathbf{w}(k) \tag{3.1}$$

$$\mathbf{y}(k) = C(k)\mathbf{x}(k) + \mathbf{v}(k) \tag{3.2}$$

where $F(k)$ is the state transition matrix and $C(k)$ is the measurement matrix of dimensions $n \times n$ and $m \times n$ respectively, $\mathbf{w}(k)$ is the n-dimensional system process noise vector and $\mathbf{v}(k)$ is the m-dimensional measurement noise vector. The best prediction of the state vector, based on the measurements up to stage k is denoted by $\hat{\mathbf{x}}(k+1/k)$. The noise vectors are assumed to be zero mean, independent processes with known covariance matrices $R_w(k)$ and $R_v(k)$ respectively. The noise $\mathbf{w}$ is assumed uncorrelated with $\mathbf{v}$ (i.e. $E[\mathbf{w}(i)\mathbf{v}(j)^{\mathrm{T}}] = 0$ for all i, j).

3.2.2 Covariance Filter

The conventional Kalman filter equations can be summarized in the following covariance form.

Time update process:

$$\hat{\mathbf{x}}(k+1/k) = F(k)\hat{\mathbf{x}}(k/k) \tag{3.3}$$

$$P(k+1/k) = F(k)P(k/k)F^{\mathrm{T}}(k) + R_w(k) \tag{3.4}$$

Measurement update process:

$$\hat{\mathbf{x}}(k/k) = \hat{\mathbf{x}}(k/k-1) - K(k)[\mathbf{y}(k) - C(k)\hat{\mathbf{x}}(k/k-1)] \tag{3.5}$$

$$P(k/k) = P(k/k-1) - K(k)C(k)P(k/k-1) \tag{3.6}$$

$$K(k) = P(k/k-1)C^{T}(k)[C(k)P(k/k-1)C^{T}(k) + R_v(k)]^{-1} \tag{3.7}$$

where P represents the (positive definite) covariance matrix such that:

$$P(k/k) = E[(\mathbf{x}(k) - \hat{\mathbf{x}}(k/k))(\mathbf{x}(k) - \hat{\mathbf{x}}(k/k))^{T}] \tag{3.8}$$

$$P(k+1/k) = E[(\mathbf{x}(k+1) - \hat{\mathbf{x}}(k+1/k))(\mathbf{x}(k+1) - \hat{\mathbf{x}}(k+1/k))^{T}]$$

Figure 3.1 shows the calculation ordering of the Kalman covariance equations for one iteration.

3.2.3 Information Filter

The filter can also be implemented to propagate P^{-1} (which can be easily found by applying the matrix inversion lemma (Anderson and Moore, 1979) to (3.5) to (3.7)), the so-called *information matrix* which relates to the recursive least-squares nature of filtering. Although the covariance and information filters are algebraically equivalent, the numerical properties can differ significantly. The

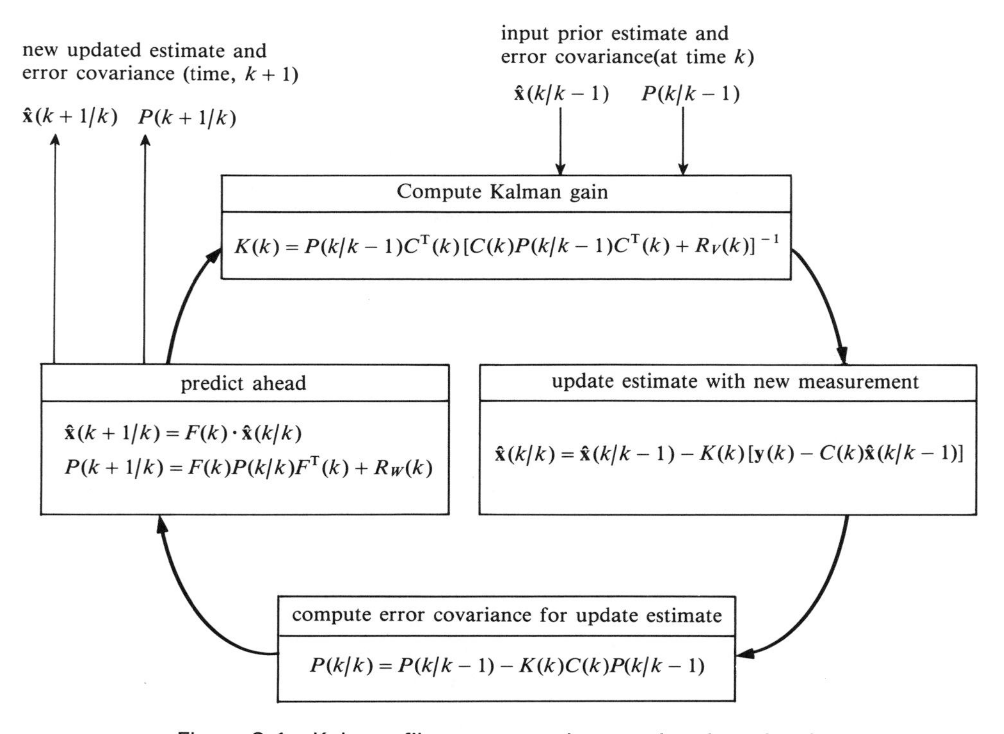

Figure 3.1 Kalman filter computations at time iteration k

suitability of one form or the other depends on the nature of the a priori information, and the dimensionality of the particular problem. The covariance form is the most popular because of its relative computational simplicity. The information filter on the other hand, has been successfully applied when either very poor or no a priori information leads to initialization problems in the covariance filter. To initialize the information filter all elements of P^{-1} are set to zero. The information filter measurement update equations can be found in Anderson and Moore (1979).

3.2.4 Numerical Properties of the Kalman Filter

A number of issues arise in the implementation of the covariance Kalman equations given in (3.3)–(3.7). Firstly the inverse of $[C(k)P(k/k-1)C^{\mathrm{T}}(k) + R_v(k)]$ must be computed. If the problem dimension is large then this could be a very time-consuming and unstable computation. In general, large matrix inversions are to be avoided. Also the propagation of error through the filter can result in a P that is *not positive definite*. Generally in the literature it is agreed that the Kalman filter is better performed by the matrix square root filtering algorithms (see Section 3.6) rather than a direct covariance approach (Anderson and Moore, 1979; Bierman, 1977; Morf and Kailath, 1975).

Numerical difficulties have occurred with the Kalman filter in problems with perfect or near perfect measurements and with singular or near singular covariance estimates. One approach to these problems is to model the system in such a way that the measurements and dynamic model are sufficiently noisy to prevent singularities from appearing. A more reliable solution is to perform some or all of the computations using extra precision arithmetic. This solution has obvious drawbacks in terms of additional processing required, which may affect the possibility of a real-time solution.

Another solution is to use computational methods that are better numerically conditioned. One such method is the *matrix square root Kalman filter*. This formulation of the Kalman filter has better numerical stability and accuracy than the standard filter (Bierman, 1977). (These algorithms found widespread and successful use in early space programs (Apollo lunar missions, Mariner 9 Mars.)) In general, however, matrix square root algorithms have received a cold reception probably owing to the conception that the techniques are too complicated to be implemented, especially in real-time. It is therefore appropriate to investigate parallel processing implementations of these algorithms. In Section 3.6 parallel implementations of the matrix square root Kalman filter will be investigated, and in Section 3.5 a least-squares based Kalman implementation (which is actually a formulation of the matrix square root information filter (Paige and Saunders, 1979)) will be presented.

3.3 DIRECT PARALLEL KALMAN FILTER

A first step in parallelizing the Kalman filter is to look directly at the covariance Kalman equations, and investigate what, if any, parallelism is achievable.

There are a number of ways of exploring parallelism of an algorithm. The simplest method is to study the algorithm using a dependency graph, where the computational tasks are shown in the order in which they must be performed (Katsikas *et al.*, 1991; Kung, 1988). The dependency of the computations of the Kalman filter equations, as shown in Figure 3.1, can be further detailed down to the matrix computation level as shown in Figure 3.2. Given this computation ordering of the various parts of the Kalman algorithm, an ordering schedule for the steps 1–15 can be realized. The operations to be performed are all matrix algebra, i.e.:

1. Matrix–vector multiplication (nodes 6, 8, 9).
2. Matrix–matrix multiplication (nodes 1, 2, 5, 11, 13, 14, 15).
3. Matrix or vector addition (nodes 3, 7, 10, 12).
4. Matrix inversion (node 4).

3.3.1 Covariance Kalman Latency

By the very nature of the Kalman filter, it is clear that it is a recursive algorithm. In Figure 3.2 there is clearly feedback from node (15) to (9), and from node (8) to (1). Therefore pipelining of the operations (1) to (15) is not possible. It is, however, possible to perform some of the operations concurrently, assuming the hardware is available. For example, nodes (9) and (10) can be performed independently from nodes (1) to (5) (note that node (6) is dependent on node (10)). The *critical* path for a dependence graph can be found by finding the node path, from input to output, that will require the longest execution time (latency). For Figure 3.2, the critical path is (1), (2), (3), (4), (5), (6), (11), (12), (13), (14), (15). The latency of this path defines the minimum execution speed possible for one iteration of the Kalman filter.

3.3.2 Parallel Array Implementation

It is obviously not practical to have 15 different arrays for the 15 different matrix algebra steps of the Kalman filter as the cost would be prohibitive, and no pipelining is possible. Therefore one parallel array of processors should share the 15 operations which are performed in sequence. (Note that to achieve maximum execution speed, two arrays would be required, one to execute the critical path computations, and the other to perform the remaining computations.)

The three matrix algebra operations to be performed are matrix–matrix multiplication, matrix–matrix addition and matrix inversion. The dimensions of the results of these computations could be $n \times n, m \times m, m \times n, n \times m, n \times 1, m \times 1$.

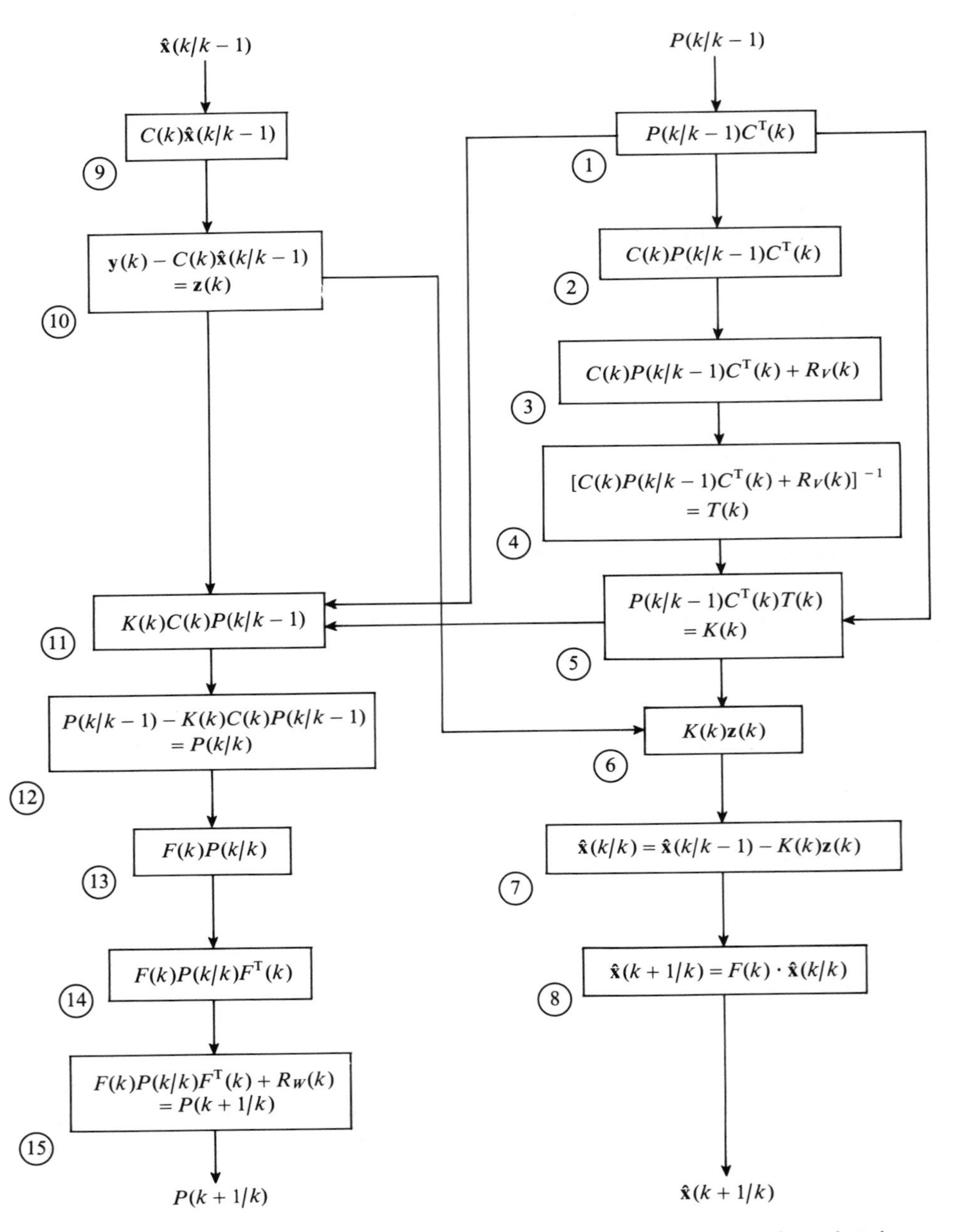

Figure 3.2 Kalman filter matrix computations dependency at time iteration k

Any array that is used must be of a large enough dimension to accommodate the largest computation. In Kung (1988) regular parallel arrays are given for the three matrix algebra operations required. Hence the regular array must be of dimension $\max(m \cdot n)$ (i.e. which ever is the largest, the m-dimensional measurement vector, or the n-dimensional state vector). A suitable parallel array is shown in Figure 3.3. Each of these processors must be capable of performing the multiply, add and division operations required in the matrix inversion, which is obviously more complex to implement than matrix multiplication or addition.

The array implementation of Figure 3.3 is not very efficient in its use of available resources. For example when a vector–vector addition, or even a matrix–vector

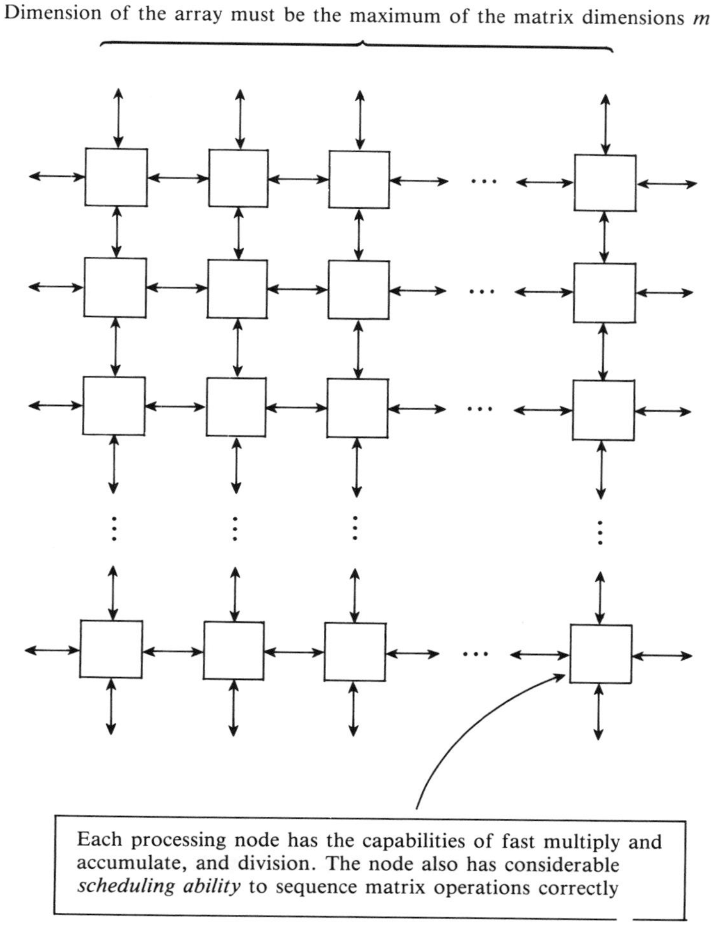

Dimension of the array must be the maximum of the matrix dimensions m or n

Each processing node has the capabilities of fast multiply and accumulate, and division. The node also has considerable *scheduling ability* to sequence matrix operations correctly

Figure 3.3 Parallel array to implement matrix equations of covariance Kalman filter

multiplication is being performed on the array, most of the processors will be idle. Furthermore, scheduling of the array will be particularly difficult; considerable inter-array communications and scheduling software will be necessary to ensure that the correct matrices are in the right place at the right time. The more nonlocal communications between processors, the less efficient and more complex the array will become.

3.3.3 Numerical Implementation

As with any numerical algorithm (parallel or sequential), round-off noise in the computer can lead to serious numerical difficulties when the number of iterations becomes very large. The simplest method of reducing round-off noise is to use high-precision arithmetic. Therefore in the array of Figure 3.3, each processor would use high-precision floating point arithmetic. Another common problem, is that the symmetric matrix P loses its symmetry. This problem can be reduced by taking the mean of elements that should be symmetric and writing the mean values back into the appropriate elements of the matrix, or simply calculating only the upper or lower half of the matrix. Note that this procedure introduces yet more nonlocal computation into the array.

It should also be pointed out that the operation of matrix inversion can be unstable and should be avoided if possible owing to the numerical instability that can occur. As the matrices to be inverted are symmetric definite, then the Cholesky method of matrix inversion can be used, which is considerably more numerically stable than straight matrix inversion. However, this once again introduces even more complexity and nonlocal computation into the parallel array.

As an alternative to the direct implementation, Yeh (1988) considered implementing the Kalman filter via the Faddeeva algorithm. This is still, however, essentially a direct implementation of the Kalman filter time and update equations. In Yeh's architecture Gaussian elimination is used which is very unstable unless some form of pivoting is used. (Golub and van Loan, 1983). Introducing pivoting, however, requires an architecture with nonlocal communication.

In conclusion the parallel Kalman filter based on the the covariance equations, although providing a speedup, is not a particularly efficient implementation, and suffers from the same numerical problems that would affect a sequential implementation, especially when the filter dimensions m and n are large. Therefore the rest of this chapter will discuss parallel Kalman filter arrays based on orthogonal matrix algebra, which is known to provide excellent numerical properties, and also has particularly *elegant* parallel array implementations.

3.4 ORTHOGONAL MATRIX ALGEBRA FOR KALMAN FILTERING

When Kalman filtering is formulated as a least-squares problem (see Section 3.5) or performed using the matrix square root approach (see Section 3.6), numerically robust and efficient parallel algorithms can be realized using orthogonal matrix (or linear) algebra techniques. In this section orthogonal matrix algebra will be reviewed and the key matrix algorithms of QR (using Givens transformations) and Cholesky decomposition will be introduced. These algorithms will then be used in Sections 3.5 and 3.6 to derive parallel regular array implementations for Kalman filtering.

The basis underlying orthogonal transforms is that they map a data set into another sparse data set that consists mostly of zeros, but is equivalent, in some sense, to the original data set. The sparse data set has the singular advantage that it is much easier to work with computationally, compared to the original data set. The equivalence of the original (dense) and transformed (sparse) data sets is that of *covariance invariance*. In other words, the covariance or correlation matrix is the same before and after orthogonal transformation of the data. For these transforms to be applicable to specific problems it is sufficient that the solution depend only on the data covariance. This is applicable in exact or recursive least-squares methods (Haykin, 1986; Sibul, 1987; McWhirter, 1983) and Kalman filtering (Kung and Hwang, 1991).

3.4.1 Least Squares Using QR Decomposition

The basic idea of least squares is that there is a set of real valued measurements $b_0, b_1, \ldots, b_{N-1}$ made at times $x_0, x_1, \ldots, x_{N-1}$, and the requirement is to construct a curve to fit these points in some optimum fashion. If the time dependence is $f(t_i) = A\mathbf{x}$ then the best fit is obtained by minimizing the sum of squares between $f(t_i)$ and x_i.

Consider the problem of finding a vector $x \in \mathscr{R}^N$ such that $A\mathbf{x} = \mathbf{b}$, where $A \in \mathscr{R}^{M\times N}$ and $\mathbf{b} \in \mathscr{R}^M$. The problem of least squares differs from general linear systems solvers in that there are more equations than unknowns. In this case $(M > N)$ and the system $A\mathbf{x} = b$ is overdetermined and has no exact solution, since b must be an element of the space spanned by the rows of A. The general approach to solve a problem of this form is to compute an $N \times 1$ vector of regression coefficients, $\mathbf{x}$, in order to minimize the sum of the squares of the elements of the $M \times 1$ residual vector $\mathbf{r}$, defined by:

$$\mathbf{r} = A\mathbf{x} - b \tag{3.9}$$

Minimizing the 2-norm of (3.9) is the least-squares approach and has exactly one solution:

$$\min_{\mathbf{x}} \| (A\mathbf{x} - \mathbf{b}) \|_2 \tag{3.10}$$

where (3.10) is a continuously differentiable function. The solution to (3.10) is well known to satisfy the normal equations:

$$(A^{\mathrm{T}}A)\mathbf{x} = A^{\mathrm{T}}\mathbf{b} \tag{3.11}$$

$$\mathbf{x} = (A^{\mathrm{T}}A)^{-1}A^{\mathrm{T}}\mathbf{b} \tag{3.12}$$

However, actually forming these equations numerically can often lead to $A^{\mathrm{T}}A$ being ill-conditioned, thus yielding an answer that may be inaccurate (Golub and van Loan, 1983). Again explicitly calculating the matrix inverse (3.12) is to be avoided for numerical reasons (Golub and van Loan, 1983).

Using the QR decomposition (3.10) can be reformulated to yield a more numerically stable and simpler set of equations to solve. An important property of the 2-norm is that it is invariant with respect to orthogonal transformations, i.e. for all orthogonal matrices Q and P (of appropriate dimension):

$$\| QAP \|_2 = \| A \|_2 \tag{3.13}$$

By premultiplying A by an orthogonal matrix Q^{T} such that:

$$A = QR \tag{3.14}$$

where R_N is an upper triangular square matrix:

$$R = \begin{bmatrix} R_N \\ 0 \end{bmatrix} \begin{matrix} N \\ M-N \end{matrix} \tag{3.15}$$

and Q is an orthogonal $M \times M$ matrix (i.e. $QQ^{\mathrm{T}} = I_M$), (3.10) can then be rewritten as:

$$\min_{\mathbf{x}} \| ((Q^{\mathrm{T}}A)\mathbf{x} - Q^{\mathrm{T}}\mathbf{b}) \|_2 \tag{3.16}$$

From (3.14), (3.16) can be simplified to:

$$\min_{\mathbf{x}} \| (R\mathbf{x} - \mathbf{b}') \|_2 = \min_{\mathbf{x}} \| (R\mathbf{x} - c) \|_2 + \| \mathbf{d} \|_2 \tag{3.17}$$

where

$$\mathbf{b}' = Q^{\mathrm{T}}\mathbf{b} = \begin{bmatrix} \mathbf{c} \\ \mathbf{d} \end{bmatrix} \begin{matrix} N \\ M-N \end{matrix} \tag{3.18}$$

Since the second term in (3.17) is constant, the least-squares estimate is therefore given by:

$$\mathbf{x}_{LS} = R^{-1}\mathbf{c} \tag{3.19}$$

which can be easily solved by back-substitution, since R is upper triangular.

Clearly this triangular system is solved for $\mathbf{x}$, without the need for forming the cross-product matrix, and with an accurate decomposition better answers are obtained. (Note in fact that R is the Cholesky factor (square root) of the matrix $A^{\mathrm{T}}A$ (cf. (3.11) and (3.14)) and noting the orthogonality of Q, i.e. $A^{\mathrm{T}}A = R^{\mathrm{T}}Q^{\mathrm{T}}QR = R^{\mathrm{T}}R$).

3.4.2 Givens Transformations

The QR algorithm for triangularization of an $M \times N$ matrix, A, using Givens transformations, performs a series of ordered plane rotations nullifying the subdiagonal elements of A, and finally reducing to an upper triangular form in least-squares solution (Golub and van Loan, 1983). For a nonsingular matrix A, the upper triangular matrix R can be obtained as follows:

$$Q^{T} \cdot A = R \tag{3.20}$$

$$Q^{T} = Q_{(M-1)} Q_{(M-2)} \dots Q_{(1)} \tag{3.21}$$

and

$$Q_{(n)} = Q_n^{(n)} Q_{n+1}^{(n)} \dots Q_{M-1}^{(n)} \tag{3.22}$$

where $Q_m^{(n)}$ is the Givens rotation to zero the $(m+1)$th row and nth column element by rotation with the mth row, and is of the form:

$$Q_m^{(n)} = \begin{bmatrix} 1 & \cdots & & & \cdots & 0 \\ \vdots & \ddots & & & & \vdots \\ & & c & s & & \\ & & -s & c & & \\ \vdots & & & & \ddots & \vdots \\ 1 & \cdots & & & \cdots & 0 \end{bmatrix} \begin{matrix} \\ \\ m \\ m+1 \\ \\ \\ \end{matrix} \tag{3.23}$$

(columns m, $m+1$)

Note that the matrix agrees with the identity matrix except in a principal 2×2 submatrix.

Examining only the updated mth and $(m+1)$th rows, and reducing Q to the appropriate 2×2 submatrix, the orthogonal rotation to zero the $(m+1, 1)$th element of matrix A is:

$$\begin{bmatrix} c & s \\ -s & c \end{bmatrix} \begin{bmatrix} a_{m,1} & a_{m,2} & a_{m,3} & \cdots & a_{m,N} \\ a_{m+1,1} & a_{m+1,2} & a_{m+1,3} & \cdots & a_{m+1,N} \end{bmatrix} \rightarrow \begin{bmatrix} \alpha_{m,1} & \alpha_{m,2} & \alpha_{m,3} & \cdots & \alpha_{m,N} \\ \alpha_{m+1,1} & \alpha_{m+1,2} & \alpha_{m+1,3} & \cdots & \alpha_{m+1,N} \end{bmatrix} \tag{3.24}$$

It is clear from (3.24) that to zero $a_{m+1,1}$ where the first matrix of (3.24) is orthogonal, c and s are set to:

$$c = \cos\theta = \frac{a_{m,1}}{r} \text{ and } s = \sin\theta = \frac{a_{m+1,1}}{r} \tag{3.25}$$

where

$$r = \sqrt{(a_{m,1}^2 + a_{m+1,1}^2)} \tag{3.26}$$

and the general row update equations can be written as:

$$\alpha_{m,i} = c a_{m,i} + s a_{m+1,i} \tag{3.27}$$

$$\alpha_{m+1,i} = -s a_{m,i} + c a_{m+1,i}, \qquad i = 1, \dots, N \tag{3.28}$$

$$\alpha_{jk} = a_{jk}, \qquad \text{if } j \neq m, m+1 \tag{3.29}$$

Note that

$$\alpha_{m,1} = \sqrt{(a_{m,1}^2 + a_{m+1,1}^2)} = r \tag{3.30}$$

In matrix–vector form the row update (3.27) and (3.28) can be written:

$$\begin{bmatrix} \alpha_{m,i} \\ \alpha_{m+1,i} \end{bmatrix} \leftarrow \begin{bmatrix} c & s \\ -s & c \end{bmatrix} \begin{bmatrix} \alpha_{m,i} \\ \alpha_{m+1,i} \end{bmatrix} \tag{3.31}$$

3.4.3 Parallel QR Triarray

The Givens transformation can be broken into two parts: (a) the Givens generation (GG) for the calculation of c, s and $r\ (= \alpha_{m,1})$ ((3.25) and (3.26)); and (b) the Givens rotation (GR) phase which is the rotation of other elements in the mth and $(m+1)$th rows ((3.27) and (3.28)).

Therefore in a recursive system, after A has been upper triangularized, R is augmented by a new row a:

$$\begin{matrix} x & x & x & x & c \\ & x & x & x & c \\ & & x & x & c \\ & & & x & c \\ & & & & x \\ \\ a & a & a & a & b \end{matrix} \tag{3.32}$$

The new matrix can be retriangularized by rotating the new row successively with the first, second, third and so on rows of R until the new row has entirely been eliminated. A further transform includes in the residual sum of squares whatever is left in the new row element of $\mathbf{b}$, and the transformed new row can be discarded. Note that if R is initialized to zero, then this is exactly the procedure used to upper triangularize the $M \times N$ matrix A.

By observing the independence between the transformation of the rows of the matrix A, and pipelining the operations, a parallel implementation can be realized. Gentleman and Kung (1981) first presented the parallel QR triangular topology parallel array (triarray) (which was later modified for recursive least squares by McWhirter, 1983). The triarray shown in Figure 3.4 uses two types of processor: the diagonal nodes of the array perform the GG, and all other nodes the GR computations. This triarray will form the key processing element of the orthogonal matrix algebra based Kalman implementations in Sections 3.5 and 3.6.

3.4.4 Square Root Free Givens Transformations

A number of authors have adopted square root free algorithms, reasoning that computation will be faster as the 'slow' square root computation has been circumvented. This has been prevalent in parallel Kalman filtering designs, where many authors have noted the desire to avoid square root computations in their calculations (Jover and Kailath, 1986; Bierman, 1977; Chisci and Mosca, 1987). As discussed in Section 3.4.5 this motivation is entirely unfounded, and is currently one of the great misinterpretations of the control and signal processing communities!

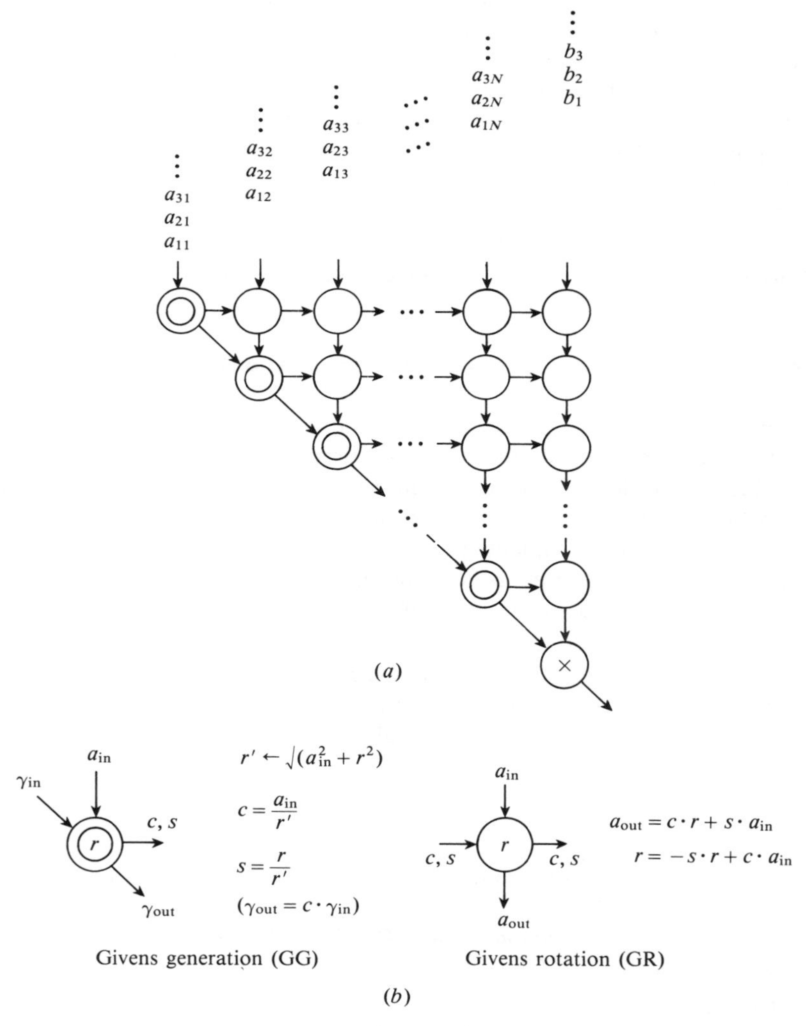

Figure 3.4 (a) QR triarray for least-squares computation; (b) GR and GG processor nodes

Gentlemen (1973) suggested a square root free (SRF) Givens transformation. The SRF Givens transformations are essentially a rearrangement of standard Givens, such that for each rotation, no arithmetic square roots are to be performed. Applying the SRF Givens transformation actually never explicitly forms matrix **R** in (3.14). In fact a diagonal matrix D, and a unit diagonal upper triangular matrix $\hat{R}$ such that:

$$R = D^{1/2}\hat{R} \tag{3.33}$$

are formed. $D^{1/2}$, however, is never explicitly formed: it is only necessary to calculate D and there is no need to use arithmetic square roots. The solution to the least-squares problem posed in (3.10) can be readily formed from the SRF Givens (Gentleman, 1973). There are two *popular* forms of Givens square root free transformations, which will be referred to as Givens I and Givens II. Givens II has fewer computations than Givens I, but at the expense of reduced numerical integrity (see Table 3.1).

The topology of the QR triarray in Figure 3.4 can also be used for square root free Givens rotations. The actual processor operations for the GG and GR operations will obviously be different (Gentleman, 1973).

3.4.5 Numerical Properties of Givens Transformations

In the development of a parallel Kalman filter array, it is vital that the most numerically stable and robust algorithms are used. Both standard and SRF Givens are amenable to parallel architectures, and the decision regarding which algorithm to use depends on the relative numerical attributes of the algorithms, and which algorithms are cheaper to implement.

It has been shown that standard Givens and SRF Givens I are numerically stable (Hammarling, 1974; Stewart *et al.*, 1989), although SRF Givens II is not. In addition to (and distinct from) stability, the problem of element growth leading to overflow and underflow must also be analyzed. Although an algorithm can often show a favourable error analysis, it must also have good scaled numerical properties, i.e. overflow/underflow of calculated values does not occur anywhere in the calculation.

Table 3.1 Arithmetic complexity of some Givens transformations

Algorithm	*Multiplications* GR	GG	*Divisions* GR	GG	*Square roots* GR	GG	*Stable*	*Overflow/ underflow*
Standard Givens	4	2	0	2	0	1	Yes	No
Square root free Givens I	3	3	0	2	0	0	Yes	Yes
Square root free Givens II	2	3	0	2	0	0	No	Yes

When using SRF Givens transformations I, II elements of the matrix D can easily double in magnitude on just one iteration. However, that largeness of elements in D is exactly counteracted by small magnitudes in other elements of the computation. Hence although the system shows a favourable error analysis, exponential element growth is happening at each iteration. This possibility of element growth requires that D is monitored throughout the computation, and scaling or pivoting must be performed in order to avoid overflow (Golub and van Loan, 1983). Clearly therefore all square root free Givens transformations (without scaling or pivoting) suffer from severe overflow problems. Note from (3.25)–(3.29) that no such overflow occurs with standard Givens (that utilizes the square root). Table 3.1 can be compiled showing the relative arithmetic complexities of standard Givens, and SRF Givens I and II. The table also summarizes (relative) numerical stabilities, and susceptibility to overflow/underflow.

In Stewart and Chapman (1990) and Stewart *et al.* (1989) it is shown that square roots can be implemented almost twice as fast, and in half of the silicon area of analogous divide operations. This was well known back in the early 1970s (Guild, 1970; Majithia and Kitai, 1971) long before the recent VLSI revolution. Many parallel application specific array designers have yet to realize this fact, and need to be made aware of it when choosing between standard and SRF Givens. One of the reasons for the general consensus that divisions are much simpler than square roots is that microprocessor benchmarks usually state the computation time for a square root, as up to N times that for a divide, where N is the word length. (Most people also have an ability to do mental arithmetic for divisions, but have little idea where to start for mental arithmetic square roots. Hence the conclusion that square roots are more complex than divides.) The reason for this is simply that to date there has not been a great demand for fast square roots on microprocessors, and hence their execution time has not been optimized. This, however, is changing; benchmarks for the new Texas TMS320C40 parallel processing DSP chip gives comparable execution times for both division and square root.

From Table 3.1 standard Givens transformations have the best numerical performance, and knowing that square roots are *cheaper* than divides, they have much simpler implementational complexity and have the better numerical integrity. SRF Givens, although once a useful algorithm, seems to have no relevance to parallel array processors.

3.4.6 Cholesky Decomposition

Cholesky decomposition is a very numerically well-conditioned algorithm and is very useful in the development of the least-squares parallel Kalman filtering array. This algorithm can also be shown to be numerically well behaved. Frequently in control and signal processing a linear system can be expressed such that $A \in \mathscr{R}^{N \times N}$ is a symmetric positive definite matrix. Symmetry is when $A^{\mathrm{T}} = A$ and positive definite is the condition that:

$$\mathbf{x}^{\mathrm{T}} A \mathbf{x} > 0 \tag{3.34}$$

for any arbitrary non-zero vector $\mathbf{x}$. Intuitively, symmetric matrices appear in problems whose laws are *fair*, i.e. each action has an equal and opposite reaction and the entry of i on to j is matched by j on to i. In signal processing covariance matrices derived from a stationary stochastic process are almost always symmetric positive definite (Papoulis, 1984).

If $A \in \mathscr{R}_{N \times N}$ and is symmetric positive definite, then there exists a lower triangular matrix $L \in \mathscr{R}^{N \times N}$ with positive diagonal elements such that:

$$A = LL^{\mathrm{T}} \tag{3.35}$$

This is called the Cholesky decomposition of A. (L is often referred to as the *matrix square root* of A.) The Cholesky algorithm is known to be very numerically stable (Golub and van Loan, 1983). From this decomposition the solution of $A\mathbf{x} = \mathbf{b}$ is obtained by solving the triangular system of equations:

$$\begin{aligned} Ly &= \mathbf{b} \\ L^{\mathrm{T}}\mathbf{x} &= \mathbf{y} \end{aligned} \tag{3.36}$$

for y and then x by using backsubstitution (Golub and van Loan, 1983).

Cholesky decomposition is also very useful for finding the inverse of symmetric matrices. (The backsubstitution process used above is implicitly forming the inverse.) To find A^{-1}:

$$A^{-1} = (LL^{\mathrm{T}})^{-1} = L^{-\mathrm{T}} L^{-1} \tag{3.37}$$

where L^{-1}, the inverse of a lower triangular matrix, is easily calculated (Golub and van Loan, 1983; Kung, 1988).

3.5 LEAST-SQUARES BASED KALMAN FILTER PARALLEL ARRAYS

This section presents a solution for Kalman filtering, based on a triangular array (triarray) architecture. By expressing the Kalman filtering problem in a least-squares formulation, orthogonal matrix decompositions can be used to solve the estimation problem. The design is suitable for both white and coloured noise cases, and has advantages in both numerical accuracy and computational efficiency.

The least-squares Kalman approach was proposed by Paige and Saunders (1979) and is essentially an expanded matrix iteration of the state-space representation of (3.2). Chen and Yao (1987) proposed a trapezoidal systolic array for both the measurement and time updates. Using a formulation similar to that of Gentleman and Kung (1981), this section presents a triangular array (triarray). This formulation is more efficient than the previous designs and was first presented by Kung and Hwang (1991).

3.5.1 Least-Squares Formulation

For the case of coloured noise, the noise covariance matrices $R_w(k)$ and $R_v(k)$ are not identity matrices. In order to use the least-squares formulation pre-whitening of the system and measurement noise vectors is necessary. By expressing the covariance matrices as $R_w^{-1}(k) = W(k)^T W(k)$ and $R_v^{-1}(k) = V(k)^T V(k)$, the upper triangular whitening operator matrices $W(k)$ and $V(k)$ can be realized. These matrices can be obtained by reverse Cholesky decomposition of $R_w(k)$ and $R_v(k)$ (Kung and Hwang, 1991).

By applying the pre-whitening operators and then collating the consecutive vectors of (3.2) up to the current stage, k, and the measurement vectors up to stage $k-1$, two accumulated vectors $\mathbf{X}(k)$ and $\mathbf{Y}(k)$ can be formed and the least-squares formulation obtained (Chen and Yao, 1987; Paige and Saunders, 1979):

$$\tilde{U}(k) = \tilde{A}(k)\mathbf{X}(k) + \tilde{\mathbf{Y}}(k) \tag{3.38}$$

where

$$\mathbf{X}(k) = [\mathbf{x}^T(1)\mathbf{x}^T(2) \ldots \mathbf{x}^T(k)]^T \tag{3.39}$$

$$\tilde{\mathbf{U}}(k) = [\tilde{\mathbf{w}}^T(0)\tilde{\mathbf{v}}^T(1)\tilde{\mathbf{w}}^T(1)\tilde{\mathbf{v}}^T(2) \ldots \tilde{\mathbf{v}}^T(k-1)\tilde{\mathbf{w}}^T(k-1)]^T \tag{3.40}$$

$$\tilde{\mathbf{Y}}(k) = [0\tilde{\mathbf{y}}^T(1)0\tilde{\mathbf{y}}^T(2) \ldots \tilde{\mathbf{y}}^T(k-1)0]^T \tag{3.41}$$

and

$$\tilde{A}(k) = \begin{bmatrix} W(0) & & & & & \\ \tilde{C}(1) & & & & & \\ \tilde{F}(1) & W(1) & & & & \\ & \tilde{C}(2) & & & & \\ & \tilde{F}(2) & W(2) & \cdots & & \\ & & \vdots & \ddots & & \\ & & & & W(k-2) & \\ & & & & \tilde{C}(k-1) & \\ & & & & \tilde{F}(k-1) & W(k-1) \end{bmatrix} \tag{3.42}$$

where $\tilde{F}(k) = -W(k)F(k)$, $\tilde{C}(k) = V(k)C(k)$, $\tilde{\mathbf{y}}(k) = -V(k)\mathbf{y}(k)$, $\tilde{\mathbf{w}}(k) = W(k)\mathbf{w}(k)$ and $\tilde{\mathbf{v}}(k) = -V(k)\mathbf{v}(k)$, which ensures that the noise covariance matrices are identity matrices. It is also assumed that the initial conditions are such that $\mathbf{x}(0) = \mathbf{0}$ and therefore $\tilde{\mathbf{w}}(1) = W(1)\mathbf{x}(1)$.

3.5.2 Recursive Least-squares Formulation

As a consequence of the whitening operators the vector $\tilde{\mathbf{U}}(k)$ in (3.38) now has a zero mean identity covariance matrix formulation. Hence the best predicted state vector $\hat{\mathbf{x}}(k/k-1)$ is the solution of (3.38) formulated as a least-squares estimation problem in the same way as (3.10):

$$\min_{\mathbf{x}} \| \tilde{A}(k)\mathbf{X}(k) + \tilde{\mathbf{y}}(k) \| \tag{3.43}$$

Applying an orthogonal matrix operator Q of dimension $((k-1)m+kn)\times((k-1)m+km)$ to realize the QR decomposition of $\tilde{A}(k)$ at stage k yields:

$$Q\tilde{\mathbf{U}}(k) = Q\tilde{A}(k) + Q\tilde{\mathbf{y}}(k) \tag{3.44}$$

where

$$[Q\tilde{A}(k) \mid Q\tilde{\mathbf{Y}}(k)] = \begin{bmatrix} R_{11} & R_{12} & & & & \mathbf{b}_1 \\ & R_{22} & R_{23} & \cdots & & \mathbf{b}_2 \\ & & \vdots & \ddots & & \vdots \\ & & & R_{k-1,k-1} & R_{k,k-1} & \mathbf{b}_{k-1} \\ & & & & R(k) & \mathbf{b}_k \\ 0 & 0 & 0 & \cdots & 0 & \mathbf{r}_1 \\ \vdots & \vdots & \vdots & \ddots & \vdots & \vdots \\ 0 & 0 & 0 & \cdots & 0 & \mathbf{r}_{k-1} \end{bmatrix} \tag{3.45}$$

Note that after the orthogonal transformation the noise vector $Q\tilde{\mathbf{U}}(k)$ remains white, and therefore the optimal predictor depends only on the vector $\mathbf{b}_k$:

$$\hat{\mathbf{x}}(k/k-1) = -R^{-1}(k)\mathbf{b}_k \tag{3.46}$$

Since $R(k)$ is upper triangular then (3.46) can be solved by back substitution. It has been shown (Paige and Saunders, 1979) that the matrix $R(k)$ is in fact the square root of the inverse error covariance matrix of the Kalman filter. This type of formulation is therefore equivalent to the *square root information filter* (SRIF) (Anderson and Moore, 1979).

At the next iteration the new measurement $\tilde{\mathbf{y}}(k)$ is available and the updated system equation for estimating $\hat{\mathbf{x}}(k+1/k)$ is given by:

$$\hat{\mathbf{U}}(k+1) = \hat{A}(k+1)\mathbf{X}(k+1) + \hat{\mathbf{Y}}(k+1) \tag{3.47}$$

where

$$\hat{\mathbf{U}}(k+1) = \begin{bmatrix} \mathbf{Q}\tilde{\mathbf{U}}(k) \\ \cdots \\ \tilde{\mathbf{v}}(k) \\ \tilde{\mathbf{w}}(k) \end{bmatrix} \tag{3.48}$$

and $[\hat{A}(k+1) \mid \hat{\mathbf{Y}}(k+1)] =$

$$\begin{bmatrix} R_{11} & R_{12} & & & & & \mathbf{b}_1 \\ & R_{22} & R_{23} & \cdots & & & \mathbf{b}_1 \\ & & \vdots & \ddots & & & \vdots \\ & & & R_{k-1,k-1} & R_{k,k-1} & & \mathbf{b}_{k-1} \\ & & & & R(k) & & \mathbf{b}_k \\ 0 & 0 & 0 & \cdots & 0 & 0 & r_1 \\ \vdots & \vdots & \vdots & \ddots & \vdots & \vdots & \vdots \\ 0 & 0 & 0 & \cdots & 0 & 0 & \mathbf{r}_{k-1} \\ - & - & - & - & - & - & - \\ 0 & 0 & 0 & \cdots & \tilde{C}(k) & 0 & \tilde{\mathbf{y}}(k) \\ 0 & 0 & 0 & \cdots & F(k) & W(k) & 0 \end{bmatrix} \tag{3.49}$$

To triangularize $\hat{A}(k+1)$ by QR decomposition, the matrix of interest is the $(2n+m) \times (2n+1)$ matrix consisting of $R(k)$ and the new system information, i.e. $\tilde{F}(k)$, $\tilde{C}(k)$ and $\tilde{\mathbf{w}}(k)$. Hence the QR decomposition of the submatrix yields:

$$Q_1 \begin{bmatrix} R(k) & \mathbf{0} & \mathbf{b}_k \\ \tilde{C}(k) & \mathbf{0} & \tilde{\mathbf{y}}(k) \\ \tilde{F}(k) & W(k) & \mathbf{0} \end{bmatrix} = \begin{bmatrix} \tilde{R}_{k,k} & \tilde{R}_{k,k+1} & \mathbf{b}_k \\ \mathbf{0} & R(k+1) & \mathbf{b}_{k+1} \\ \mathbf{0} & \mathbf{0} & \mathbf{r}_k \end{bmatrix} \tag{3.50}$$

where $R_{k,k}$ and $R(k+1)$ are upper triangular matrices and $\mathbf{r}_k$ is a residual (*don't care*) vector. Further, $\hat{\mathbf{x}}(k+1/k)$ can now be computed as in (3.46), i.e. $\hat{\mathbf{x}}(k+1/k) = R^{-1}(k+1)\mathrm{b}_k$.

3.5.3 A Triangular Parallel Array

A parallel array to perform the orthogonal QR decomposition in (3.50) is required. The matrix is already highly structured: $R(k)$ and $W(k)$ are upper triangular, and there are three $\mathbf{0}$ (zero) matrices. Kung and Hwang (1991) derived an $n \times n$ (a total of $n(n+1)/2$ processor nodes) triarray that could, in block sequence, process all six of the submatrices of the left-hand side of (3.50) to yield the right-hand side of this equation. The key operation of this array is QR decomposition, performed using Givens transformations as in Figure 3.4.

The general blocks of the submatrix to be nullified are shown in Figure 3.5. Rather than using a $2n \times (2n+1)$ triarray to nullify the entire matrix, the sparsity of the submatrix allows us to use only one $n \times n$ triarray by performing the overall triangularization operation in two phases:

1. Nullification of $[C^{\mathrm{T}}(k)F^{\mathrm{T}}(k)]$ by rotation with $R(k)$ on a QR triarray.
2. Post-rotation of $W(k)$ while retaining its triangular structure.

1a. *Nullifying $\tilde{C}(k)$*: the arriving matrix $\tilde{C}(k)$ is nullified by rotating with the resident triarray matrix $R(k)$ as shown in Figure 3.6.

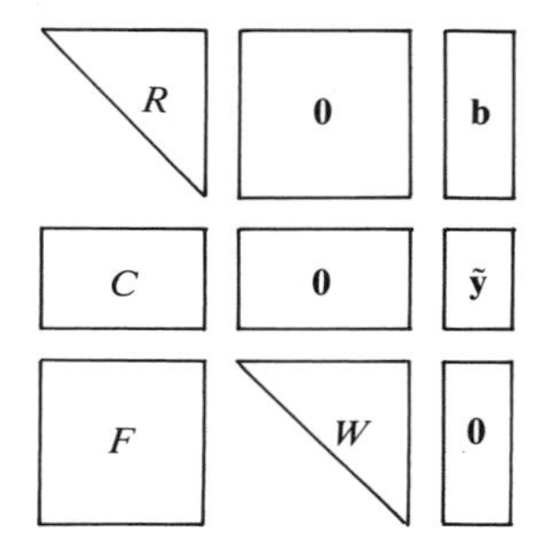

Figure 3.5 Blocks of submatrix to be triangularized

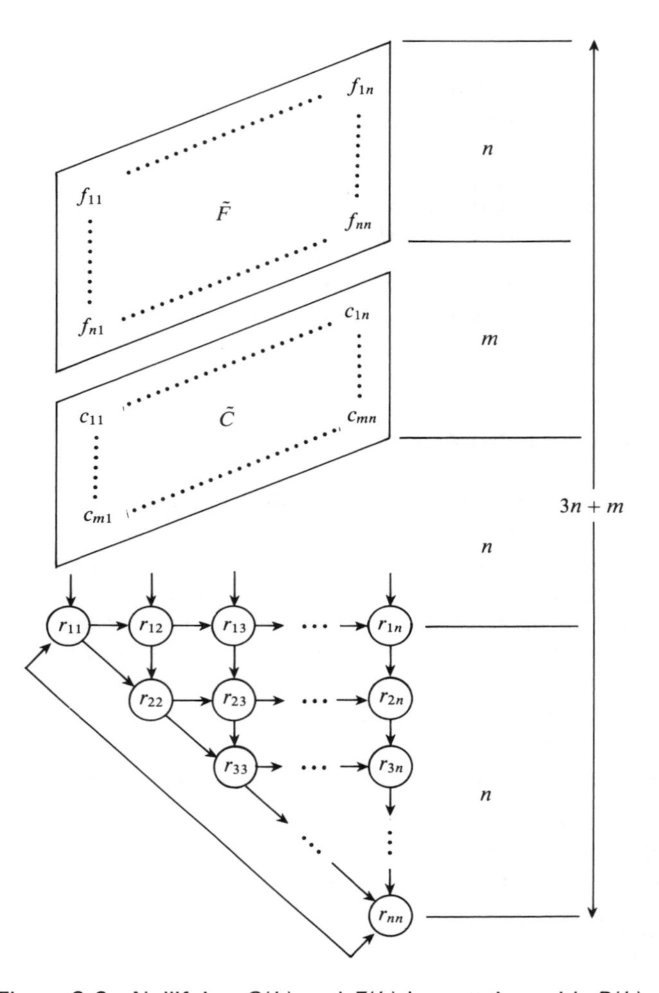

Figure 3.6 Nullifying $C(k)$ and $F(k)$ by rotation with $R(k)$

1b. *Nullifying* $\tilde{F}(k)$: after m time units the nullification of $\tilde{F}(k)$ will continue after the operation on $C(k)$. In this operation the diagonal nodes perform the Givens generation as normal and pass the angle parameters θ_{ij} on to the right neighbour node. However, the rotations of the $\tilde{F}(k)$ rows must also be applied to the rows of $W(k)$ and hence angle rotation parameters must be stored. Note from Figure 3.6 that $\tilde{F}(k)$ is nullified in a bottom-up order to ensure that $W(k)$ remains upper triangular.

2. *Post-rotation* of $W(k)$: after $3n + m$ time steps, the nullification of $\tilde{C}(k)$ and $\tilde{F}(k)$ is complete and the θ_{ij} for the post-rotation are available. At this time $W(k)$ has already been loaded into the array and the rotation

parameters θ_{ij} become available for use. There is a duality relationship between triangularization of $[R^{\mathrm{T}} F^{\mathrm{T}}]^{\mathrm{T}}$ and the post-rotation of $[O^{\mathrm{T}} W^{\mathrm{T}}]^{\mathrm{T}}$. Hence $W(k)$ matrix is now post-rotated with $Z(k)$, which is initialized to a zero matrix and realizes its value on flowing through the triarray in a total of $2n$ time steps as shown in Figure 3.7. Note that upward and rightward edges are required in this triarray.

The best prediction vector $\hat{\mathbf{x}}(k+1/k)$ can now be solved for on a back substitution array. After $4n+m$ time units the next recursion can begin (i.e. $\tilde{C}(k+1)$ can enter the triarray). Figure 3.8 shows the overall flow of data for the operation of the Kalman filter on the triarray.

3.5.4 Noise Pre-whitening

Very often the system noise $w(k)$ and the measurement noise $v(k)$ will be white, and therefore the covariance matrices $R_w(k)$ and $R_v(k)$ are simple diagonal matrices (i.e. all off-diagonal elements are zero):

$$\begin{aligned} R_w(k) &= \mathrm{diag}\,[w_1\ w_2\ \dots\ w_n] \\ R_v(k) &= \mathrm{diag}\,[v_1\ v_2\ \dots\ v_m] \end{aligned} \tag{3.51}$$

Therefore the derivation of the whitening operators $W(k)$ and $V(k)$ is simply

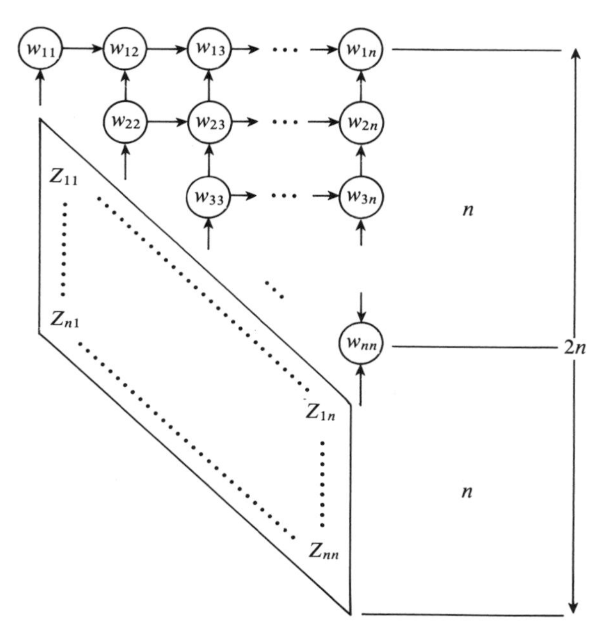

Figure 3.7 Post-rotation of $W(k)$ with $Z(k)$

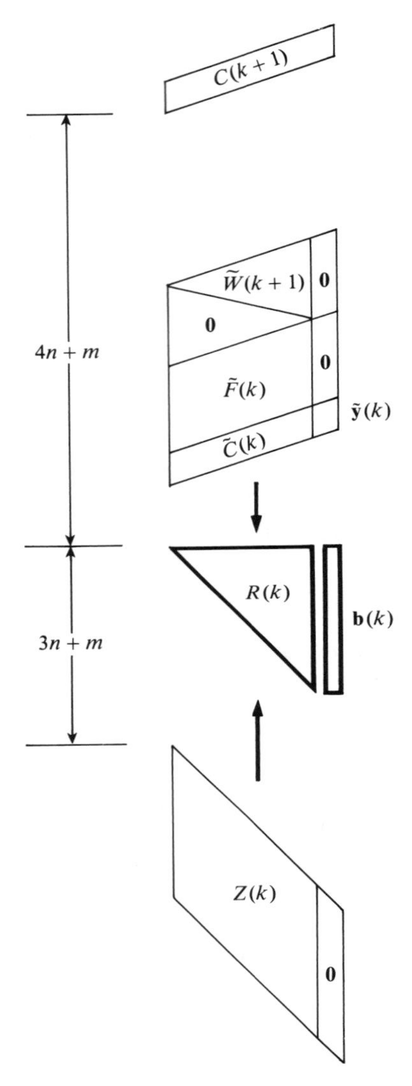

Figure 3.8 Overall scheduling for the Kalman triarray

performed by square root and division operations on each element of the diagonal:

$$W(k) = \mathrm{diag}\,[w_1{}^{1/2}\; w_2{}^{1/2}\; \ldots\; w_n{}^{1/2}] \tag{3.52}$$

$$V(k) = \mathrm{diag}\,[v_1{}^{1/2}\; v_2{}^{1/2}\; \ldots\; v_m{}^{1/2}] \tag{3.53}$$

For applications which have coloured noise, the reverse Cholesky decomposition can be performed on the triarray. Kung and Hwang (1991) detail this array.

3.5.5 Alternative Designs

Chen and Yao (1987) were first to present a parallel architecture based on a least-squares formulation of Kalman filter. The array was rectangular, and used the scalar square root free Givens II algorithm, which is actually unstable. However, their array could be easily modified to use standard (and inherently stable) Givens.

Gaston and Irwin (1989) also derived a parallel array based on the least-squares approach of Paige and Saunders (1979). Their array is also rectangular, and uses Givens transformations for orthogonal transformation. The parallel array has unidirectional data movement, but has a particularly large number of feedback loops around the array.

3.6 SQUARE ROOT KALMAN FILTERING

Another parallel array for Kalman filtering can be derived based on the *matrix square root Kalman filter* (Bierman, 1977). In many filtering problems propagation of the error covariance matrices in the Kalman filter equations can result in a matrix that is not positive semidefinite. This occurs either when a linear combination of the state vectors are known with high precision but others are virtually unobservable, or when the covariance matrix is rapidly reduced by processing very accurate measurements. Both situations can lead to numerical problems due to ill-conditioned quantities. The numerical accuracy degradation is often accompanied by a computed covariance matrix that loses its nonnegative definiteness. Reformulating the Kalman algorithm in terms of a matrix square root (or Cholesky factor) of the error covariance will preserve the nonnegative definiteness of the computed covariance. This method of propagating the matrix square root covariance matrix, rather than the actual covariance, to the next iteration is completely successful in maintaining the positive semi-definiteness of the error covariance matrix. Kaminski *et al.*'s (1971) work verified the excellent numerical characteristics of the matrix square root filter. It should be noted that the matrix square root filter does require more computation than the standard covariance filter; however, based on its numerical attributes and suitability to parallel implementation, this is still a desirable solution.

3.6.1 Numerical Properties

Propagating the matrix square root of the covariance (the Cholesky factor), $P^{1/2}(=P^{1/2}(k+1/k))$ will ensure that the covariance matrix is nonnegative definite:

$$P = P^{1/2}P^{\mathrm{T}/2} \tag{3.54}$$

where $P^{1/2}$ is a lower triangular matrix, and the product of $P^{1/2}P^{T/2}$ can never be indefinite even in the presence of round-off noise. Also the numerical conditioning of $P^{1/2}$ is generally much better that that of P. Considering the condition number of P denoted by

$$\varkappa(P) = \frac{\lambda_1}{\lambda_n} \tag{3.55}$$

where λ_1^2 is the maximum and λ_n^2 is the minimum eigenvalue of PP^T. Then when computing in m-digit arithmetic problems can be expected as $\varkappa(P)$ approaches 2^m. The real advantage of the matrix square root approach is clear from considering that:

$$\varkappa(P) = \varkappa(P^{1/2}P^{T/2}) = (\varkappa(P^{1/2}))^2 \tag{3.56}$$

i.e. the condition number of $P^{1/2}$ is the scalar square root of $\varkappa(P)$ and therefore the matrix square root formulation of a problem will not expect problems until $\varkappa(P) = 2^{2m}$. Therefore the numerical precision has been effectively doubled, and this *extra* precision will manifest itself as better numerical behaviour.

3.6.2 Cholesky Factorization of the Kalman Equations

The basic idea of the matrix square root filter is the propagation of the matrix square root covariance $P^{1/2}(k+1/k)$, to the next filter iteration. Consider again the time update arising in (3.1) and (3.4):

$$P(k+1/k) = F(k)P(k/k)F^T(k) + R_w(k) \tag{3.57}$$

The time update is solved by finding any orthogonal matrix Q_1 to triangularize the array:

$$[R_w^{1/2}(k) \quad F(k)P^{1/2}(k/k)] \tag{3.58}$$

such that

$$Q_1[R_w^{1/2}(k) \quad F(k)P^{1/2}(k/k)] = [\mathbf{0} \quad P^{1/2}(k+1/k)] \tag{3.59}$$

The existence of a suitable matrix Q_1 is well known and hence a complete proof is omitted, except to note that a basic element is

$$w = [R_w^{1/2} \quad FP^{1/2}] \begin{bmatrix} R_w^{T/2} \\ P^{T/2} \cdot F^T \end{bmatrix} \tag{3.60}$$

or

$$w = R_w + FPF^T = P_+ \tag{3.61}$$

where time indices at time k have been dropped for notational convenience (i.e. $R_w(k) = R_w$) and $P(k/k-1) = P$, $P(k/k) = P_k$, $P(k+1/k) = P_+$.

Similarly the matrix square root formulation can be found for the measurement update (3.5)–(3.7):

$$\begin{aligned} P(k/k) &= P(k/k-1) - K(k)C(k)P(k/k-1) \\ &= P(k/k-1) - P(k/k-1)C^{\mathrm{T}}(k)R_e^{-1}(k)C(k)P(k/k-1) \end{aligned} \tag{3.62}$$

where

$$R_e(k) = C(k)P(k/k-1)C^{\mathrm{T}}(k) + R_v(k) \tag{3.63}$$

Noting that the right-hand side of (3.60) is a Schur complement (Golub and van Loan, 1983), then there exists an orthogonal decomposition $\mathbf{Q}_2$ such that:

$$Q_2\begin{bmatrix} R_v^{1/2}(k) & C(k)P^{1/2}(k/k-1) \\ \mathbf{0} & P^{1/2}(k/k-1) \end{bmatrix} = \begin{bmatrix} R_e^{1/2}(k) & \mathbf{0} \\ P(k/k-1)C^{\mathrm{T}}(k)R_e^{-T/2}(k) & P^{1/2}(k/k) \end{bmatrix} \tag{3.64}$$

Again, the existence of a suitable matrix Q_2 is well known and hence a complete proof is omitted, except to note that it makes use of the following:

$$\begin{bmatrix} R_e^{1/2} & \mathbf{0} \\ PC^{\mathrm{T}}R_e^{-\mathrm{T}/2} & P_k^{1/2} \end{bmatrix} \begin{bmatrix} R_e^{\mathrm{T}/2} & R_e^{-1/2}CP^{\mathrm{T}} \\ \mathbf{0} & P_k^{\mathrm{T}/2} \end{bmatrix} = \begin{bmatrix} R_e & CP^{\mathrm{T}} \\ PC^{\mathrm{T}} & PC^{\mathrm{T}}R_e^{-1}CP^{\mathrm{T}} \end{bmatrix} \tag{3.65}$$

and

$$\begin{bmatrix} R_v^{1/2} & CP^{1/2} \\ \mathbf{0} & P^{1/2} \end{bmatrix} \begin{bmatrix} R_v^{\mathrm{T}/2} & \mathbf{0} \\ P^{\mathrm{T}/2}C^{\mathrm{T}} & P^{\mathrm{T}/2} \end{bmatrix} = \begin{bmatrix} R_v + CP^{\mathrm{T}}C & CP^{\mathrm{T}} \\ PC^{\mathrm{T}} & P_k \end{bmatrix} \tag{3.66}$$

From (3.59) and (3.64), the submatrices $P(k/k-1)C^{\mathrm{T}}(k)R_e^{-1/2}(k)$, and $R_e^{1/2}(k)$ produced by the orthogonal transformations can be used to form the Kalman update equation. Note that $R_e^{-1/2}(k)$ is required but is not explicitly produced by (3.59) and (3.64). $R_e^{1/2}(k)$ is, however, available from (3.64). Therefore as $R_e^{1/2}$ has been preserved to be upper triangular, it can be easily inverted using a triangular array (Kung, 1988), and the Kalman update equation is:

$$\begin{aligned} \hat{\mathbf{x}}(k/k+1) &= F(k)\hat{\mathbf{x}}(k/k-1) \\ &\quad - F(k)P(k/k-1)C^{\mathrm{T}}(k)R_e^{-1/2}(k)R_e^{-1/2}(k)[\mathbf{y}(k) - C(k)\hat{\mathbf{x}}(k)] \end{aligned} \tag{3.67}$$

3.6.3 Parallel Square Root Kalman Filter Array

A parallel array for square root Kalman filtering should be general purpose as it requires the ability to perform the following matrix computations:

1. Cholesky decomposition to find $R_v^{1/2}$ and $R_w^{1/2}$ for use in (3.51) and (3.59).
2. The matrix multiplication required to set up the submatrices in (3.59) and (3.64): $F(k)P^{1/2}(k/k)$, and $C(k)P^{1/2}(k/k-1)$.
3. The orthogonal decomposition to perform the measurement update of (3.64) generating $P(k/k-1)C^{\mathrm{T}}(k)R_e^{-1/2}(k)$ and $R_e^{1/2}(k)$.

4. The orthogonal rotation of (3.59) to generate the predicted error covariance $P^{1/2}(k+1/k)$.
5. The matrix inversion of the upper triangular matrix $R_e^{1/2}(k)$.
6. The matrix multiplication required in (3.67).

The above matrix computations have various data dependencies, and therefore must be performed in a specific sequence. Hence there is no advantage to having a dedicated array for each of the matrix computations. A more cost-effective solution would be to have one array performing the $(m, n) \times (m, n)$ matrix computations in parallel and pipelining operations 1 to 6 according to an efficient schedule. This array will need to be of a big enough dimension for the largest computation to be partitioned onto the array. The choice of array topology (square, triangular or trapezoidal) will have to be carefully chosen, by comparing the various dimensions of the matrices to be computed (as defined by the Kalman filter dimensions) to computation complexity and latency (i.e. QR decomposition is considerably more complex than matrix multiplication).

3.6.4 LDU Matrix Square Root Kalman Filtering

Bierman (1977) pointed out that the computation of a triangular covariance matrix requires N scalar square roots in its computation. As a motivation for developing his LDL^T (or UDU^T) matrix square root factorization, Bierman cited the very slow computation of scalar square roots compared to other arithmetic operations. In the early 1970s this fact was particularly relevant since, for example, the IBM 360 machine took 10 times as long to do a square root than a division operation. A number of other authors have also followed Bierman's line of argument to develop both serial and parallel LDL^T matrix square root Kalman filters (Jover and Kailath, 1986; Chen and Yao, 1987). (The name LDL^T matrix square root Kalman filter is somewhat of a misnomer, as the algorithm does not perform any scalar square roots; it is more correctly termed the LDL^T Kalman filter.) It is therefore worthwhile considering this algorithm, and why, despite the designs of others, it is not relevant to parallel array implementations of others.

Cholesky decomposition is a special case of the more general decomposition used to decompose a nonsingular matrix ($\in \mathscr{R}^{N\times N}$) into a lower triangular (L), diagonal (D) and upper triangular matrix (U). For any symmetric, but not necessarily positive definite, matrix A, L is a unit lower triangular matrix, $D = \text{diag}(d_1, d_2, \ldots, d_N)$ is a diagonal matrix and $A = LDT^T$. In the special case of a matrix A that is symmetric positive definite, Cholesky decomposition can be performed such that

$$A = (LD^{1/2})(D^{1/2}L^T) \tag{3.68}$$

The LDL^T factorization for the matrix square root filters can be developed by first defining (with obvious subscripts in what follows) the decompositions:

$$P = LDL^{\mathrm{T}} \tag{3.69}$$

$$R = L_R D_R L_R^{\mathrm{T}} \tag{3.70}$$

where, in comparison to (3.54), $P^{1/2} = LD^{1/2}$. The matrix square root time update of (3.59) can be written as the following LDL^{T} factorization:

$$Q_1 [FL_k \quad L_{Rw}] \begin{bmatrix} D_k & \mathbf{0} \\ \mathbf{0} & D_{Rw} \end{bmatrix}^{1/2} = [L_+ \quad \mathbf{0}] \begin{bmatrix} D_+ & \mathbf{0} \\ \mathbf{0} & \mathbf{0} \end{bmatrix}^{1/2} \tag{3.71}$$

The measurement update can be expressed in LDL^{T} form by rewriting (3.64) as:

$$Q_1 \begin{bmatrix} L_{Rv} & CL \\ \mathbf{0} & L \end{bmatrix} \begin{bmatrix} D_{Rv} & \mathbf{0} \\ \mathbf{0} & D \end{bmatrix}^{1/2} = \begin{bmatrix} L_{Re} & \mathbf{0} \\ PC^{\mathrm{T}} R_e^{-1} L_{Re} & L_k \end{bmatrix} \begin{bmatrix} D_{Re} & \mathbf{0} \\ \mathbf{0} & D_k \end{bmatrix}^{1/2} \tag{3.72}$$

For a suitably defined transform Q_D:

$$Q_D \begin{bmatrix} L_{Rv} & CL \\ \mathbf{0} & L \end{bmatrix} = \begin{bmatrix} L_{Re} & \mathbf{0} \\ PC^{\mathrm{T}} R_e^{-1} L_{Re} & L_k \end{bmatrix} \tag{3.73}$$

This approach can be directly related to the use of the scalar square root free Givens transformations to perform the Kalman update (Jover and Kailath, 1986).

Two numerical problems can be highlighted with the LDL^{T} approach. Firstly the matrix square root free orthogonal transformations will suffer from overflow/underflow and in some cases are unstable. The second problem is that the advantage of extended numerical precision is lost with the LDL^{T} matrix square root Kalman filter. One of the original reasons for realizing the matrix square root Kalman filter was that it gives an effective doubling in the number of bits of computational precision, as shown in Section 3.6.1. The LDL^{T} decomposition of a matrix P is denoted as:

$$P = LDL^{\mathrm{T}} \tag{3.74}$$

where L is unit lower triangular and D is upper triangular. The condition number of P is given by

$$\varkappa(P) = \varkappa(LDL^{\mathrm{T}}) \tag{3.75}$$

$$\leqslant \varkappa(L)\varkappa(D)\varkappa(L^{\mathrm{T}}) \tag{3.76}$$

Since the eigenvalues of a lower triangular matrix are the same as the diagonal elements (Golub and van Loan, 1983), then λ_{max} and λ_{min} of L both equal 1. Hence comparing with (3.55) the condition number of D is:

$$\varkappa(D) \geq \frac{\lambda_1}{\lambda_n} \tag{3.77}$$

where λ_1^2 is the maximum, and λ_n^2 is the minimum eigenvalue of PP^{T}. As stated above when computing in m-digit arithmetic, problems can be expected as $\varkappa(P)$

approaches 2^m. Unlike the real matrix square root factorization (Cholesky) this LDL^T computation requires the computation D which has a condition number the same as P. Therefore problems can be expected when $\varkappa(P) = 2^m$, unlike the standard matrix square root formulation (using Cholesky factors) which will not expect problems until $\varkappa(P) = 2^{2m}$. Hence the matrix square root filter is numerically superior to the LDL^T. Furthermore the standard Givens is faster to implement and more numerically robust than the square root free Givens rotations.

In Jover and Kailath (1986) a parallel architecture for the measurement update equations of the matrix square root covariance Kalman filter is presented. This paper used Bierman's LDL^T matrix square root formulation and restricted attention to the parameter estimation problem, or equivalently the measurement update portion of the Kalman filter. Although the parallel array uses orthogonal linear algebraic techniques, the unstable 2-multiply Givens II transformation, which suffers from overflow/underflow, was used.

3.7 HARDWARE PARALLEL KALMAN FILTER IMPLEMENTATIONS

To date, what is the real practicality of arithmetic level parallel arrays? Can they really be built? Is it feasible to design triarrays of, say, dimension 10, with some 50 processing elements each with multiply, divide, square root, and at least 4 bidirectional communication ports? Systolic arrays have been around now for more than 10 years. But how many designs have made the transition from elegant paper designs, or software simulations to actual hardware implementations in the true sense? The VLSI, or even ULSI technology to design these systems is not available to date: that is not to say, however, that it will not appear in the next 10 years.

3.7.1 A Dedicated Kalman Processing Element

For the orthogonal triarray Kalman designs, the core processing technique is Givens transformations. Therefore the diagonal nodes (see Figure 3.4) require the arithmetic functions of multiplication, division and square root, and high speed links to communicate with neighbouring processors. Figure 3.9 shows an *applications specific integrated circuit* (ASIC) block layout for the Kalman triarray processing nodes. As discussed in Section 3.6, the area required for an N-bit square root array is exactly half the area for the N-bit division array. The square root array also has half of the latency (Stewart *et al.*, 1989). Hence in an ASIC implementation there are no reasons at all for implementing square root free Givens I and II. These implementations would be slower, require more silicon area and, worst of all, be numerically ill-conditioned.

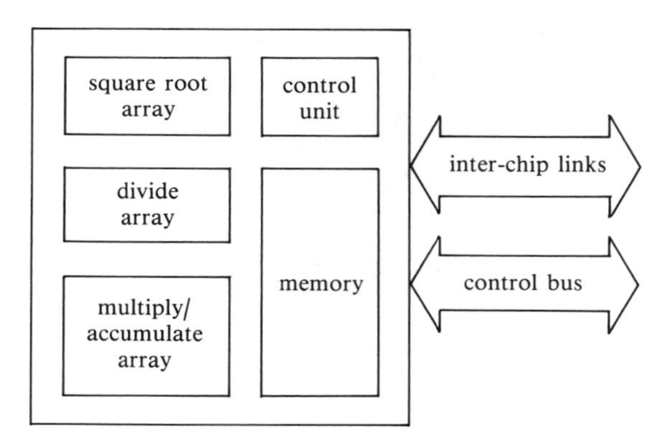

Figure 3.9 ASIC Kalman processor capabilities

3.7.2 Parallel Array Partitioning

To design a Kalman parallel array, using any of the architectures discussed above, the Kalman filter problem dimension must be mapped on to the parallel array dimension. If a particular problem has a measurement vector of length 20, then the parallel array will have the order of $20 \times 20 = 400$ processors. Clearly this is impractical, and may even be an over-provision of processing power. Therefore techniques are required to partition the paper design of the array on to a fixed size hardware array of general-purpose processors.

The techniques for *partitioning* large parallel array designs onto a smaller number of physical processors have been widely studied, but there is no general procedure to provide an efficient mapping. This is still a considerable area of parallel algorithms and architectures research. Although there are many architectures in the open literature for Kalman arrays, very few even consider the problem of partitioning. Any partitioned parallel arrays are complex in their control and no longer comply with the straightforward systolic concept of simple locally connected elements with minimal control. Nodes are now more likely to be general-purpose signal processing chips.

In Jainandunsing (1986) and Kung (1988) the concepts of locally sequential globally parallel (LSGP) and locally parallel globally sequential (LPGS) partitioning are analyzed. Simply stated, LSGP takes a parallel array or signal flow graph (SFG) and groups nodes together to be performed on a single processor, e.g. a 6×6 array could be split into blocks of 2×2 and each block implemented on a 3×3 processor array. LPGS would split the array into the same 2×2 blocks, but use an array of 2×2 processors to execute each block in sequence. Again the nodes are necessarily very general signal processors. These two strategies are illustrated in Figure 3.10. With particular relevance to the triarray in this chapter, Moldovan (1984) considered a partitioned architecture for QR decomposition.

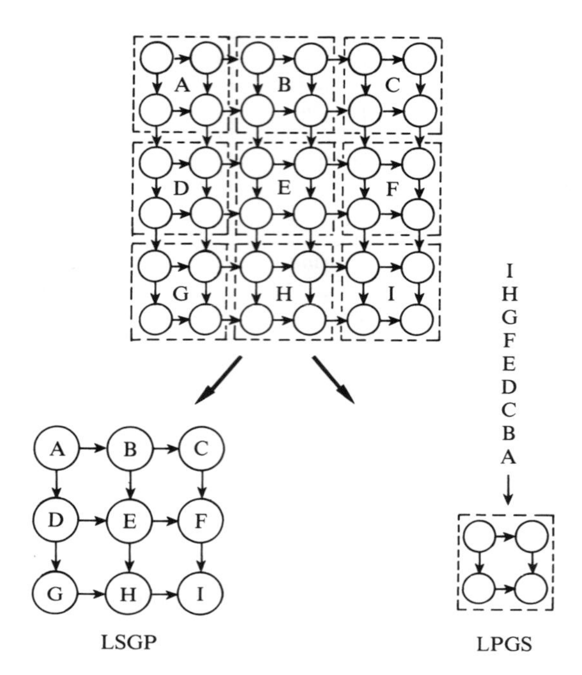

Figure 3.10 LSGP and LPGS array partitioning strategies

In Stewart (1988) a novel folding partitioning strategy was presented. The concept was part of a four stage mapping methodology, starting with deriving a dependence graph (DG) (Kung, 1988) for an algorithm, mapping this DG to an SFG, retiming as a systolic/wavefront array, and finally partitioning the (large) dimensioned problem on to a fixed array architecture. The folding can be conceptually considered as equivalent to taking a piece of paper and folding to a smaller size. A simple example for a triarray is shown in Figure 3.11 where the target architecture is a 4×4 triarray. The technique is very general and practically any sized problem can be folded down to a lower dimensioned hardware array. However, the efficiency of the mapping will depend on the actual data flow in the folded array. The folding can be performed in such a way that I/O edges of the original parallel array are preserved at the edges of the target array. An application of this technique to parallel Kalman filtering on a transputer array is given in Kung and Stewart (1988). To help ensure an efficient partitioning, communication and arithmetic processing should be completely independent.

3.8 CONCLUSIONS

This chapter has presented parallel processor implementations for Kalman filtering. The direct implementation in parallel of the covariance Kalman filter equations was

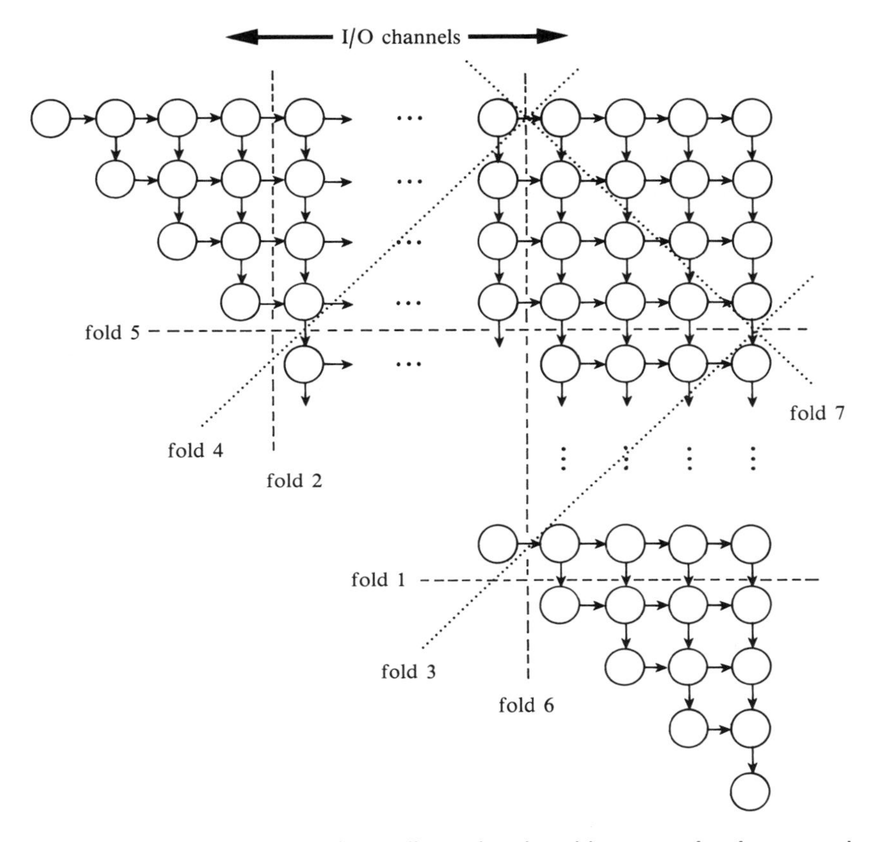

Figure 3.11 Fold partitioning a large dimensional problem to a 4×4 target triarray

first investigated. Although a parallel architecture was achievable from a consideration of the algorithm dependence graph, it was found not to be very efficient in its use of available computational resources, or to be local in its communication. It would also suffer from the same numerical problems as a sequential Kalman filter. Therefore a numerically robust least-squares formulation of the Kalman filter, based on orthogonal matrix algebra, was presented. An efficient parallel array for the implementation of the least-squares formulation was presented. ASIC processors with the capabilities of multiply, divide and square root were required for the triarray. A parallel implementation of the matrix square root Kalman filter was also presented. This filter is numerically attractive as it always preserves the nonnegative definiteness of the covariance matrix and gives other numerical advantages. The price paid for these advantages is an increase in the amount of computation over the direct Kalman covariance equations, due to the orthogonal linear algebraic calculations that are required. It was also shown that the

LDL^T (square root free) Kalman based parallel arrays recently suggested in the literature have no computational advantage over the standard formulation. In fact they have a number of numerical and implementational disadvantages. Lastly the very important consideration of parallel array partitioning was considered. In particular, for large dimensioned problems array partitioning is essential in order to realize a tractable and economic solution. This last area is still a major area of research for regular parallel array implementations.

REFERENCES

ANDERSON, B. D. O. and MOORE, J. B. (1979) *Optimal Filtering*, Englewood Cliffs, NJ: Prentice Hall.

BIERMAN, G. J. (1977) *Factorisation Methods for Discrete Sequential Estimation*, London: Academic Press.

CHEN, M. J. and YAO, K. (1987) 'On realisations least-squares implementation and Kalman filtering by systolic arrays' in *Proc. 1st International Workshop on Systolic Arrays, Oxford* (eds Moore, W. *et al.*), Bristol: Adam Hilger, pp. 161–70.

CHISCI, L. and MOSCA, E. (1987) 'Parallel architectures for RLS with directional forgetting' *Int. J. of Adaptive Control and Signal Processing*, **1**, 69–88.

GASTON, F. M. F. and IRWIN, G. W. (1989) 'Systolic approach to square root information Kalman filtering', *Int. J. Control*, **50** (1), 225–48.

GENTLEMAN, W. M. (1973) 'Least squares computation by Givens transformations without square roots', *J. Inst. Math. Appl.*, **12**, 329–36.

GENTLEMAN, W. T. and KUNG, H. T. (1981) 'Matrix triangularization by systolic arrays', *Proc. SPIE*, **289**, 19–26.

GOLUB, G. H. and VAN LOAN, C. F. (1983) *Matrix Computations*, Baltimore: John Hopkins Press.

GUILD, H. H. (1970) 'Cellular logical array for nonrestoring square root extraction', *Electron. Lett.* , **6**(3), 66–7.

HAMMARLING, S. (1974) 'On modifications to the Givens plane rotations', *J. Inst. Math. Appl.*, **13**, 215–18.

HAYKIN, S. (1986) *Adaptive Filter Theory*, Englewood Cliffs, NJ: Prentice Hall.

JAINANDUNSING, K. (1986) 'Optimal partitioning schemes for wavefront/systolic array processors', in *Proceedings of the IEEE International Conference on Circuits and Systems*, pp. 940–3.

JOVER, J. M. and KAILATH, T. (1986). 'A parallel architecture for Kalman filter measurement update and parameter estimation', *Automatica*, **22**(1), 43–57.

KALMAN, R. E. (1960) 'A new approach to linear filtering and prediction problems', *J. Basic Eng.*, **82D**, 35–45.

KAMINSKI, P. G., BRYSON, A. E. JR and SCHMIDT, S. F. (1971) 'Discrete square root filtering: a survey of current techniques' *IEEE Trans. Auto Contr.*, **AC-16**, 727–35.

KATSIKAS, S. K., LIKOTHANSSIS, S. D. and LAINIOTIS, D. G. (1991) 'On parallel implementation of the linear Kalman and Lainiotis filters and their efficiency, *Signal Processing*, **25**, 289–305.

KUNG, S. Y. (1988) *VLSI Array Processors*, Englewood Cliffs, NJ: Prentice Hall.

KUNG, S. Y. and HWANG, J. N. (1991) 'A systolic array processor for Kalman filtering', *IEEE Trans. Signal Processing*, **39**(1), 171–82.

KUNG, S. Y. and STEWART, R. W. (1988) 'Kalman filtering on a fixed transputer array' in *Proceedings of EUSIPCO Conference, Grenoble*, pp. 347–50.

McWHIRTER, J. G. (1983) 'RLS minimisation using a systolic array', *Proc SPIE*, **43**(1), 415–31.

MAJITHIA, J. C. and KITAI, R. (1971) 'A cellular array for the nonrestoring estimation of square roots', *IEEE Trans. Computers*, 1617–18.

MOLDOVAN, D. I. (1984) 'Partitioned QR algorithm for systolic arrays' in *VLSI Signal Processing* (eds S.Y. Kung *et al.*), Los Angeles: IEEE Press, 350–62.

MORF, M. and KAILATH, T. (1975) 'Square-root algorithms for least squares estimation', *IEEE Trans. Auto. Contr.*, **AC-20**(4), 487–97.

PAIGE, C. C. and SAUNDERS, M. A. (1979) 'Least squares estimation of discrete linear systems using orthogonal transforms', *SIAM J. Num. Algebra*, **14**(2), 180–93.

PAPOULIS, A. (1984) *Probability, Random Variables and Stochastic Processes*, New York: McGraw Hill.

SIBUL, S. (ed.) (1987) *Adaptive Signal Processing*, New York: IEEE Press.

SORENSON, H. W. (1970) 'Least squares estimation: from Gauss to Kalman', *IEEE Spectrum*, **7**, 63–8.

STEWART, R. W. (1988) 'Mapping signal/image processing algorithms to fixed architectures', in *Proceedings of the IEEE International Conference on Acoustics, Speech and Signal Processing, New York*, pp. 2037–40.

STEWART, R. W. and CHAPMAN, R. (1990) 'Fast stable Kalman filter algorithms utilising the square root, in *Proceedings of the IEEE International Conference on Acoustics, Speech and Signal Processing, Albuquerque*, pp. 1815–19.

STEWART, R. W., CHAPMAN, R. and DURRANI, T. S. (1989) 'The square root in signal processing', in *SPIE Real-Time Signal Processing* XII.

YEH, H.-G. (1988) 'Systolic implementation of Kalman filters', *IEEE Trans. ASSP*, **36**(9), 1514–17.

Part II

Architectures for Intelligent Control

Intelligent control is a general area which is receiving increasing attention by the general control systems community. In general, the implementation of these schemes can impose a heavy computational load and this, coupled with their inherent parallelism, requires a parallel treatment. Amongst the various candidate architectures, artificial neural networks (ANNs) are currently the subject of much research effort and, in effect, provide a self-learning capability and ultra-high processing speed by massive parallelism. Typically, an ANN consists of a very large number of processing elements – the neurons – interconnected by weights which can be updated, or trained, during operation. The actual network topology varies according to the choice of various models and activation rules.

In the context of the work reported here, the well-known back-propagation model can be interpreted as a 'partitioned complete graph' with each partition termed a layer. Further, every neuron within a layer is run in parallel and every layer is run in a pipeline. Alternatively, in the Hopfield model all neurons are interconnected to each other with no topological hierarchy.

These networks can be implemented in either hardware or software. Further, their self-learning property, in particular, offers the promise of powerful techniques, in comparison to existing techniques such as adaptive schemes, for solving nonlinear control problems. The three chapters in this section are a cross-section of current 'state of the art' research on the application of ANNs to control problems.

Chapter 4 critically examines the role of ANNs in the important general area of process control with emphasis on currently difficult problems in modelling, estimation and prediction. Their use as 'software sensors' is also examined. Chapter 5 develops the multivariate B-spline based ANN for adaptive or learning nonlinear controllers involving least-squares estimation, stochastic approximation and nonlinear time series prediction.

The work of Chapter 5 can also be viewed as extracting useful concepts from fuzzy logic based systems and a comparison between these two types of systems is also included. This leads naturally on to Chapter 6 where the group method of data handling is mapped onto a multilayer perceptron type configured parallel network, and a detailed treatment of some related self-organizing systems and fuzzy logic based control schemes is given.

4

Artificial Neural Networks: A Possible Tool for the Process Engineer

M. J. Willis, G. A. Montague, A. J. Morris and M. T. Tham

4.1 INTRODUCTION

It has been claimed that artificial neural networks (ANNs) are a representation that attempts to mimic the functionality of the brain. For several decades scientists have been trying to emulate the real neural structure of the brain, believing that the human process of learning might be reproduced by an algorithmic equivalent. The principal motivation behind this research is the desire to realize the sophisticated level of information processing that can be achieved by the brain. Unfortunately, the broad-based allure of the concept has resulted in hype playing a significant role in the history of artificial neural network research.

The origins of the subject can be traced back several decades; indeed a paper by McCulloch and Pitts (1943) is often considered to be the initiator. However, it was not until 1957 that the first real success can be found with Rosenblatt's Mark I Perceptron. Widrow and Hoff's 'adaline' processor followed, the concept of which is still in widespread use today. Considerable success was achieved by workers during the early 1960s; however, studies were primarily experimentally based with little theoretical foundation. Nevertheless as a result of initial successes the literature began to postulate wildly and even such concepts as the 'artificial brain' appeared. This was clearly not to the benefit of researchers. The lack of rigour in neural network analysis and failure to substantiate these assertions resulted in significant loss in confidence and consequently funding. It was not until the mid 1980s that there was a resurgence in interest: a key driving force being US Defense Agency funding. The much wider availability and power of computing systems, together with new research studies, resulted in a far greater 'market' for the technology.

It is apparent, however, that present research aims are not directed at emulating the sheer complexity of the brain. Indeed, biological parallels would seem to be somewhat dubious at the levels of complexity that are normally considered. So what is the attraction of this technology? Two quite different reasons are often put forward in support of artificial neural networks. From the viewpoint of computer science, the primary argument is based upon the parallel nature of computation which is claimed to provide extremely fast processing facilities. This, unfortunately, has yet to be justified. Indeed whilst experience with present day multiprocessor machines has indicated that some performance improvements can be gained during training it should be noted that the topology (structure) of the network and the training methods adopted can significantly modify these findings. For example, in some situations it has been found that the optimal number of processors on which to simulate a multilayer, multineuron network is one! Ultimately, however, it will be essential to make use of parallel processing techniques for which the material in this chapter is essential background from, in particular, a process control standpoint.

The second reason for the high level of interest in ANNs would appear to be because of the ability to develop a nonlinear model of a system. Indeed, a specific niche for a neural network appears to be where the replacement of a conventional model may accrue significant benefits. These benefits generally arise due to the ability of the network to provide a generic 'cost-effective' modelling tool; a potentially apt philosophy for often notoriously complex chemical/biochemical processes. Historically such systems often necessitate the devotion of considerable time in order to develop a realistic mechanistic model. Moreover, simplifying assumptions have to be made in many instances to enable a tractable solution to the modelling problem. A first principle model will, therefore, often be very costly to construct and will be subject to inaccuracies due to the assumptions made during the development. A desirable objective is therefore the application of a technique which possesses generality of model structure (facilitating rapid and cheap development) but which could also be capable of learning and expressing the process nonlinearities and complexities. The ANN appears to offer this possibility. Indeed, recent studies (for example, Bhat *et al.*, 1989; Birky and McAvoy, 1989; Di Massimo *et al.*, 1990; Willis *et al.*, 1991) have demonstrated the utility and flexibility of the concept within the domain of process engineering. Such practical applications complement theoretical research on ANNs and go some way to demonstrating the claimed capabilities. However, the obvious question still remains as to whether the pattern of the mid-1960s is going to reoccur, will artificial neural networks be a passing phase with few benefits gained, or are real industrial applications going to result?

4.2 PROCESS MODELLING VIA ARTIFICIAL NEURAL NETWORKS

Although a number of ANN architectures have been proposed (see Lippmann, 1987), the 'feedforward' ANN (FANN) is by far the most widely applied. Cybenko (1989) claimed that any continuous function can be approximated arbitrarily well on a compact set by a FANN, comprising two hidden layers and a fixed continuous nonlinearity. This result essentially states that a FANN could be used to model a wide range of nonlinear relationships. The implications of this statement are therefore considerable. In view of this, subsequent discussions will therefore be restricted to FANNs.

A FANN is made of neuron-like elements, called nodes. These nodes are organized in layers as shown in Figure 4.1 (where the circles represent the 'neurons'). Each interconnection has associated with it a weight which acts to modify the signal strength. Scaled data are fed into the network at the input layer. The signals are then 'fed-forward' through the network via the connections to the output layer. The 'neurons', except those in the input layer, perform two functions. First, they act as a summing point for the input signals and second they propagate the input through a sigmoidal processing function. In order to extract a system model the neural network has to be trained on appropriate system data.

4.2.1 Network Training

During training it is required to search for an extremum (maximum or minimum) of a real function f of n variables (the network weights) each of which can take on any value. The extremum can either be global, in which case it is truly the highest or lowest value of the function, or it can be local, in which case it will only be the highest or lowest value in a finite neighbourhood. Obviously, the ultimate objective is to find the global extrema $\mathbf{w}^*$ of the objective function generally specified as:

$$\mathrm{f}(\mathbf{w}) = \sum_{i=1}^{p} E_p \tag{4.1}$$

where $E_p = (Y_\mathrm{d} - Y)^2$ and Y is the output of the network and Y_d is the desired output of the network.

The minimization of the error can be achieved by the application of a suitable minimization routine, the most popular being that of back-error propagation

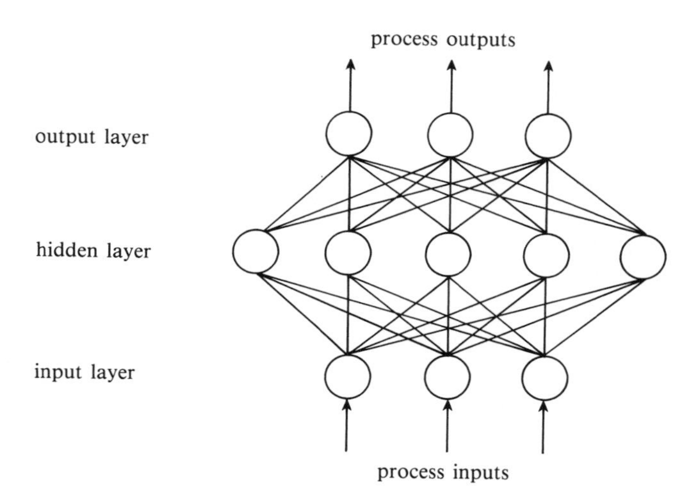

Figure 4.1 Feedforward network structure

(Rumelhart and McClelland, 1986). This technique is based upon one of the more simplistic optimization techniques known as the method of steepest descent. The method creates sequences of successive approximations that converge to points that satisfy local optimality criteria. In its simplest form the technique may be described as follows. The partial derivatives of the function $f(w_1, w_2, \ldots, w_n)$ denoted $\nabla f(\mathbf{w})$ are first found with respect to a single training example. For minimization, the search direction is the negative of the gradient and hence at the kth iteration, the transition from the weight vector $\mathbf{w}^k$ to the new vector $\mathbf{w}^{k+1}$ is given by:

$$\mathbf{w}^{k+1} = \mathbf{w}^k - b^k \nabla f(\mathbf{w}) \tag{4.2}$$

where b^k is a scalar that determines the step length in the direction $-\nabla f(\mathbf{w})$. Thus, the negative gradient allows the determination of the direction of minimization; however, it does not give any information about the size of the step (b^k) that has to be taken. Generally, for implementation simplicity when training networks the step length is prespecified and this, or a heuristically amended version, is used during the training phase.

Recently, there have been several attempts to improve the back-propagation algorithm. A common and easily implemented modification to the standard back-propagation philosophy is based upon the concept of 'batching' (Rumelhart and McClelland, 1986). Recall that in the back-propagation technique discussed above the gradients $\nabla f(\mathbf{w})$ of the objective function were calculated with respect to the squared error associated with a single training example. It should be noted therefore, that this is not the direction of steepest descent with respect to the overall objective function (4.1). The direction of steepest descent with respect to the overall objective function can be obtained by taking the sum of the gradients of the individual training examples, i.e.:

$$L(\mathbf{w}) = \Sigma \ \nabla f(\mathbf{w})_p \tag{4.3}$$

where the subscript 'p' represents the pth input–output data set. Using this gradient the weight change can be determined according to:

$$\mathbf{w}^{k+1} = \mathbf{w}^k - b^k L(\mathbf{w}) \tag{4.4}$$

It is quite well known that this philosophy can provide a consistent improvement to the sequential approach. As with the sequential approach it is still necessary to define the step size b_k heuristically. However, as it is obvious that a fixed step size is undesirable (a step of arbitrary length in a local direction does not necessarily result in a reduction of the objective function, moreover, a fixed b^k can cause cycling about a minimum), the step-size is determined at each iteration via a line search. This is used to ensure that the objective function is minimized, in the direction of search, at each iteration. However, it should be noted that the steepest descent method (and as a direct result, back-propagation) is based upon local linearization of the function $f(\mathbf{w})$, i.e. the method of steepest descent can be interpreted as fitting a tangent plane to the objective function. Thus, when using this minimization technique all information about the function is lost except that contained in the slope. In an attempt to improve the rate of convergence to an extrema, so-called second-order optimization techniques incorporate additional information, namely the curvature of the slope. This information can be obtained by utilizing the second partial derivatives of f(**w**) with respect to the network weights. This leads to a slight modification of (4.4), yielding a Newton optimization algorithm:

$$\mathbf{w}^{k+1} = \mathbf{w}^k - b^k[H(\mathbf{w}^k)]^{-1}L(\mathbf{w}^k) \tag{4.5}$$

where $[H(\mathbf{w}^k)]^{-1}$ is the inverse of the Hessian matrix $H(\mathbf{w}^k)$ of f(**w**), i.e. the matrix of second partial derivatives with respect to **w** calculated at $\mathbf{w}^k$. An obvious drawback in employing this methodology, especially for large networks, is the inversion of a large matrix. Thus, whilst convergence properties could be improved, significantly more calculations per epoch would be required. Hence, the recourse generally is to approximate the Hessian matrix. In order to implement this modification it is necessary to modify Newton's method to read:

$$\mathbf{w}^{k+1} = \mathbf{w}^k - b^k\mathbf{x}(\mathbf{w}^k)L(\mathbf{w}^k) \tag{4.6}$$

where **x** is referred to as a direction matrix which is continuously updated using a formula of the form:

$$\mathbf{x}^{k+1} = \mathbf{x}^k + \Delta\mathbf{x}^k \tag{4.7}$$

where **x** eventually approximates $[H(\mathbf{w}^k)]^{-1}$. Usually, the method is initiated with $\mathbf{x}^{\mathrm{O}} = I$ (identity matrix), thus again the technique is initially identical to that of steepest descent and hence back-error propagation. Subsequent iterations, however, use the direction matrix, which is updated from one iteration to the next. Further, to ensure a monotonic reduction in the objective function an optimized procedure for step size can be implemented. It should be noted, that there are many of these quasi-Newton techniques available in the literature. Watrous (1987) investigated Davidon–Fletcher–Powell (DFP) and Broyden–Fletcher–Goldfarb–Shanno (BFGS) variants and compared these to back-error propagation with an optimized learn rate. It was shown that both these algorithms took fewer iterations than back-error propagation; however, it was pointed out that the Hessian updates in DFP and BFGS incurred significant computational overheads. Indeed, because of this, Watrous (1987) concluded that second-order methods may not be advantageous for large networks.

With the desire to improve the rate of convergence and yet minimize computational burden, Leonard and Kramer (1990) suggested the use of a conjugate directions routine. Quasi-Newton methods are usually more rapidly convergent and more robust than conjugate gradient methods; however, they require significantly more storage. An advantage of the conjugate gradient method is that it relinquishes the need for second derivatives of the objective function whilst retaining convergence properties of second-order techniques. A conjugate gradient methodology is thus a well-established contender for problems with a large number of variables, i.e. training of an artificial neural network.

The basic philosophy is to generate a conjugate direction as a linear combination of the current steepest descent direction and the previous search direction. With this technique minimization is initiated as with steepest descent, i.e.

$$\mathbf{w}^{k+1} = \mathbf{w}^{k} - b^{k}\mathbf{s}^{k} \tag{4.8}$$

where $\mathbf{s}^{k} = L(\mathbf{w})$ is the initial search direction; after this iteration a new direction of search is required. This direction ($\mathbf{s}^{k+1}$) is chosen so that it is conjugate with $\mathbf{s}^{k}$ and it can be shown that this is given by:

$$\mathbf{s}^{k+1} = \frac{-L(\mathbf{w}^{k+1})1 + \mathbf{s}^{k}[L^{\mathrm{T}}(\mathbf{w}^{k+1})L(\mathbf{w}^{k+1})]}{[L^{\mathrm{T}}(\mathbf{w}^{k})L(\mathbf{w}^{k})]} \tag{4.9}$$

Minimization then proceeds in the direction defined by (4.9). It should be noted, however, that when this technique is applied to nonquadratic functions, the exact minimum will not be found in a finite number of steps. Practical experience suggests that resetting the algorithm to the steepest descent direction every n iterations (where n is the number of network weights) is superior to the repeated use of the conjugate gradient method.

For the purposes of this contribution we use (and compare) three optimization techniques: conjugate gradients, DFP quasi-Newton technique and batched back-error propagation (BP), all of which use an optimized learn rate. In order to compare the algorithms, under known experimental conditions a simple nonlinear function will be utilized. Obviously, any comparison of the techniques described above will be highly dependent upon the network topology chosen to be appropriate to approximate the nonlinear function. Thus, rather than use an ad hoc choice of network topology (as often appears to be the situation in most evaluation studies appearing in the literature), the evaluation is conducted with the use of an 'automatic' network topology selection procedure.

4.2.2 Topology Selection Procedure

A technique for selecting the number of neurons per hidden layer in a static feedforward neural network has been proposed by Wang *et al.* (1992). A link has been established between the number of neurons in the hidden layers and the dimension of the topological spaces in which the outputs of the neurons in the hidden layers reside. Indeed, it is claimed that the artificial neural network can be regarded as an approximate canonical decomposition of the nonlinear function to be estimated. Consequently, the technique may be used during the training process in order to specify the number of neurons required by each hidden layer.

The details of network mapping are as follows. Let $\mathbf{u}(t)$ and $\mathbf{y}(t)$ be the input and output data respectively. By definition the functional transformation of each layer of a network J can be represented by the composition of an affine transformation (weighted sum of the inputs to the layer, W) with a nonlinear mapping (usually the sigmoidal function, S). Let 1, 2 and o denote the first and second hidden layers and the output layer respectively. Thus,

$$\begin{aligned} \mathbf{h}_1(t) &= J_1(\mathbf{u}(t)) = S \cdot W_1(\mathbf{u}(t)) \\ \mathbf{h}_2(t) &= J_2(\mathbf{h}_1(t)) = S \cdot W_2(\mathbf{u}(t)) \\ \mathbf{y}(t) &= J_o(\mathbf{h}_2(t)) = S \cdot W_o(\mathbf{u}(t)) \end{aligned} \tag{4.10}$$

where $\mathbf{h}_1(t)$, $\mathbf{h}_2(t)$ and $\mathbf{y}(t)$ are the output vectors of the layers and $\cdot$ denotes the composition operator. An approximation to the original nonlinear mapping by a neural network with two hidden layers is therefore given by:

$$\mathbf{y}(t) = J_o \cdot J_2 \cdot J_1(\mathbf{u}(t)) \tag{4.11}$$

To ascertain the number of neurons required in each hidden layer it is necessary to determine the dimension of the maximal subspaces of the input space and the output space so as to provide the appropriate spaces for the canonical decomposition. This can be achieved by checking the independence of the neuron outputs of each hidden layer.

Suppose $[h1_1(t), h1_2(t), \ldots, h1_{N1}(t)]$ is the vector output sequence of the first hidden layer, where $t = 1, 2, \ldots, L$, $N1$ is the number of neurons in the first hidden layer and L is the number of training data samples. The independence of the outputs of the layer can be verified by computing in the case when it is a square, the determinant of a matrix $\mathbf{M}$ formed from the vector output sequence of the layer, i.e.

$$M = \begin{bmatrix} h1_1(1), h1_2(1), \ldots, h1_{N1}(1) \\ h1_1(2), h1_2(2), \ldots, h1_{N1}(2) \\ h1_1(L), h1_2(L), \ldots, h1_{N1}(L) \end{bmatrix}^{\mathrm{T}} \tag{4.12}$$

In general, however, M is a rectangular matrix and hence one option is to calculate the determinant of the square matrix $M^{\mathrm{T}}M$. This procedure can be performed for each layer in the network. To summarize, the implementation procedure for a k hidden layer network is as follows:

Step 1. Initialize $n_i (i = 1 \text{ to } k)$ to a small integer.
Set $i = 1$.
Step 2. Start training and let the FANN converge.
Step 3. Calculate:
$d_i = \det(\mathbf{M}_i^T \mathbf{M}_i)$
Step 4. If $d_i > \delta$ then add a neuron to the ith hidden layer and go to step 2.
Else:
Step 5. $i = i + 1$ Go to step 2.

δ is typically set to 0.001. It should be noted that the use of this procedure can be computationally excessive, hence our desire to use improved training paradigms.

4.2.3 A Simple Comparison of the Training Paradigms

In the results that follow all topology selections were based upon training via the method of conjugate directions. It should be noted, however, that the selection procedure should be independent of the choice of algorithm, hence this represents an arbitrary choice.

In the experiment the 'simple' nonlinear function $f(\mathbf{x}) = \mathbf{x}^2$ was to be approximated using a FANN. Input–output data were generated in the range $[-0.9, 0.9]$ and then utilized to train the network. The topology selection procedure was then used to generate an appropriate network topology to represent the function. Starting from a small topology, say 1–2–2–1 (i.e. a network that has one neuron in the input layer, two hidden layers with two neurons in each layer and an output layer with one neuron) or 1–3–3–1, the network was trained using the conjugate gradient routine until the topology selection criteria were satisfied. Once convergence of the training for a particular network topology had occurred, the determinants of the $M^{\mathrm{T}}M$ matrices were checked to see if they were below the threshold δ which was set to 0.001. A summary of the results is given in Table 4.1.

It may be noted that the topology selection procedure automatically augments the number of neurons in the two hidden layers. The procedure stops when the determinants of the $M^{\mathrm{T}}M$ matrices of the two hidden layers are below the threshold value. At this point the network contains a neuron, the presence of which gives no significant additional contribution to the learning of the mapping between input and output data. Thus, we step back one stage in the selection procedure and the 1–4–4–1 topology is chosen as being the appropriate topology for the nonlinear approximation.

Once an appropriate topology had been selected the three candidate training algorithms were compared. Table 4.2 illustrates the results obtained. It can be

Table 4.1 Topology selection for the function $f(x) = x^2$

Topology	d_1	d_2	E^2
1–2–2–1	3.558	0.836	0.514
1–3–3–1	2.8×10^{-2}	1.6×10^{-3}	0.416
1–4–3–1	5.0×10^{-4}	2.02×10^{-3}	0.1735
1–4–4–1	1.05×10^{-4}	3.01×10^{-3}	2.1×10^{-2}
1–4–5–1	1.1×10^{-9}	2.9×10^{-12}	1.0×10^{-2}

Table 4.2 Approximation of $f(x) = x^2$ with 1–4–4–1 FANN

	Batched BP (+ line search)	*DFP*	*Conjugate gradient*
Iterations	1029	72	197
E^2	6.68×10^{-2}	1.21×10^{-2}	2.1×10^{-2}

observed that both the DFP and the conjugate gradients technique out-perform the batched BP with optimized learn rate. Computationally, the DFP and the conjugate gradient techniques are slightly more demanding than the batched BP; however the order of magnitude reduction in the number of iterations more than compensates for this difference. Moreover, since the topology selection procedure requires repeated training it is desirable to utilize efficient learning paradigms to enhance the viability of the technique.

4.3 DYNAMIC MODELLING USING A FANN

The results presented above indicate the potential of a FANN to model nonlinear systems. The result is a steady-state (albeit nonlinear) model of the process. Unfortunately, if the technique is to achieve widespread application (and it is application where our emphasis is placed) then it must be capable of modelling dynamic systems. The most straightforward way to move from steady-state neural network models to dynamic is to adopt an approach similar to that taken in linear ARMA (auto-regressive moving average) modelling. In this approach a time series of past process inputs and outputs are used to predict the present process outputs. Obviously important process characteristics such as system delays can be accommodated by utilizing only those process inputs beyond the dead-time of the process. Additionally, any uncertainty in process time delay can be taken into consideration by using an extended time history of process inputs. Adopting this philosophy when using a neural network inevitably necessitates a significant number of network inputs. An alternative methodology would be to modify the neuron processing to incorporate dynamics inherently within the network. Thus, in addition to the sigmoidal processing of nodes, the neurons (or transmission between neurons) can be given dynamic characteristics. Clearly the simplest approach here is to incorporate a first-order transfer function, i.e. to assume a first-order dynamic response. Whilst this can eliminate the need to utilize a time history of process variables, any uncertainty in system delays may still warrant a limited history of variables. Even this can be avoided by the incorporation of delays with the dynamics of the neuron, yielding a neuron processing function as shown in Figure 4.2, where a second-order Pade approximation is used to represent the time delay

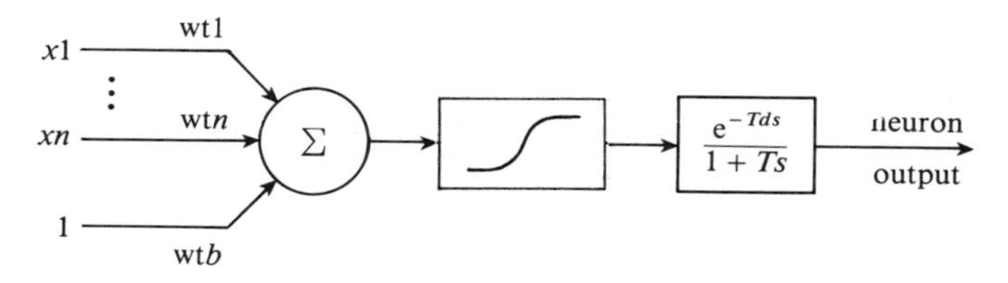

Figure 4.2 Processing at the neuron level

(Stephanopoulos, 1984). The advantage here is the avoidance of the need for a time series of inputs. This reduces network dimension and thus the number of process 'parameters' to be determined a priori. Recent studies have shown that the incorporation of dynamics into the network in this manner is highly beneficial in many real process application studies (*e.g.* Montague *et al.*, 1991).

If the ARMA modelling philosophy is adopted then particular care is needed when training and assessing the quality of the resulting model. Willis *et al.* (1991) demonstrate the problems that can occur using industrial examples; however, a very simple example can serve more effectively as a warning. Taking a simple first-order model (gain of one, time constant of 10 seconds) sampled at one second intervals with a time delay of 2 seconds, two neural networks were developed to capture the process dynamics; one using the time series approach, the other using the dynamic neuron concept. The simple system was subjected to step changes in input (u_t) and the output (y_t) monitored.

Using the 'time series' approach four network inputs were specified – these were $u_{t-2}, u_{t-3}, u_{t-4}$ and y_{t-1}, i.e. delayed values of process input and output – in an attempt to predict process output, y_t. Note that because of an inherent delay due to sampling, u_{t-1} has no effect on y_t, hence it was not utilized as a network input. The topology of the network was selected as one that had one hidden layer with

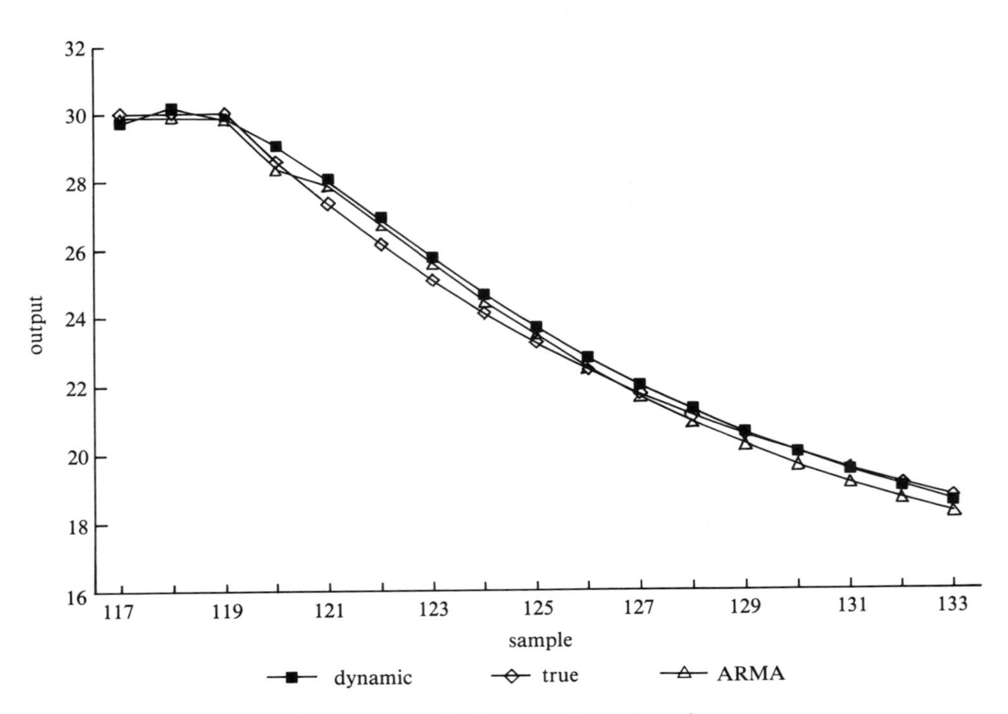

Figure 4.3 Single step ahead learning

three neurons in this layer, thus a 4–3–1 network was employed. Although the topology selection approach has as yet not been proven on a dynamic network, intuitively the network dimension will be smaller than that of the time series representation. Hence, a 1–2–1 network was utilized with the network input being u_{t-1}. Moreover, it should be noted that in both cases a single hidden layered network was employed as this was found to be sufficient. Clearly, however, whilst the automatic topology selection procedure discussed above can indicate the number of neurons per layer it does not reveal the number of layers that are necessary. This remains an arbitrary facet of an otherwise promising technique.

The parameters of the two networks were then determined, and the results based upon training the network to minimize the current output prediction error gave a high quality of fit. The portion of the response shown in Figure 4.3 demonstrates the dynamic neuron approach; both this and the time series approach exhibit reasonable behaviour.

It would appear that both the network models have captured the essential process characteristics. If, however, the models are used to predict the output of the process further into the future, as may be required for control and optimization, it is obvious that the alternative modelling approaches are not comparable. Figure 4.4 shows a 15 step ahead prediction of system output. Clearly, the time series network

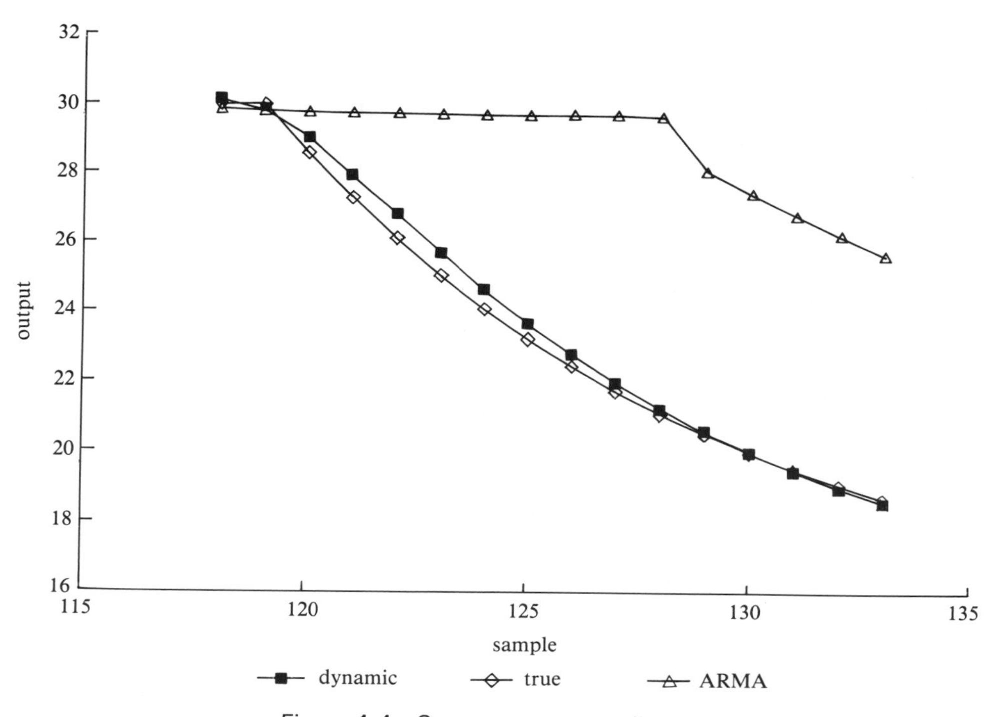

Figure 4.4 System output prediction

model has not captured the process dynamics. The reason for this lies in the auto-regressive nature of the approach. In order to predict further into the future, past estimates must be used as network inputs. Thus, any errors in the past estimates accumulate as the prediction horizon increases. Since the dynamic network model is not auto-regressive this problem does not arise and the predictions of the process output are thus reasonable. The problems of using the time series approach can be overcome by minimizing the network prediction error not just for the current output but also for all output predictions to a prediction horizon. Nevertheless, a lesson to be learnt is that when using model fitting techniques that involve minimization of error, a careful assessment of the quality of the resulting model must be made. Here, for example, a relatively small one step ahead prediction error does not necessarily indicate a close resemblance to the true process characteristic.

4.4 INDUSTRIAL APPLICATION RESULTS

If artificial neural networks are capable of capturing nonlinear process relationships and modelling dynamic nonlinear systems then obviously the range of applications for which they could be considered is large. Such diverse areas as data interpretation for process plant design, the training of process operators and the on-line optimization of plant performance have been suggested (as well as the obvious applications in nonlinear process modelling). Unfortunately, reports of real applications are rare. Industrial confidentiality or bureaucracy hindering the publication of recent work could be a factor here, as it is our contention that interest in neural networks is not confined to the 'ivory-towered' institutes of higher learning and research establishments.

4.4.1 Artificial Neural Networks in Estimation

A problem of present and increasing industrial interest where artificial neural networks may be of benefit is in providing a means by which to improve the quality of on-line information available to plant operators for plant control and optimization. Major problems exist in the chemical and biochemical industries (paralleled with the food processing industries) concerned with the on-line estimation of parameters and variables that quantify process behaviour. In a nutshell, the problem is that the key 'quality' variables cannot be measured at a rate which enables their effective regulation. This can be due to limited analyzer cycle times or a reliance upon off-line laboratory assays. An obvious solution to such problems could be realized by the use of a model along with secondary process measurements, to infer product quality variables (at the rate at which the secondary variables are available) that are either very costly or impossible to measure on-line. Hence, if the relationship between quality measurements and on-line process variables can be captured then the resulting model can be utilized within a control scheme to enhance process regulation. The concept is known as inferential estimation. Historically, with varying degrees of success, linear models, adaptive models and process-specific mechanistic models have been used to perform this task. It is suggested, however, that the use of a neural network model, because it is not process-specific and has the ability to capture nonlinear process characteristics, may be beneficial for these applications. Indeed, in this area results from industrial evaluations have been promising (see, for example, Willis *et al.*, 1991). Whilst such schemes operating in 'open-loop' can be used to assist process operators, it should be noted that with the availability of fast and accurate product quality estimates, the possibility of closed-loop inferential control becomes instantly feasible.

Here the inferred estimates of the controlled output are used for feedback control. The effective elimination of a time delay caused by the use of an on-line analyzer or the need to perform off-line analysis affords the opportunity of tight product control through the use of standard industrial controllers. Consequently reduction in product variability caused by process disturbances, and hence reduction in off-spec product, can be achieved. Whilst such specific process applications allow short term benefits, it should be noted that in the long term, the concept of inferential control of quality variables could remove the need for anything but rudimentary analysis.

Many industrial systems subject to measurement limitations could be considered. However, the complexity of the penicillin fermentation process operated by SmithKline Beecham is an ideal example to highlight the possible benefits of applying the techniques. Owing to industrial confidentiality a thorough description of the fermentation process is not permissible. Suffice to say that the biomass levels inside a fermenter need to be regulated to achieve good performance. Unfortunately, the only method available to determine biomass is by off-line laboratory analysis. The delay induced by this sampling procedure and the frequency at which samples can be taken reduce the effectiveness of control. A complex nonlinear relationship is known to exist between on-line measured variables, such as off-gas concentrations, and biomass levels in the fermenter. Thus the objective was to model this relationship using the neural network modelling procedure. A series of experimental trials were performed and the results were utilized to provide training data linking off-gas to biomass levels. The current on-line measurement of oxygen uptake rate (OUR) was used as one of the inputs to the neural network. Additionally, since the characteristics of the fermentation are also a function of time, the batch time was also considered a pertinent input. A FANN was then trained to capture the complex relationship and thus provide estimates of biomass at a much increased frequency. The training procedure adopted is identical to that discussed above. The results of the topology selection procedure are given in Table 4.3. The results indicate that an appropriate topology for this process is a 2–4–3–1 network. Using this topology the three candidate training methodologies are compared in Table 4.4. Here again, it may be observed

Table 4.3 Topology selection for batch fermenter

Topology	d_1	d_2	E^2
2–2–2–1	0.1625	3.9×10^{-2}	3.09
2–3–3–1	1.3×10^{-3}	1.3×10^{-7}	3.09
2–4–3–1	4.2×10^{-3}	2.6×10^{-4}	2.8×10^{-2}
2–5–3–1	3.2×10^{-7}	3.3×10^{-4}	2.4×10^{-2}

Table 4.4 Comparison of three candidate training methodologies

	Batched BP (+ line search)	*DFP*	*Conjugate gradient*
Iterations	228	44	42
E^2	3.10×10^{-2}	1.89×10^{-2}	2.8×10^{-2}

that the performance of both the DFP algorithm and the conjugate gradient technique are superior to the batched BP methodology.

Figure 4.5 indicates the typical performance of the estimator applied to the industrial system at two different process operating conditions. Whilst the quality of the neural network model is apparent it is particularly interesting to compare the model development time (approximately one month) with the many months spent deriving a less representative first principles model. The neural network based estimator is now successfully operating on-line to provide information to the process operators. Experience gained in operation has suggested other areas where neural networks may contribute. For example, when an unlikely event such as contamination occurred the neural network model detected process changes before they were apparent to operators using conventional techniques. A potential to provide early warning to process faults is therefore apparent.

4.4.2 Artificial Neural Networks in Control

If a FANN model is of sufficient accuracy, then it should be possible to employ the model directly within a model-based control strategy. A potentially useful algorithm may be one which minimizes future output deviations from set-point, whilst taking suitable account of the control sequence necessary to achieve this objective. Being common to most predictive control algorithms, this concept is not new (see, for example Clarke *et al.*, 1987). However, the attraction of using the neural network instead of other model forms within the control strategy is the ability to represent

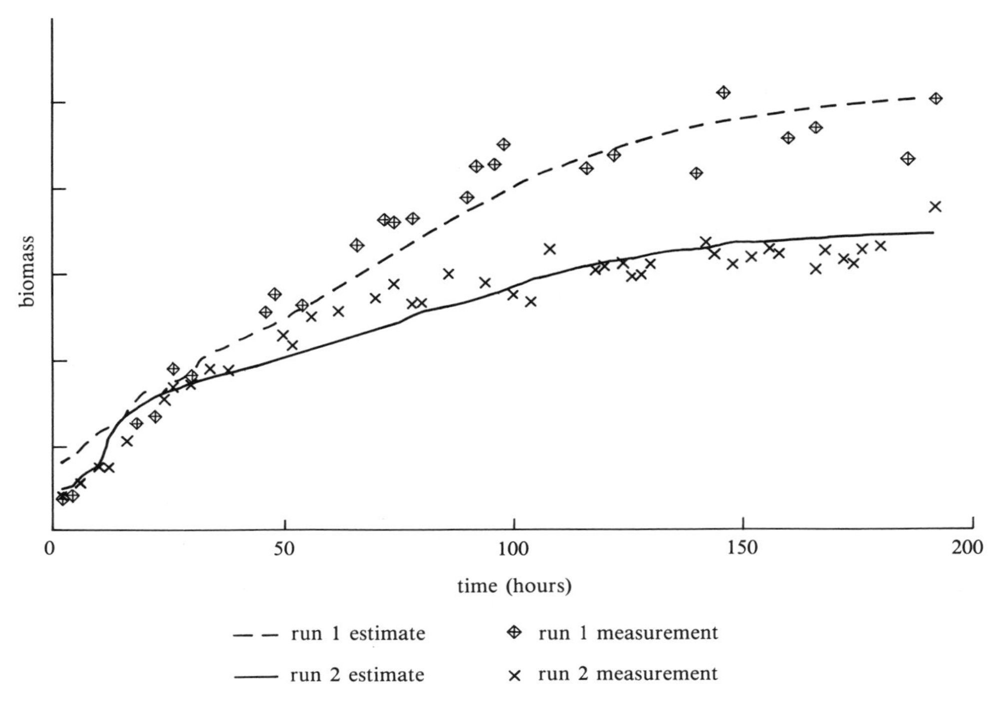

Figure 4.5 Biomass estimation in penicillin fermentation

complex nonlinear systems effectively. Thus, the resulting controller may prove to be more robust in practical situations, where the structure of the nonlinearity is usually unknown: a common cause of failures in current approaches to nonlinear control system design. The predictive control algorithm developed in this contribution is centred around an iterative solution of the following cost function:

$$J = \sum_{i=1}^{N_L} \left\{ \sum_{n=N_{1,i}}^{N_{2,i}} [w_i(t+n) - y_i(t+n)]^2 + \sum_{i=0}^{N_{u,i}} [\lambda_i \, \Delta u_i(t+i)]^2 \right\} \tag{4.13}$$

where $y_i(t)$, $u_i(t)$ and $w_i(t)$ are the controlled output, manipulated input and set-point sequences of control loop 'i' respectively; $N_{1,i}$ and $N_{2,i}$ are the minimum and maximum output prediction horizons; $N_{u,i}$ is the control horizon; and λ_i is a weighting which penalizes excessive changes in the manipulated input of loop 'i', and N_L is the number of control loops.

In the cost function $y_i(t+n)$, $n = N_1, \ldots, N_2$ represent a 'sequence' of future process output values for each loop $i = 1, N_L$, which are unknown. Thus, in order to minimize the cost function, the sequence of process outputs has to be replaced by their respective n step ahead predictions.

n step ahead neural network predictor

For a linear discrete representation, it is easy to show that prediction equations may be developed implicitly (e.g. Peterka, 1984; Albertos and Ortega, 1989). With nonlinear systems, this is not possible and an explicit approach is required. At each sampling instant, (4.11) can be used to develop a one step ahead prediction of the process output $\hat{\mathbf{y}}(t+1 \mid t)$. Thus, by successive substitutions of the previous predictions a sequence of output predictions, $\mathbf{y}(t+n \mid t)$, may be obtained in a recursive manner, i.e.

$$\hat{\mathbf{y}}(t+n \mid t) = J_o \cdot J_2 \cdot J_1[\mathbf{u}(t+n), \hat{\mathbf{y}}(t-1+n)] \tag{4.14}$$

Note that when generating the predictions, known values of $\mathbf{y}$ and $\mathbf{u}$ will be used where appropriate.

Solution of the objective function

With the ability to predict the future outputs, $\hat{\mathbf{y}}(t+n \mid t)$, and with known future set-points then the future controls which will minimize the cost function (4.13) may be determined. In common with most predictive control strategies, an assumption is made concerning future control moves. Beyond the control horizon, $N_{u,i}$, it is assumed that the control action remains constant. The optimization altorithm therefore searches for $N_{u,i}$ control values in order to minimize the cost function. Since the neural network model is nonlinear, an analytical solution of the cost function is difficult, if not impossible. However, adapting a numerical optimization approach enables a solution to the cost function.

As noted earlier, most optimization algorithms employ some form of search technique to scan the feasible space of the objective function until an extremum point is located (e.g. Luenberger, 1973; Edgar and Himmelblau, 1989). The search is generally guided by calculations on the objective function and/or the derivatives of this function. The various procedures available may be broadly classified as either 'gradient based' or 'gradient free'. In an on-line situation, where process measurements are often corrupted by noise, the use of gradient based methods may not be feasible due to their susceptibility to discontinuities. With gradient free methods, Fletcher (1980) showed that the most efficient technique was due to Powell (1964). This method located the extremum of a function using a sequential unidirectional search procedure. Starting from an initial point, the search proceeds according to a set of conjugate directions generated by the algorithm until the extremum is found. This technique was therefore chosen as the basis of the nonlinear predictive control philosophy advocated herein.

Ensuring offset free response

It is also required that set-points are tracked with zero errors. Since control is based upon the prediction of future outputs obtained from a nominal model of the process, offsets may occur due to disturbances and plant–model mismatch. Following the internal model control (IMC) concept the discrepancy between model and process responses can be estimated at each sample instant as:

$$d_i(t) = y_i(t) - \hat{y}_i(t \mid t-1) \tag{4.15}$$

where $y_i(t)$ is the actual process output of loop 'i' at time t and $\hat{y}_i(t \mid t-1)$ is the corresponding process model output. In circumstances where the process noise is significant, the estimate of prediction offset can be filtered in order to reduce the effect of noise and enhance stability. This estimate is then used to 'correct' the predictions obtained from the model, i.e.

$$\hat{y}_{ic}(t+n \mid t) = \hat{y}_i(t+n \mid t) + d_i'(t) \qquad n = N_1, \ldots, N_2 \tag{4.16}$$

where the subscript 'c' denotes a corrected value and $'$ refers to a filtered value. It is this sequence of corrected predictions that are then used in the cost function 4.13.

Summary of algorithm

The implementation procedure of an unconstrained predictive control algorithm within the proposed framework is summarized below:

Step I. For each respective control loop: sample process output, $y_i(t)$, and calculate the error, $d_i(t)$, between $y(t)$ and $\hat{y}(t \mid t-1)$. Filter the error to give $d_i'(t)$.

Step II. The process model is used to generate predictions of process outputs $\hat{y}_i(t+n \mid t)$, $n = N_1, \ldots, N_2$; $i = 1, \ldots, N_L$.

Step III. Correct the predictions from step II using the result of step I, to obtain $\hat{y}_{ic}(t+n \mid t), n = N_1, \ldots, N_2; i = 1, \ldots, N_L$.

Step IV. Formulate cost function (4.12).

Step V. An optimization algorithm is used to determine the sequence of controls: $u_i(t+i_o)$, $i_o = 0, \ldots, N_{u,i}$; $i = 1, \ldots, N_L$.

Step VI. Implement $u_i(t), i = 1, \ldots, N_L$. Wait until next sample and repeat procedure from step I.

The procedure is summarized schematically in Figure 4.6.

4.4.3 Multi-input Single-output (MISO) Dynamic Network Control

The process under consideration is a nonlinear simulation of the 10 stage pilot plant column, installed at the University of Alberta, Canada. It separates a 50–50 wt% methanol–water feed mixture which is introduced at a rate of 18.23 g s^{-1} into the column on the fourth tray. The column is modelled by a comprehensive set of dynamic heat and mass balance relationships (Simonsmeir, 1977). Both the column and the nonlinear model have been used by many investigators to study different advanced control schemes (see, for example Morris *et al.*, 1981). A column

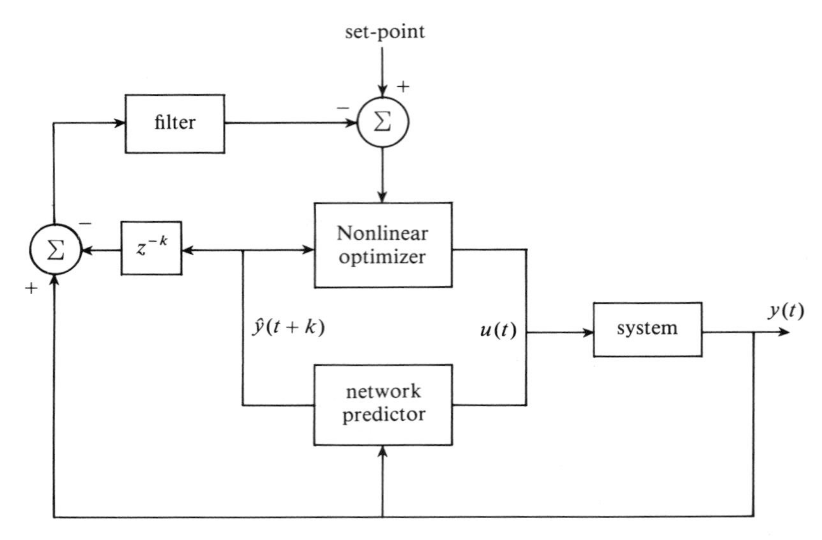

Figure 4.6 Network-based predictive control

schematic is shown in Figure 4.7. The top composition loop is controlled by manipulating reflux flow rate, using a proportional plus integral (PI) controller. The objective is to control the bottom product composition using the steam flow rate to the reboiler as manipulated input. For the purposes of this evaluation study, it was assumed that the column composition measurements were not subject to any analyzer delay (alternatively one could assume perfect inferential estimation). The performance of the proposed predictive control scheme, in conjunction with a neural network based model, was then compared with the performance of conventional PI control and a linear predictive control strategy where the sample time for each controller was 1 minute.

First, the relationships between bottom product composition and steam flow rate were established by the use of both a linear model and a FANN. Both descriptions also took into account the effects of reflux changes (interaction) and feed flow rate (disturbance) on bottom product composition. In developing the linear model, simple first-order plus dead-time transfer functions were used to describe the dynamics of each input–output pair, whilst the neural network model (NNM) used had a (3–4–4–1) structure. After identification of the respective models, they were incorporated within the proposed control scheme to provide, respectively, a linear and a nonlinear (FANN based) predictive controller. Note that the linear predictive controller is basically a variant of the generalized predictive control (GPC) algorithm of Clarke *et al.* (1987). However, the cost function (4.13) is solved using an iterative optimization procedure instead of an analytical approach. The output prediction horizons for both predictive controllers were set as $N_1 = 1$ and $N_2 = 7$ (i.e. the output horizon was set to predict a significant

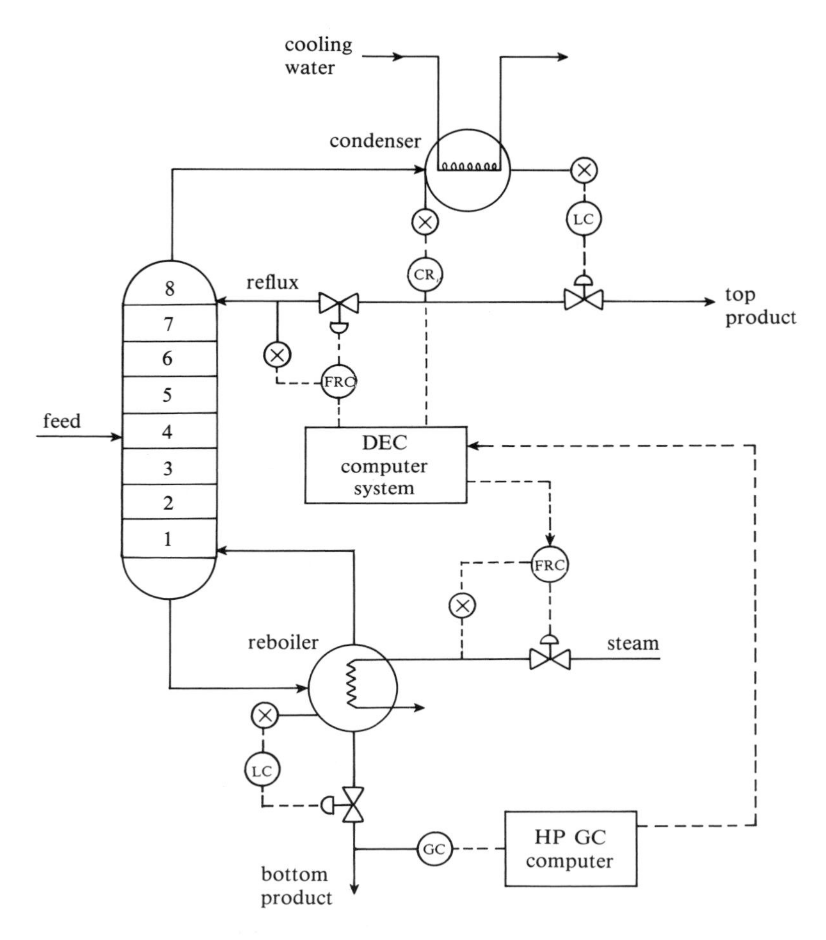

Figure 4.7 Distillation column schematic: AR analyzer recorder, FRC flow recorder/ controller, GC gas chromatograph, LC level controller

proportion of plant rise time), while the control horizons were chosen to be $N_u = 1$. A set of n step ahead output predictions were obtained following the procedure outlined above.

The disturbance rejection properties of both predictive algorithms as well as a 'well-tuned' conventional PI controller were to be compared and evaluated. The control objective was to maintain bottom product composition at 5 wt% methanol, when subject to step disturbances in feed flow rate. The integral of the absolute error (IAE) was used to quantify the performance characteristics of the three controllers.

Figure 4.8 shows the performance of the above controllers. It can be observed

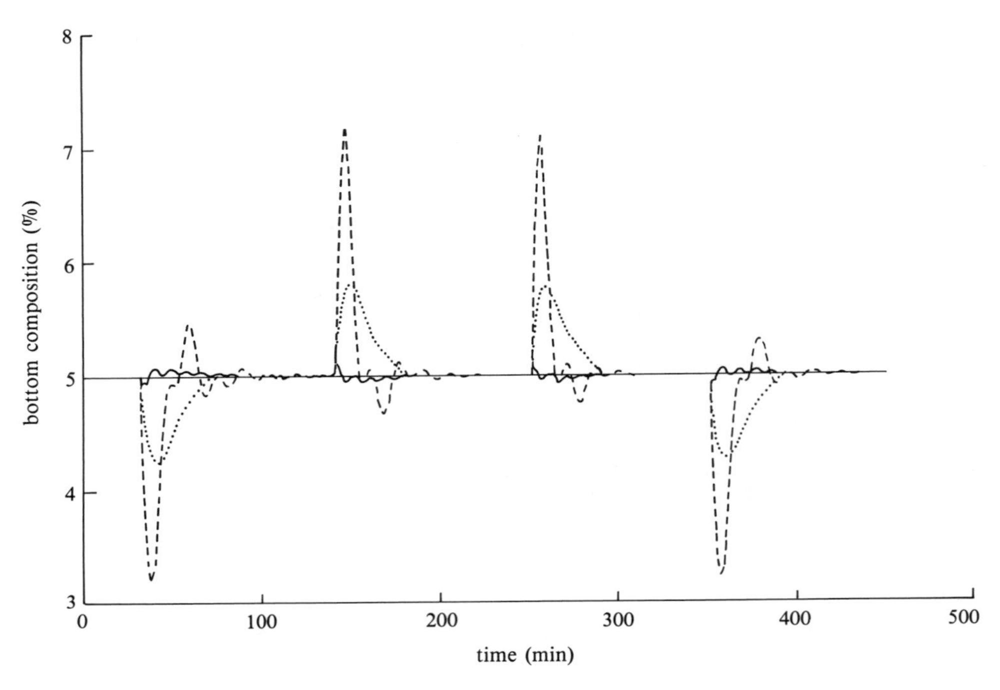

Figure 4.8 Distillation column control: $N = 7$, $\nu = 1$, $\lambda = 1.4$. — net model IAE = 5.7, ··· linear model IAE = 62.1, --- PI control IAE = 78.1

that the PI algorithm resulted in the worst control performance, yielding an IAE of 78.1. This is primarily because the algorithm does not provide feedforward compensation against the feed flow rate changes. Better disturbance rejection characteristics were observed when the linear predictive control strategy was employed. However, bearing in mind that the nominal model used in formulating the linear predictive controller also accounts for the effects of disturbance and loop interactions, the recorded IAE value of 62.1 does not represent a significant improvement over the performance of the well-tuned PI strategy. The best regulatory performance was obtained when the nonlinear predictive control strategy was applied as is evident from Figure 4.8. The corresponding IAE value in this application was 5.7, showing an order of magnitude improvement in closed-loop performance. This superior disturbance rejection behaviour is mainly attributed to the ability of the neural network model to represent the dynamic characteristics of the process more accurately. Since the performance of model-based controllers is directly dependent on the accuracy of the nominal model used in controller synthesis, superior control is achieved.

4.4.4 Multi-input Multi-output (MIMO) Neural Network Based Control

The last section has shown the regulatory performance improvements possible with a MISO neural network predictive controller. However, in obtaining these results from the distillation system, the top composition PI controller was designed in order to reduce the loop interaction effects which have been observed in earlier studies (see, for example, Montague (1987)). The development of a multivariable controller that takes such loop interactions into account offers the promise of a further improvement in the overall control performance. Thus this section investigates the performance of a MIMO neural network predictive control algorithm when applied to the column simulation.

As in the MISO case, the first step in the implementation procedure is the determination of the process model. The neural network model used had eight inputs, two hidden layers with six neurons in each layer and two outputs, i.e. (8–6–6–2). The inputs to the networks were steam flow rate at times $t-1, t-2$ and $t-3$ with a corresponding time history for reflux values. In addition, both top and bottom product composition at time $t-1$ were included. The outputs from the

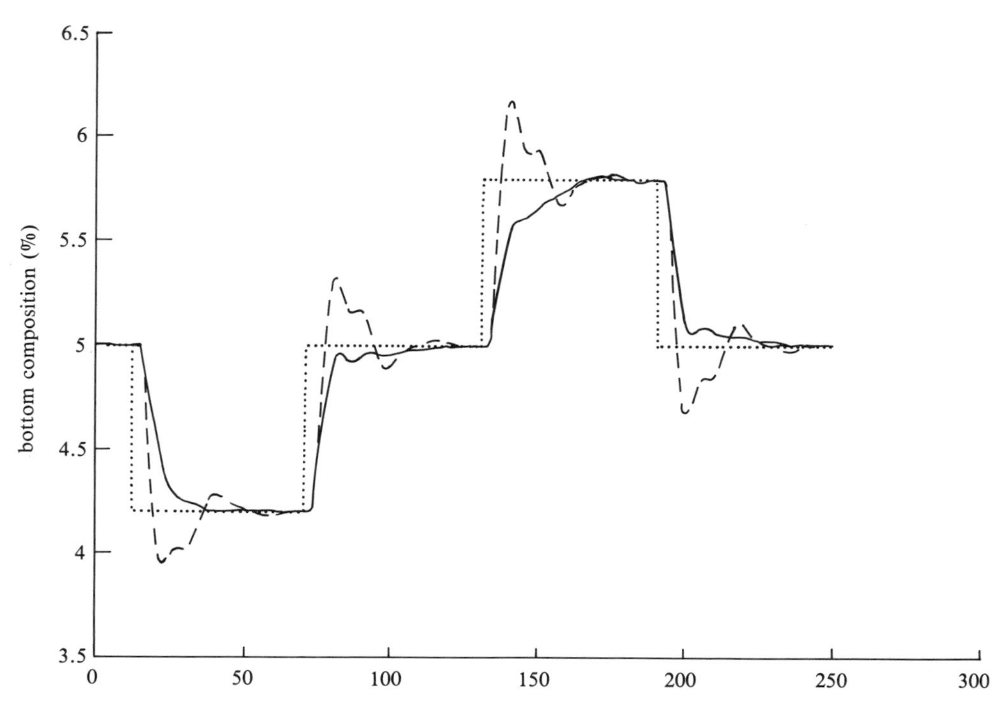

Figure 4.9 Servo control of bottom product composition (MIMO controller): — net control, --- PI control, ··· set point

network were estimates of top and bottom product composition at time t. Having trained the network the model was then utilized within the nonlinear predictive control strategy. Here, the values used for both loops were $N_1 = 1$ and $N_2 = 6$, while the control horizons were chosen to be $N_u = 1$. A small λ weighting was also used for both loops to reduce excessive control activity. The servo response characteristics of the bottom composition loop and the resulting disturbance rejection properties of the top composition loop were investigated. The nonlinear predictive controller was compared to a 'well-tuned' conventional PI algorithm. The control objective was to follow a series of set-point changes in the bottom loop whilst simultaneously maintaining top product composition at 95 wt% methanol. With the nonlinear predictive control philosophy, an IAE of 3.66 was achieved for the top loop and 28.45 for the bottom loop as compared to 11.52 for PI control of the top loop and 31.58 for the bottom loop. For bottom composition control, the IAEs obtained are similar although, as may be observed in Figure 4.9, closed-loop response is less oscillatory. The differing control profiles are a characteristic of the respective control types. However, the significant aspect of these results is due to the decoupling capabilities of the multivariable neural network based controller. The ability of the network model to predict the effects of bottom loop disturbances on

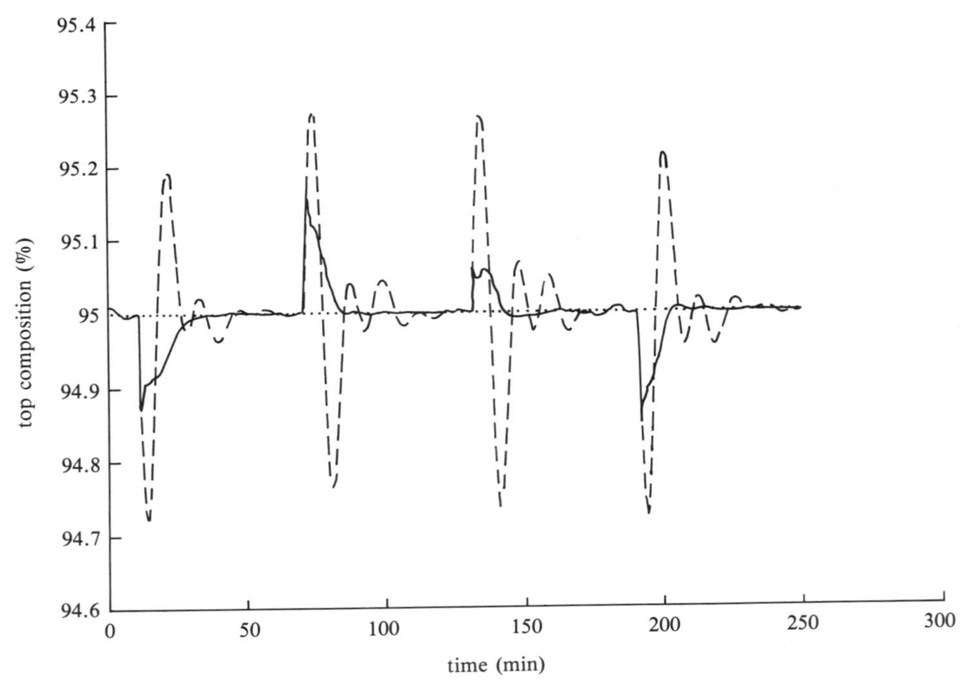

Figure 4.10 Regulatory control of top product composition (MIMO controller): — net control, --- PI control, ··· set point

the output of the top loop enables more effective regulation of the top product composition (Figure 4.10).

4.5 CONCLUSIONS

In this chapter, the concept of artificial neural networks has been briefly reviewed with particular emphasis on their suitability for solving some currently difficult process engineering modelling problems. Application to industrial process data revealed that given an appropriate topology the network could be trained to characterize the behaviour of the system considered. It is our contention, therefore, that ANNs exhibit good potential as 'software-sensors', i.e. sensors that are based upon software rather than hardware alone. Moreover, note that the additional benefits obtained from accurate output prediction, using secondary variables, allows for the effects of load disturbance compensation in a feedforward sense. In addition to inferential estimation, it was also demonstrated that significant improvements in process regulation could be achieved when a neural network was utilized as the basis for a model based predictive control algorithm.

Thus, there is evidence that ANNs can be a valuable tool in alleviating many current process engineering problems. However, it is stressed that the field is still very much in its infancy and many questions still have to be answered. For example, one disadvantage of network models is that, in concert with their linear ARMAX and nonlinear NARMAX counterparts, inspection of the identified structure does not provide any mechanistic information about the system being modelled. The information held is distributed across all the network parameters and bears no relationship to the physical nature of the system. It is therefore difficult to interpret what the network has characterized and as a result predict its abilities and limitations in handling new unseen data. This is particularly important, especially since the problems escalate rapidly with large complex systems. In addition, the generation of training data sets capable of providing all the information necessary for the identification of large dimensioned networks, and the network training itself, could become prohibitive.

Conversely, mechanistic models can impart structure and meaning to the representation. Even though the difficulty in developing such models is often cited as a major reason for using a neural network approach, the use, association or even incorporation of known process knowledge into a network structure is very appealing. Yet another important issue that has to be addressed before the full potential of artificial neural networks can be realized is a method for determining robustness and stability. Nevertheless, given the resources and effort that are currently being infused into both academic and commercial research in this area, it is anticipated that, within the decade, neural networks will have been well established as a valuable tool. Hype, of course, will remain an unwelcome accompaniment; indeed it would be interesting to know just how much interest the approach would have generated if it had not been called 'neural' networks. Despite such reservations, in the short term ANNs may indeed provide some interesting, useful and pragmatic solutions for solving process engineering problems. A panacea they will not be.

ACKNOWLEDGEMENTS

The authors would like to thank the members of the Artificial Neural Network research groups at the Universities of Newcastle and Maryland and the Industrial Neural Network Club members.

REFERENCES

ALBERTOS, P. and ORTEGA, R. (1989) 'On generalised predictive control: two alternative formulations', *Automatica*, **25**(5), 753–5.

BHAT, N., MINDERMAN, P. and McAVOY, T. J. (1989) 'Use of neural nets for modelling of chemical process systems', *Preprints IFAC Symposium Dycord + 89, Maastricht*, pp. 147–53.

BIRKY, G. J. and McAVOY, T. J. (1989) 'A neural net to learn the design of distillation controls' *Preprints IFAC Symposium. Dycord + 89, Maastricht*, pp. 205–13.

CLARKE, D. W., MOHTADI, C. and TUFFS, P. S. (1987) 'Generalised predictive control: Part I: the basic algorithm', *Automatica*, **23**, 137–48.

CYBENKO, G. (1989) 'Approximations by superpositions of a sigmoidal function', *Math. Cont. Signal Systems*, **2**, 303–14.

DI MASSIMO, C., WILLIS, M. J., MONTAGUE, G. A., THAM, M. T. and MORRIS, A. J. (1990) 'On the applicability of neural networks in chemical process control', AIChE Annual Meeting, Chicago.

EDGAR, T. F. and HIMMELBLAU, D. M. (1989) *Optimization of Chemical Processes*, London: McGraw-Hill.

FLETCHER, R. (1980) *Practical Methods of Optimization*, New York: Wiley.

LEONARD, J. A. and KRAMER, M. A. (1990) 'Improvement of the back-propagation algorithm for training neural networks', *Computers Chem. Eng.*, **14**, 337–41.

LIPPMANN, R. P. (1987) 'An introduction to computing with neural nets', *IEEE ASSP Magazine*, April.

LUENBERGER, D. G. (1973) *Linear and Nonlinear Programming*, Reading, MA: Addison-Wesley.

McCULLOCH, W. S. and PITTS, W. (1943) 'A logical calculus of the ideas immanent in nervous activity' *Bull. Math. Bio.*, **5**, 115–33.

MONTAGUE, G. A. (1987) 'Inferential self-tuning control of the fed-batch penicillin fermentation, PhD Thesis, University of Newcastle, UK.

MONTAGUE, G. A., WILLIS, M. J., MORRIS, A. J. and THAM, M. T. (1991) 'Artificial neural network based multivariable predictive control', in proceedings of IEE International Conference ANN 91, Bournemouth, 119–23.

MORRIS, A. J., NAZER, Y., WOOD, R. K. and LIEUSON, H. (1981) 'Evaluation of self-tuning controllers for distillation column control', in *IFAC Conference on Digital Computer Approach to Process Control, Dusseldorf*, Oxford: Pergamon, pp. 345–54.

PETERKA, V. (1984) 'Predictor based self-tuning control', *Automatica*, **20**, 39–50.

POWELL, M. J. D. (1964) 'An efficient method for finding the minimum of a function of several variables without calculating the derivatives', *Comput. J.*, **7**, 155–62.

RUMELHART, D. E. and McCLELLAND, J. L. (1986) *Parallel Distributed Processing: Explorations in the Microstructure of Cognition. Vol. 1: Foundations*. Cambridge, MA: The MIT Press.

SIMONSMEIR, U. F. (1977) 'Modelling of a nonlinear binary distillation column', MSc Thesis, University of Alberta, Edmonton, Canada.

STEPHANOPOULOS, G. (1984) *Chemical Process Control*, Englewood Cliffs, NJ: Prentice Hall.

WANG, Z., THAM, M. T. and MORRIS, A. J. (1992) 'Multilayer feedforward neural networks: A canonical form approximation of nonlinearity', *Int. J. Cont.* **56**(3), 655–72.

WATROUS, R. L. (1987) 'Learning algorithms for conectionest networks: Applied gradient methods of non-linear optimisation', in *Proceedings of the International Conference on Neural Networks*, II, pp. 619–27.

WILLIS, M. J., DI MASSIMO, C., MONTAGUE, G. A., THAM, M. T. and MORRIS, A. J. (1991) 'Artificial neural networks in process engineering', *Proc. IEE, Pt D*, **138**(3), 256–66.

5

The B-spline Neurocontroller

M. Brown and C. J. Harris

5.1 INTRODUCTION

Intelligent controllers determine their structure and consequent actions in response to external commands from the observed input/output behaviour of the plant with minimal reference to a mathematical or model based description of the plant. It should be noted that few restrictions are placed upon the linearity, time invariance, deterministic character or complexity of the plant; rather it is for processes which are nonlinear, time varying, stochastic and complex that intelligent control is aimed. It is the authors' view that there is little to be gained in applying intelligent control to linear time invariant systems: classical control theory suffices for such systems. Two approaches dominate the intelligent control field: (a) self-organizing fuzzy logic intelligent controllers (SOFLICs) (Harris *et al.*, 1992) and (b) neural net based controllers (neurocontrollers).

SOFLICs are relatively new, using experiential evidence to determine a relational matrix (in terms of fuzzy-like production rules) in which the fuzzy input/output information is stored and updated by new information. The type and amount of generalization or interpolation are determined by the size, shape and overlap of the fuzzy sets defined on each of the inputs. As with conventional adaptive control, there are two approaches to SOFLICs: (a) the *direct* SOFLIC which utilizes a fixed fuzzy performance index in order to adapt the fuzzy control rules in response to the performance of the plant (Sutton and Jess, 1991), and (b) the *indirect* SOFLIC (Moore, 1991) which generates a fuzzy relational matrix of the plant and then inverts this model in order to find the control which realizes the desired next state. The indirect method has the advantage that the system's response can be changed by simply supplying a new desired next-state rule base, rather than retraining the whole fuzzy controller. However, unlike the direct SOFLIC it is not yet multivariate. Benchmark experience on SOFLICs (Moore, 1991) indicates that they are extremely robust with respect to parametric variations, external disturbances, system nonlinearities and system faults. Equally they have very rapid training or learning cycles and recently their mathematical learning convergence properties have been established (Brown and Harris, 1991a).

However, this discussion will concentrate upon a new class of neural networks, designed for control applications, the B-spline networks (Basis-spline nets) which can be shown to have a one-to-many relationship with fuzzy production rules and are similar to the more common neural architectures: the Albus CMAC, radial basis function networks and the Kanerva memory model.

Artificial neural networks (ANNs) originate from the work performed in the 1950s and 1960s; indeed the first neurocontroller was developed by Widrow in 1963 which was taught to control an inverted pendulum. ANNs have many desirable properties which make them suitable for intelligent control applications:

1. ANNs learn by experience rather than by programming.
2. ANNs have the ability to generalize, that is the networks map similar inputs to similar outputs.
3. ANN architectures are generally distributed, inherently parallel and potentially real-time.
4. ANNs can form arbitrary continuous nonlinear maps and hence *few* a priori assumptions about the plant dynamics need be made.

ANNs are typically constructed from parallel layers of simple computational nodes, with weighting elements between nodes that are adjusted (by some optimization procedure) to yield the appropriate input/output mapping. The most popular ANNs in neurocontrol are the multilayer perceptron (MLP) (Narendra and Parathasarathy, 1990), functional link networks (FLN) (Pao, 1989), associative memory networks such as the Albus CMAC (Miller, 1987) and radial basis function (RBF) (Chen and Billings, 1992), and the B-spline network which is discussed below. Each layer in the MLP forms a weighted sum of the input variables followed by a nonlinear transformation. The weights are then adjusted in order to minimize some error criterion. The strength *and* the weakness of the MLP, trained using a gradient descent learning algorithm such as backwards error propagation (BEP), is that nonlinear optimization is being performed, generating a complex cost function with many local minima which traps the gradient descent learning rules. In contrast associative memory ANNs first perform a nonlinear transformation of the input space (usually to a much higher dimensional space), followed by a single *linear* layer which forms a weighted combination of the transformed input. The first layer is generally fixed in which minimal a priori knowledge about the plant can be incorporated, whereas the weights in the second layer are adaptive, resulting in a convex cost function for which there exists a single global minima.

In this paper an adaptive or learning nonlinear controller is developed based upon multivariate B-splines (more commonly found in numerical interpolation and graphical applications). However, before introducing this new neurocontroller, general aspects of adaptive intelligent control architectures, cost functions and associated learning algorithms, with implications for the popular BEP algorithm, are discussed.

5.2 INVERSE PLANT MODELLING

Inverse plant modelling, as shown in Figure 5.1, is a common structure in control theory and signal processing. Initially we are concerned with single-input single-output plants, and controllers which are linear in their parameter vector **w** (this includes such nonlinear associative memory algorithms as the Albus CMAC, B-splines and radial basis functions, as well as conventional linear controllers) and the effect upon adapting the control parameter vector **w** through the minimization of different cost functions. A measure of how well the controller is performing can be specified by the cost functions

$$J_u = \frac{1}{2T} * \sum_{t=1}^{T} [\hat{u}(t) - u(t)]^2 \tag{5.1}$$

$$J_y = \frac{1}{2T} * \sum_{t=1}^{T} [\hat{y}(t) - y(t)]^2 \tag{5.2}$$

where $u(t)$ is the control signal at time t, $y(t)$ is the plant output and ^ denotes the desired value of the appropriate signal.

When **w** is trained off-line and the desired control signal is available, as shown in Figure 5.2, the cost function given in (5.1) is generally chosen. However, for online adaptive control only $\hat{y}(t)$ is generally available, so training **w** is performed

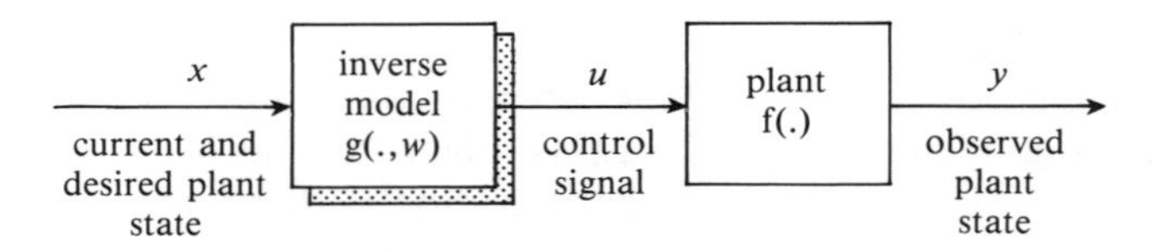

Figure 5.1 Inverse plant modelling

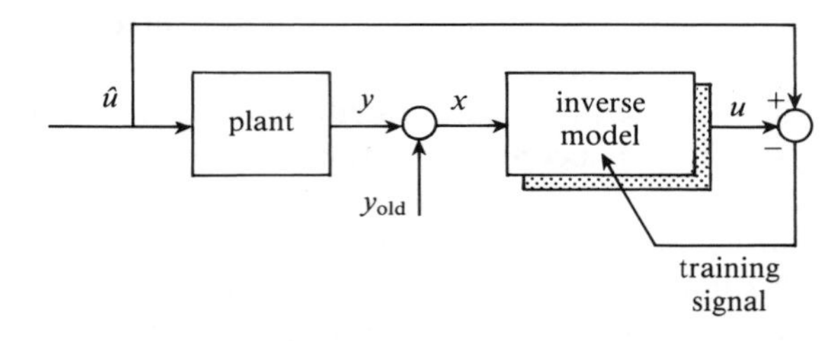

Figure 5.2 Off-line controller modelling

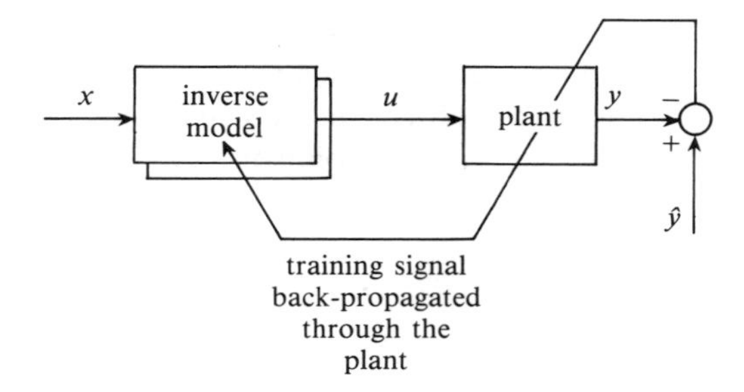

Figure 5.3 On-line controller modelling

using the cost function given in (5.2) and the plant output error can be 'back-propagated through the plant' as illustrated in Figure 5.3.

The combination of these two training algorithms for performing general and then specialized learning was first proposed in Psaltis *et al.* (1987), where it was experimentally shown that the learning rates for the algorithms were substantially different, as well as noting that little benefit was obtained by performing off-line learning in order to produce an initial parameter vector for use in the on-line learning algorithm. In the following sections this phenomenon is explained by comparing the cost functions (5.1) and (5.2), showing that they have, in general, different minima and comparing two different means of back-propagating the plant output error through the plant to provide a signal with which to train the controller's parameter vector.

For these two on-line learning algorithms, it is necessary to constrain the plant to be either smooth and monotonic or invertible. In particular, one such 'plant' is the sigmoid function which is employed in many of the neural network algorithms and was fundamental in the development of the BEP algorithm for MLPs (Rumelhart *et al.*, 1986). In this context it is possible to explain the slow convergence rates experienced when using the batch training BEP algorithm and to generate an initial value for the learning gains (based upon knowledge about the derivative of the sigmoid) which offers a substantial increase in the rate of learning convergence.

5.2.1 Parameter Adaptation

Given a performance cost function J (typically error squared), such as (5.1) or (5.2), and a time varying control coefficient vector $\mathbf{w}(t)$ (where the time index has been made explicit), then gradient based adaptation laws update $\mathbf{w}(t)$ according to

$$\Delta w_i(t-1) = -\delta_i \frac{\delta J(t)}{\delta w_i(t-1)} \tag{5.3}$$

where $\Delta w_i(t-1)$ is defined to be $w_i(t) - w_i(t-1)$, and δ_i is the ith learning rate.

If J is linear in the parameter vector $\mathbf{w}$, then the update rule given in (5.3) performs gradient descent on an error surface which has a single global minimum at $\hat{\mathbf{w}}$ and convergence is assured, subject to the usual constraints of a sufficiently exciting and bounded input signal, etc. (Johnson, 1988). This holds for a linear controller and the cost function J_u.

Off-line adaptation

The difference $\varepsilon_u(t) = [\hat{u}(t) - u(t)]$ provides an instantaneous measure of how well the controller is performing (note that for on-line learning the desired control signal $\hat{u}(t)$ is generally unavailable). For off-line training, Figure 5.2, a persistently exciting control signal (across the relevant domain) is applied to the plant and the previous plant state as well as the current plant state is recorded and transformed to form the control input vector $\mathbf{x}(t)$. Hence the controller's output, for an input $\mathbf{x}(t)$, can be compared with the desired control signal $\hat{u}$, and this error signal can be used to train the controller's coefficient vector $\mathbf{w}(t)$ using an instantaneous gradient based least-mean square (LMS) learning algorithm such as

$$\Delta w_i(t-1) = \delta_i \varepsilon_u(t) x_i(t) \tag{5.4}$$

This technique can also be used for on-line training, when the plant's state, due to the previous input, forms the desired plant state relative to the previous plant state. The controller's output for this new input can then be compared against the actual control signal, which is the desired control signal. This technique, however, will not enable the plant to explore new regions of the state-space: in particular if the coefficient vector is identically zero, both the desired and actual control signal will be zero, hence the coefficient vector will remain unaltered.

On-line adaptation

On-line adaptation is generally necessary in order to cope with plants which have time varying parameters or unmodelled dynamics. Here the desired control signal $\hat{u}(t)$ is generally unavailable; however, the difference between the desired and actual plant behaviour $\varepsilon_y(t) = [\hat{y}(t) - y(t)]$ can be employed as the performance measure. Hence gradient based LMS algorithms can be employed which attempt to minimize the cost function J_y given in (5.2). At time t, the *instantaneous* estimate of the partial derivative of $J_y(t)$ with respect to $\mathbf{w}(t-1)$ can be obtained by differentiating (5.2) using the chain rule (Jordan, 1989):

$$\frac{\delta J_y}{\delta w_i} = \frac{\mathrm{d} J_y}{\mathrm{d} y} \frac{\mathrm{d} y}{\mathrm{d} u} \frac{\delta u}{\delta w_i} \tag{5.5}$$

which evaluates to:

$$\frac{\delta J_y}{\delta w_i} = -\varepsilon_y(t)\frac{\mathrm{d}y(t)}{\mathrm{d}u(t)}\, x_i(t) \tag{5.6}$$

where $\mathrm{d}y/\mathrm{d}u$ is the plant Jacobian.

Alternatively, the relation for *small differences:*

$$\varepsilon_u(t) \approx \frac{\mathrm{d}u(t)}{\mathrm{d}y(t)}\, \varepsilon_y(t) \tag{5.7}$$

could be used to minimize J_u directly. So to a first approximation:

$$\frac{\delta J_u}{\delta w_i} = -\varepsilon_y(t)\, \frac{\mathrm{d}u(t)}{\mathrm{d}y(t)}\, x_i(t) \tag{5.8}$$

where $\mathrm{d}u/\mathrm{d}y$ is the inverse plant Jacobian.

Minimizing the cost functions J_y and J_u has produced very similar update rules, i.e. (5.6) and (5.8). Their only difference is that (5.6) is multiplied by the plant Jacobian (transposed) where (5.8) is multiplied by the inverse plant Jacobian. For plants with low (high) gains and employing unity learning rates, using (5.6) instead of (5.8) can result in algorithms which are slow to converge (unstable).

It should be noted that if the plants are linear ($\mathrm{d}y/\mathrm{d}u$ is constant) or if $\mathrm{d}y/\mathrm{d}u$ is constant for the training examples (consider the XOR problem) then the learning rules derived from (5.6) and (5.8) are equivalent because they can be scaled by the appropriate learning rate δ in (5.3). Alternatively if the plant is nonlinear and the controller can learn to provide an exact control signal (i.e. $\hat{u}(t) \equiv u(t) \forall\, t$) then the global minimum for the two learning algorithms is identical. However, these assumptions are unrealistic and so to provide insight into the difference between these two learning algorithms, (5.1) may be approximated by

$$\begin{aligned} J_u &= \frac{1}{2T} * \sum_{t=1}^{T} [\hat{u}(t) - u(t)]^2 \\ &= \frac{1}{2T} * \sum_{t=1}^{T} \left(\frac{\mathrm{d}u(t)}{\mathrm{d}y(t)} [\hat{y}(t) - y(t)]\right)^2 \end{aligned} \tag{5.9}$$

That is the cost function J_u forms an approximation to J_y where the plant errors $[\hat{y}(t) - y(t)]$ are weighted according to the value of the inverse Jacobian evaluated at the point $y(t)$. In general, the plant is not linear and the controller cannot realize an exact control signal over the whole range, so the value of the weight vector which provides the minimum MSE will be different. The appendix contains an example for which the 'optimal' weight vector is different owing to controller modelling mismatch.

Previous training strategies

Within the neurocontrol field several research groups have recently proposed on-line learning control algorithms (Psaltis *et al.*, 1987; Miller, 1987; Saerans and Soquet, 1989; Chen 1989). Psaltis *et al.* originally proposed using a combination of general off-line learning and specialised on-line learning where the plant error was propagated back through the plant, (5.6). It was noted that 'they follow different paths to the minimum' and 'though there may be some benefit to performing generalized training prior to specialised training, these simulations show no clear advantage in doing so.' These differences in the convergence rates and high initial errors when specialized (on-line) training was employed is due to the poor error signal provided by (5.6) and the weighted relationship in (5.9) which shows that minimizing the two cost functions is *not* equivalent.

Saerens and Soquet proposed using the sign of the plant Jacobian (obtained a priori) as the multiplying factor in (5.6) in order to minimize J_y. Similarly, in Miller (1987) and Chen (1989), a coarse, constant estimate of the inverse Jacobian was used in (5.8) to provide the training signal. In approximating the plant Jacobian/inverse Jacobian by a constant, (5.6) and (5.8) are equivalent and the rate of convergence will be dependent upon how closely they approximate the inverse Jacobian as this provides for the most rapid training. However, it should be emphasized that an approximation to the cost function J_u is being minimized rather than J_y.

5.3 BACKWARDS ERROR PROPAGATION FOR MULTILAYER PERCEPTRONS

Here the training rule for a single layered network is discussed, where the network contains a single output sigmoid nonlinearity, Figure 5.4. The reason for the slow convergence of BEP is discussed and initial approximate values for the learning rates of the individual layers are derived.

5.3.1 The Sigmoid

In Section 5.2.1 it was implicity assumed that the plant was invertible in deriving (5.8). One of the simplest such functions is the sigmoid given by

$$y = f(u) = \frac{1}{1 + \mathrm{e}^{-u}} \in (0, 1) \qquad (5.10)$$

whose derivative is given by

$$\frac{\mathrm{d}y}{\mathrm{d}x} = \mathrm{e}^{-u} \frac{1}{(1 + \mathrm{e}^{-u})^2} = y(1 - y) \in (0, 1/4) \qquad (5.11)$$

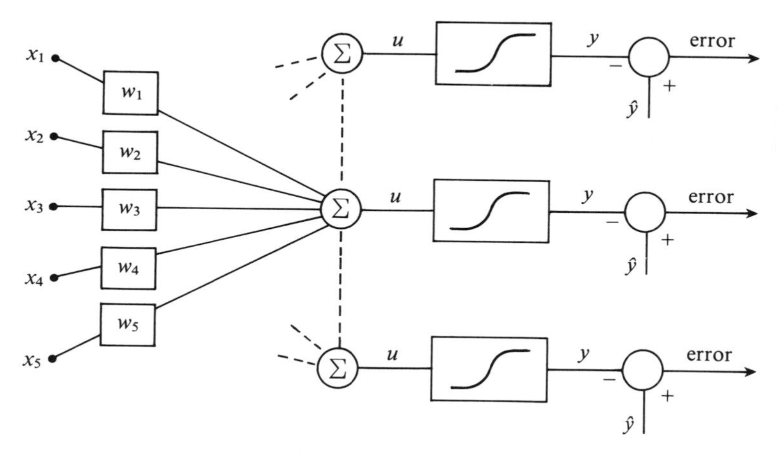

Figure 5.4 Single layer neural network

Now suppose an error in $\mathbf{w}(t)$ produces an error $\varepsilon_u(t)$, which results in an error $\varepsilon_y(t)$ so that they are related by (5.7). Then the BEP algorithm multiplies $\varepsilon_y(t)$ by the forward derivative and updates $\mathbf{w}(t)$ according to

$$\Delta w_i(t-1) = \delta_i \mathrm{f}'(u(t))\varepsilon_y(t)x_i(t) \tag{5.12}$$

or, to a first approximation,

$$\Delta w_i(t-1) = \delta_i \mathrm{f}'(u(t))^2\varepsilon_u(t)x_i(t) \tag{5.13}$$

Here the error $\varepsilon_u(t)$ is multiplied by the square of the plant jacobian which lies in the interval (0, 1/16] with an expected value of 1/30. This factor is one of the reasons for the slow convergence of error back-propagation. If instead of minimizing the cost function J_y directly, the algorithm attempted to minimize J_u instead, then the following training rule would be derived

$$\Delta w_i(t-1) = \delta_i \frac{1}{\mathrm{f}'(u(t))} \varepsilon_y(t)x_i(t) \tag{5.14}$$

and so employing the relationship given in (5.7)

$$\Delta w_i(t-1) = \delta_i \varepsilon_u(t)x_i(t) \tag{5.15}$$

Hence for small $\varepsilon_u(t)$, the training signal is of the correct magnitude, as illustrated in Figure 5.5. It should again be emphasized that the cost function J_u is being minimized rather than J_y and in the case of model mismatch the two algorithms given in (5.13) and (5.16) will converge to different optimal weight vectors (see the appendix).

An attempt to encode this information into the original BEP algorithm was performed in Rigler *et al.* (1991). Here the learning rate for the output layer was set to be 6, and the learning rates for the hidden and input layers, j layers from the

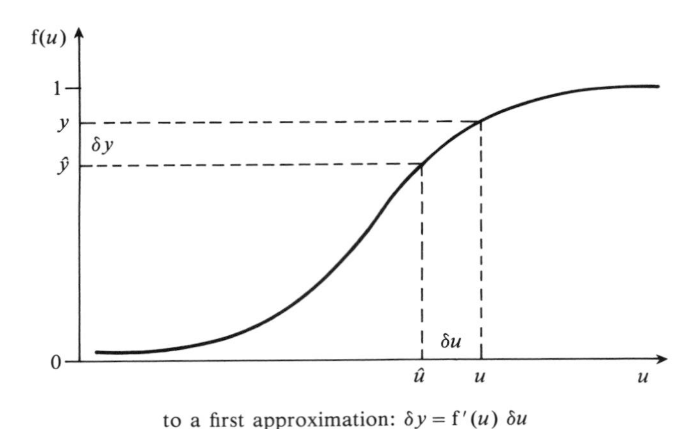

to a first approximation: $\delta y = f'(u)\ \delta u$

Figure 5.5 Sigmoid input–output error relationship

output, was chosen to be 6^j. This factor of 6 is used because the expected value of $f'(u)$ is 1/6 and so an attempt is made to 'normalize' the BEP algorithm. However even these initial learning rates may be underestimated. Comparing (5.13) and (5.15), they differ in magnitude by the square of the inverse Jacobian. So it would seem reasonable to set the initial learning rate to be the inverse of the expected value of the square of the forward Jacobian.

5.3.2 Multilayer Networks

Consider now the training algorithms for an n-layer multilayer network. Suppose that there exists an error $\varepsilon_u(t)$ which is then passed through n sigmoids to obtain the output (i.e. the output error in the previous passage must be passed through one sigmoid, the first hidden layer must be passed through two sigmoids, etc.). Then the 'error' propagated back by BEP includes a scaling factor which is given by

$$\prod_{l=1}^{n} f'(u(l))^2 \tag{5.16}$$

Now as the expected value of $f'(u)^2$ is 1/30, the expected value of (5.16) is approximately $1/30^n$ and so this is one reason for the slow convergence in MLPs trained using BEP. However, this observation can be used to provide an initial approximation to the learning gains. It should also be noted that the weight update should be normalized by the expected size of the inputs.

If the 'optimal' learning rates were used for each layer then each layer would receive a full correction, causing instability in networks with more than one layer. Instead each layer should be scaled by a nonnegative factor, with the sum over all the factors summing to a value less than or equal to 1. These factors would then indicate how plastic each layer was, for instance a simple updating strategy would be to keep the weights in the hidden layer fixed and only adapt the weights in the output layer (this optimization strategy then has a single, global minima). Finally, in order to filter out modelling and measurement noise, the learning rates should tend to zero as time increases (Section 5.5.3).

5.4 B-SPLINES FOR NONLINEAR MODELLING

In recent years ANNs have received a lot of attention due to their improved architectures and nonlinear learning algorithms. These improvements enable the user to model arbitrary continuous nonlinear mappings, learning to extract the underlying functional relationship from the supplied training set. However, these learning algorithms are (necessarily) slow, convergence cannot be established in any sense, and increasing the state-space covered by the training set results in the whole network having to be retrained in order to form the relevant (internal) concepts. Finally the relationship formed by the network is generally not very apparent to the designer. Recent research has resulted in improved (but more complex) training algorithms, but the problems of forming a complete cost function which will contain (recent) training data from the whole of the state-space, the convergence issue, and the opaque relationship extracted by the network still remain. Additionally for the purposes of neurocontrol any ANN based learning algorithm must be *temporally stable*, which refers to the ability to absorb new information (plasticity) whilst retaining knowledge (or rules) previously encoded across the network (stability). The dilemma is, how does an intelligent self-organizing system resolve the dichotomy between plastic and stable modes in order to achieve plasticity without random or chaotic learning and stable learning in order to incorporate unexpected changes in the process, plant or environment? MLPs are temporally unstable, changing all the weights in response to outputs errors, have no global convergence proofs, they are not generally real-time and are therefore inappropriate for neurocontrol.

To overcome these problems, B-spline ANNs are proposed to be used in areas of on-line adaptive modelling and control and static off-line design. A B-spline net is formed from a linear combination of basis functions which are piecewise polynomials of order k (Cox, 1990):

$$y(t) = \mathbf{a}^T(t)\mathbf{w}(t-1) \tag{5.17}$$

where $y(t)$ is the output of a B-spline network at time t (without loss of generality, single-output networks are considered), $\mathbf{a}(t)$ is the basis function output vector and $\mathbf{w}(t-1)$ is the weight vector.

Hence the output of the network is a piecewise polynomial of order k and so can approximate continuous nonlinear functions arbitrarily well. The basis functions nonlinearity transform the input into a higher dimensional space in which the desired mapping is approximately linear. This transformation is such that only a small number of the variables a_i will be nonzero (or equivalently the basis functions have compact support). For adaptive modelling, this map is fixed a priori so the adaptation only occurs for the linear set of parameters (weights) $\{w_i\}_{i=1}^{p}$. Hence on-line adaptive algorithms such as least-mean square (LMS) or normalized least-mean square (NLMS) can be used, and convergence can be established subject to the usual conditions of a sufficiently exciting input signal, bounded input, etc. These algorithms run in real-time because only a small number of the variables a_i will be nonzero, and are temporally stable – dissimilar inputs will map to a completely different set of nonzero a_i and hence the updated weight set will be completely different. For off-line design these networks have the advantage of being transparent (due to the compact support property of the basis functions), with the dependence of the network's output on a particular weight being clear.

5.4.1 Polynomial Basis Functions

From (5.17) the output of a B-spline network is a linear combination of the outputs of the (piecewise) polynomial basis functions of order k (specified by the designer). These polynomial basis functions map the network's input $\mathbf{x}(\in R^n)$ into an alternative representation $\mathbf{a}(\in [0, 1]^p$, where p is the number of basis functions) in which only a small number $\rho(=k^n)$ of the variables will have a nonzero value. In order to describe the form of these basis functions, the domain on which they are defined must first be specified.

A quantized input space

Consider one of the network's inputs x_i which is defined on the finite interval I_i:

$$I_i = \{x_i : x_i^{\min} \leqslant x_i \leqslant x_i^{\max}\} \tag{5.18}$$

and an associated partition of this interval given by

$$x_i^{\min} < \lambda_{i,1} \leqslant \lambda_{i,2} \leqslant \cdots \leqslant \lambda_{i,R_i-1} < x_i^{\max} \tag{5.19}$$

The point $\lambda_{i,j}$ is termed the jth *interior knot* of the ith coordinate. As well as these interior knots, a set of *exterior knots*, $\lambda_{i,j} : j \leqslant 0$ or $j \geqslant R_i$, must also be defined, of which there are k at each end, in order to complete the knot set λ_i:

$$\ldots\lambda_{i,-1} \leqslant \lambda_{i,0} \leqslant x_i^{\min},\ x_i^{\max} \leqslant \lambda_{i,R_i} \leqslant \lambda_{i,R-i+1}\ldots \tag{5.20}$$

Generally it is assumed that all the left-hand exterior knots are placed at the point $x_i^{\min}$ and all the right-hand exterior knots are placed at $x_i^{\max}$. However, this can cause slower rates of convergence when using LMS based adaptation laws (see Section 5.5) and, in the case of approximating a periodic function, it may be advantageous for the exterior knots to coincide with the interior knots. If two interior knots coincide then it is termed a *multiple* or *coincident* knot, similarly if a knot occurs at a unique location it is termed a *simple* knot. This knot set divides the interval I_i up into a set of R_i subintervals $I_{i,j}: 0 < j < R_i - 1$, defined by

$$I_{i,j} = x_i : \begin{cases} x_i \in [\lambda_{i,j}, \lambda_{i,j+1}] & \text{if } j = 0, 1, \ldots, R_i - 2 \\ x_i \in [\lambda_{i,j}, \lambda_{i,j+1}] & \text{if } j = R_i - 1 \end{cases} \tag{5.21}$$

Note that the subinterval $I_{i,j}$ is empty if $\lambda_{i,j} = \lambda_{i,j+1}$.

For a one-dimensional input space a series of R_1 subintervals (possibly empty) have been defined upon a closed and bounded interval of the real line. For an n-dimensional input space, these knots define an n-dimensional lattice contained within a closed and bounded hyper-rectangle $\subset R^n$.

It is these spaces on which the basis functions are defined. It has been implicitly assumed that the knot set will be supplied by the designer. As the output of the network depends nonlinearly upon the set of parameters, nonlinear adaptive rules have been developed (Cox *et al.*, 1990; Cadzow, 1990). However, this reintroduces problems associated with global convergence and temporal stability and so it has been assumed in the remainder of this chapter that the knot set is given and fixed.

Univariate basis functions

Denote the output (or membership) of a univariate basis function of order k by $N_j^k(x)$ (so that $a_j \equiv N_j^k(x)$), which is defined upon the knots $\lambda_{j-k}, \lambda_{j-k+1}, \ldots, \lambda_j$ (in this section the subscript i has been suppressed). If these knots are coincident then the output is identically zero, otherwise it can be calculated from the recurrence relationship

$$N_j^k(x) = \left(\frac{x - \lambda_{j-k}}{\lambda_{j-1} - \lambda_{j-k}}\right) * N_{j-1}^{k-1}(x) + \left(\frac{\lambda_j - x}{\lambda_j - \lambda_{j-k+1}}\right) * N_j^{k-1}(x)$$

for $k > 1$, otherwise

$$N_j^k(x) = \begin{cases} 1 & \text{if } x \in I_{j-1} \\ 0 & \text{otherwise} \end{cases} \tag{5.22}$$

This relationship provides an efficient and stable relation for calculating the output of the basis functions. Also the compact support property of the basis functions means that for any input lying in the interval I, at most k of the basis functions will have a nonzero output. The outputs of these basis functions are always nonnegative, and positive on the open interval on which they are defined (for $k = 1$, the interval may be closed at either or both ends), hence their support is compact. For any $x \in I$, the sum over all the basis functions is unity, which results in them sometimes being called normalized basis splines. Finally $N_j^k(x) \in C^{k-2}(x^{\min}, x^{\max})$ which means that (for simple knots) $N_j^k(x)$ and its derivatives up to the $(k-2)$th are continuous over I. In Figure 5.6, basis functions of orders 1, 2 and 3 are shown.

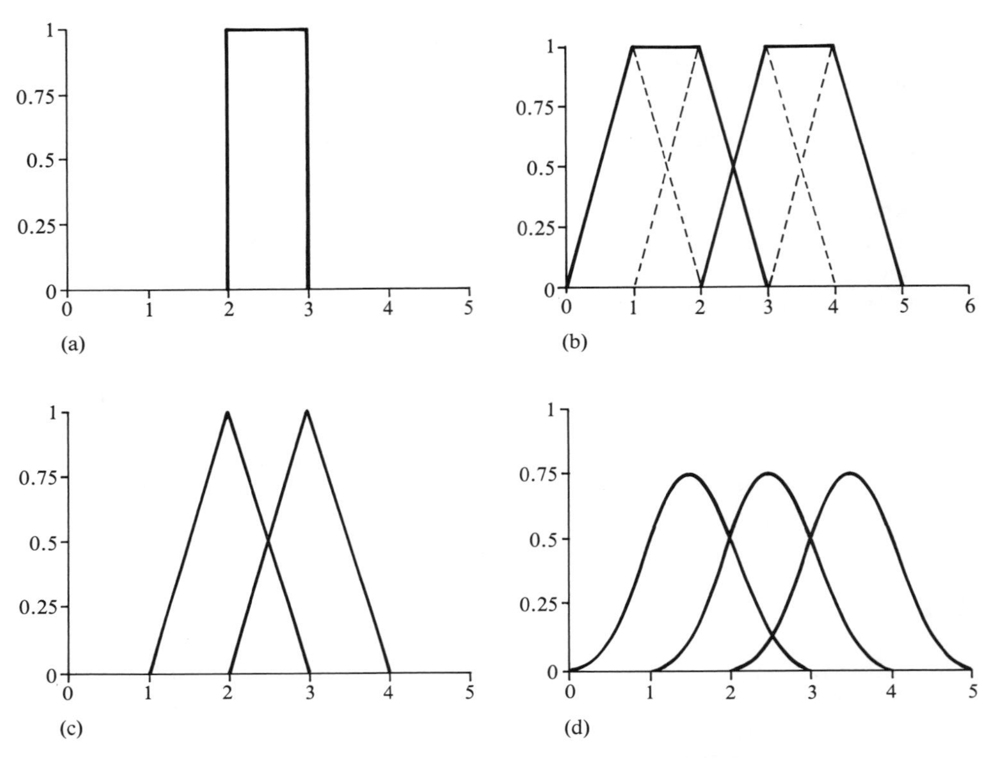

Figure 5.6 Splines of orders 1–3: (a) order 1 – piecewise constant basis spline; (b) order 2 – trapezoidal basis splines (disjunction); (c) order 2 – piecewise linear basis splines; (d) order 3 – quadratic basis splines

It should be noted that an alternative representation of the basis function over an interval I_i is simply a polynomial of order k. B-splines provide an efficient representation of this form while ensuring the continuity constraints between neighbouring intervals.

Multivariate basis functions

Consider a multivariate input $\mathbf{x}(\in R^n)$, a lattice defined across the input space as defined above and k denoting the order of the basis functions (it is possible for basis functions of different orders to be defined on the different input variables). Then the n-dimensional multivariate basis functions are formed from the product, over every possible combination, of the univariate basis functions. Hence to find the output of a particular multivariate basis function, the output of each univariate basis function is calculated, and then combined using the product operator (similar to the fuzzy AND operator). Hence the output of the multivariate basis function is nonnegative, and positive iff all of the univariate basis functions are positive, and so retains the compact support property. Also because the product operator is chosen, the multivariate basis functions are normalized (sum to unity) and, for simple knots, are $C^{k-2}(x^{\min}, x^{\max})$. 2D multivariate basis functions of orders 1, 2 and 3 are illustrated in Figure 5.7.

Again it should be emphasized that a B-spline network of order 1 is simply a look-up table and a B-spline network of order 2 generalizes piecewise linear interpolation. However, this is not a criticism of the technique; rather the algorithm provides an efficient method for generalizing these basic concepts to higher-order interpolants which generalize.

5.4.2 Computational Cost

The memory requirements of the algorithm are proportional to the number of basis

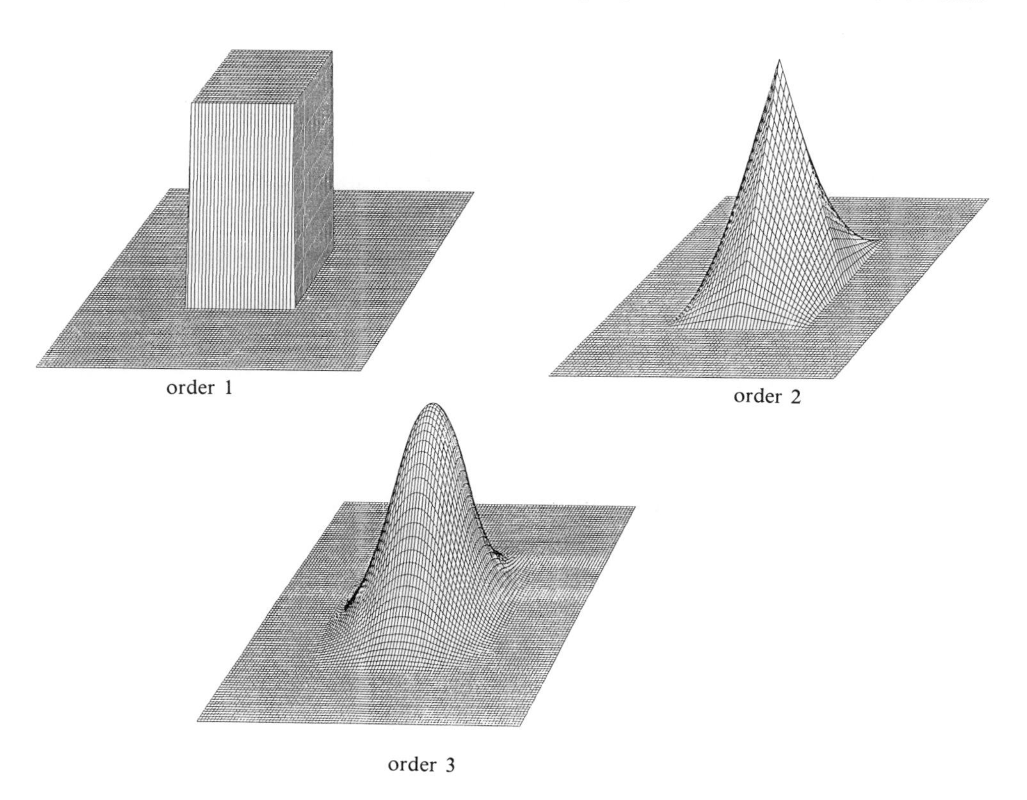

Figure 5.7 Multivariate basis splines

functions p where

$$p = \prod_{i=1}^{n} (R_i + k - 1) \tag{5.23}$$

which grows exponentially in n. Hence without memory hashing, this algorithm is suitable only for small to medium sized input spaces.

The arithmetic cost of the algorithm can be calculated by noticing that each input maps to $\rho = k^n$ multivariate basis functions, whose output is formed from multiplying together n univariate basis functions. Once the subinterval in which the input is lying can be determined, the output can be formed using approximately $1.5k^2$ floating point multiplications/divisions. Hence the total cost is

$$1.5nk^{n+2} \text{ floating point operations} \tag{5.24}$$

Again the dominant term (k^n) is exponential in n.

Whilst these expressions are highly prohibitive, it should be noted that the algorithm is highly parallel, and the required memory is local, which divides evenly across (up to) k^n processors, where an overlay (or more) would be assigned to each processor (see Section 5.4.4).

It has been assumed that the (univariate) subintervals in which the input lies are available. In fact they can be obtained using a sequential or a binary search through the respective knot sets λ_i. These take approximately $\log_2(R_i)$ comparisons, where the arithmetic cost of a comparison is a lot less than a floating point operation and so this part of the algorithm can be neglected. Again this operation could be distributed across n processors.

5.4.3 Input Space Metrics

It is interesting to investigate the various metrics defined across the input space for basis functions of small orders.

Initially consider basis splines of order 1, which results in a multivariate look-up table. Each cell in the look-up table corresponds to an n-dimensional cell defined by the knot lattice on the original input space. Hence measuring distance across the lattice and defining the centre c_j of each basis function to occur at the centre of the cell, then the output of the jth multivariate basis function can be represented by

$$N_j^1(x) = \begin{cases} 1 & \text{if } \| c_j - \mathbf{x} \|_\infty < 0.5 \\ 0 & \text{otherwise} \end{cases} \tag{5.25}$$

with an appropriate rule to assign the inputs which lie on the lattice itself.

For basis functions of order 2 (chapeau/hat functions) defined on a 2D input space, the contour plot is shown in Figure 5.8. When the input is close to the multivariate set centre, the polynomial can be approximated by a linear polynomial (in the n variables) in each of 2^n cells defined by the lattice. Hence in each cell, for

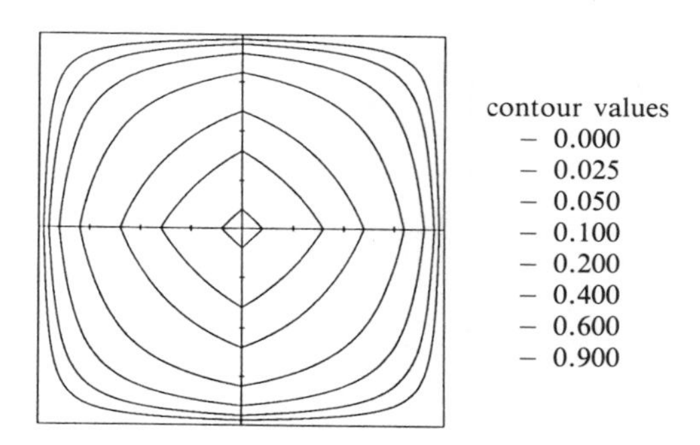

Figure 5.8 A 2D contour map for triangular basis splines formed using the product operator

(scaled) inputs close to the multivariate set centre, the output can be approximated by

$$N_j^2(\mathbf{x}) = 1 - \| c_j - \mathbf{x} \|_1 \tag{5.26}$$

When the inputs are close to the cell's edge, the output is approximately proportional to

$$N_j^2(\mathbf{x}) \propto \| c_j - \mathbf{x} \|_\infty \tag{5.27}$$

and in between the higher-order polynomial terms provides a continuous transformation between the two.

A similar result is obtained when basis functions of orders 3 and 4 are used to generate the contour plot, Figures 5.9 and 5.10, respectively. In these cases the increased order of the basis functions introduce extra degrees of smoothness into the contour plots. Indeed as k increases, the distance measure appears to approximate the Euclidean norm for inputs close to the function centre. Forming linear combinations of basis functions which measure locally w.r.t. the Euclidean norm

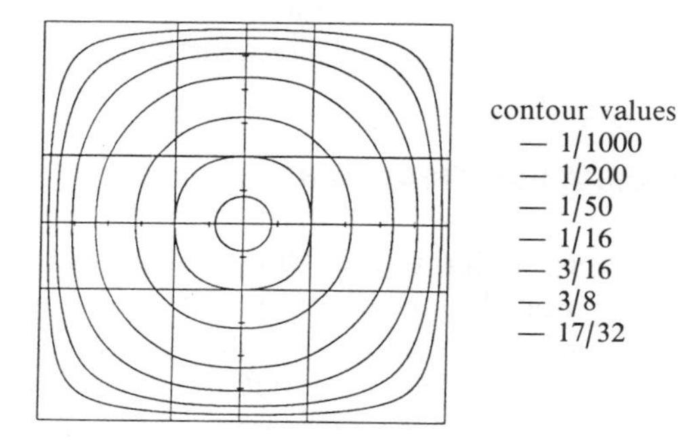

Figure 5.9 A 2D contour map for quadratic basis splines formed using the product operator

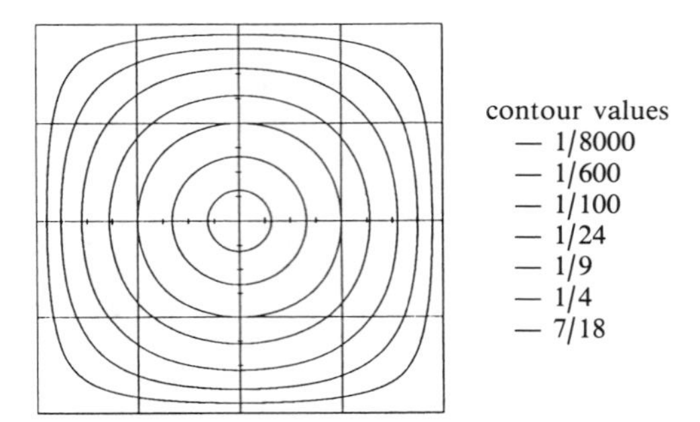

Figure 5.10 A 2D contour map for cubic basis splines formed using the product operator

is very similar to radial basis functions and the superspheres proposed in An *et al.* (1991) for the CMAC algorithm.

5.4.4 Parallel Implementation

One way of visualizing the previously described algorithm is to imagine the support of the multivariate basis functions forming $\rho(=k^n)$ overlays. Each overlay is just sufficiently large to cover the input lattice and is formed from the adjacent, non-overlapping supports (volume k^n) of the basis functions. The first overlay contains the support $(\lambda_{i,0}, \lambda_{i,k})$ for each i and then the subsequent overlays are positioned so that every possible combination of knots in this first hypercube of volume k^n is covered. Then each overlay is unique and each input will map to one and only one basis function in each overlay. This provides an efficient breakdown for parallel processing with up to ρ processors, with an overlay (or set of overlays) being assigned to each processor. If ρ processors are available, then the computational cost is dominated by approximately $1.5nk^2$ floating point operations, which is linear w.r.t. n. However the local memory requirements are still exponential w.r.t. n and so memory hashing algorithms may be required for larger dimensional spaces.

5.4.5 Similarity with the Albus CMAC and Fuzzy Logic

This network structure is very similar to the Albus CMAC where the basis functions are binary (output 1 if the input lies in the support, 0 otherwise). More specifically a generalization parameter ρ is supplied by the user, which specifies the number of nonzero basis functions mapped to for each input and an n-dimensional lattice must also be given. The algorithm then forms ρ overlays, each consisting of the supports of the basis functions which are of volume ρ^n. The first overlay is positioned as before, with the remaining $(\rho-1)$ overlays being distributed in order to achieve good generalization (An *et al.*, 1991), which is illustrated in Figure 5.11. It is possible to use higher-order basis functions, rather than binary; however, whilst this will give (in general) a continuous output, the type of surface will vary in relation to the position of the input on the network's lattice. This is because the number of basis functions mapped to is independent of the size of the input space, n, whereas the number of basis functions mapped to for the B-spline net increases exponentially in n.

Finally, it is worth noting that the B-spline network is very similar to fuzzy logic networks (Brown and Harris, 1991a). The univariate B-splines illustrated in Figure 5.6 obviously resemble fuzzy sets and can be adopted as a methodology for their

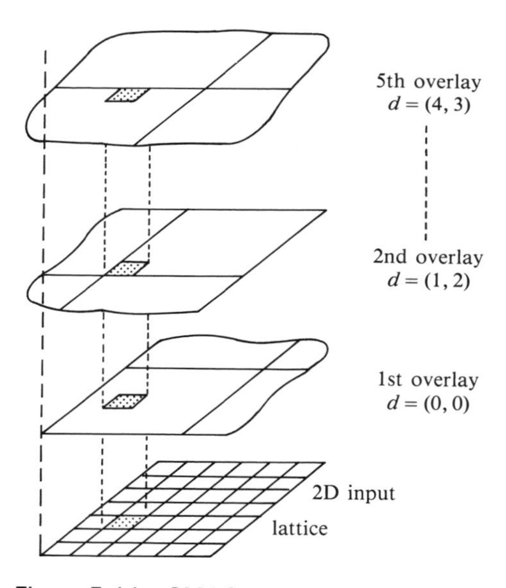

Figure 5.11 CMAC receptive fields, $\rho = 5$

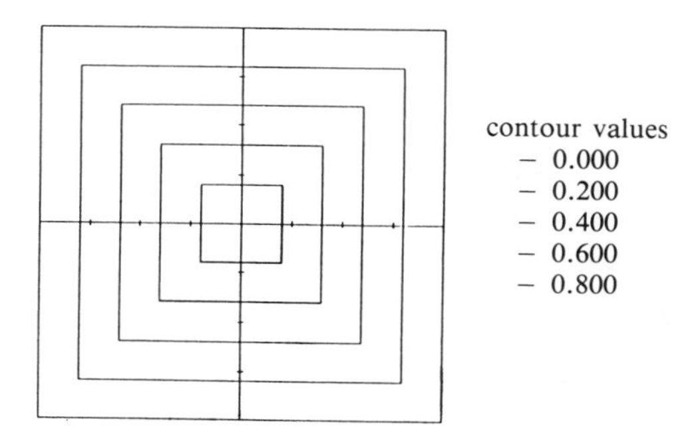

Figure 5.12 A 2D contour map for triangular fuzzy sets formed using the min operator

design. Similarly, the product operator, which is used to form multivariate basis functions, and the min operator, used to represent the fuzzy AND, can be interchanged. For completeness, a 2D contour plot for two triangular fuzzy sets (basis functions of order 2) joined using the min operator is shown in Figure 5.12. Note that the same contour plot would result if the distance between the input and the set centre had been calculated using the infinity norm and then subtracted from 1 (over the relevant domain). It can also be shown that, if the remaining fuzzy logic operators are chosen appropriately (modified centre of area defuzzification and the product operator implementing IMPLICATION), in a forward-chaining mode both networks are equivalent.

5.5 WEIGHT ADAPTATION

The output of the B-spline network is formed from a linear combination of the basis function's outputs. Hence the network's output is linearly dependent upon this set of adjustable parameters, or weights, $\mathbf{w}(t)$.

For on-line modelling, this weight vector must adapt in response to observed plant input/output mappings. There are many standard adaptation rules, five of which will be discussed in this section: recursive least squares (RLS), least-mean squares (LMS), normalized least-mean squares (NLMS), stochastic approximation least-mean squares (SALMS) and stochastic-approximation normalized least-mean squares (SANLMS). All of these rules attempt to minimize the plant output mean square error (MSE) given a set of plant input/output mappings.

5.5.1 Recursive Least Squares

Given a set of observed plant input/output mappings $\{\hat{y}(t), \mathbf{x}(t)\}_{t=1}^{T}$ (where $T \geqslant p$), the optimal weight vector in the MSE sense is given by (pseudo) inverting the equation

$$A\mathbf{w} = \hat{\mathbf{y}} \tag{5.28}$$

where $\hat{\mathbf{y}}$ is a desired output vector of size $(T \times 1)$, $\mathbf{w}$ is the weight vector of size $(p \times 1)$ and A is the matrix of size $(T \times p)$, whose tth row corresponds to the basis functions outputs for input $\mathbf{x}(t)$.

Hence the optimal weight vector $\hat{\mathbf{w}}$ is given by

$$\hat{\mathbf{w}} = (A^{\mathrm{T}}A)^{-1}A^{\mathrm{T}}\hat{\mathbf{y}} \tag{5.29}$$

provided that $A^{\mathrm{T}}A$ is invertible.

However, in order to incorporate a new sample, a full matrix pseudo-inversion must be performed as in (5.29). What is required is an algorithm for recursively updating the weight vector in order to incorporate new samples. This is called the

RLS estimator and is given by

$$\begin{aligned}\hat{\mathbf{w}}(t) &= \hat{\mathbf{w}}(t-1) + K(t)[\hat{y}(t) - \mathbf{a}^{\mathrm{T}}(t)\hat{\mathbf{w}}(t-1)]\\ K(t) &= P(t)\mathbf{a}(t) = P(t-1)\mathbf{a}(t)[I - \mathbf{a}^{\mathrm{T}}(t)P(t-1)\mathbf{a}(t)]^{-1}\\ P(t) &= [I - K(t)\mathbf{a}^{\mathrm{T}}(t)]\,P(t-1)\end{aligned} \tag{5.30}$$

where $\hat{\mathbf{w}}(t)$ is the optimal weight vector at time t.

Whilst this algorithm has many appealing features – the current optimal weight vector for the data set is generated at each update, long term random noise filtering, etc. – the computational cost is dominated by the updating of $K(t)$ and $P(t)$ which require $O(p^2)$ floating point operations (see (5.23)). Another disadvantage is that at least p input/output data pairs must be generated from the state-space, ensuring that every basis function has a nonzero output for at least one input. Finally, in order to obtain an exponential convergence rate, a forgetting factor would have to be introduced into (5.30) so that a higher importance was attached to the samples most recently collected. This would cause serious problems in updating the weights corresponding to those areas of the state-space not recently visited.

It would not be practical to implement a RLS algorithm for this particular problem.

5.5.2 Least-mean Squares and Normalized Least-mean Squares

LMS and NLMS weight adaptation rules use an instantaneous estimate of the current value of the MSE cost function in order to update the weight vector (Johnson, 1988). Unlike the RLS algorithm, however, convergence to the MSE will only occur in the limit. The LMS algorithm is given by

$$\mathbf{w}(t) = \mathbf{w}(t-1) + \delta[\hat{y}(t) - \mathbf{a}^{\mathrm{T}}(t)\mathbf{w}(t-1)]\,\mathbf{a}(t) \tag{5.31}$$

where δ is the constant weight updating factor (>0). Note that the vector $K(t)$ in (5.30) is being estimated by $\delta\mathbf{a}(t)$ in (5.31).

Despite the fact that convergence will only occur in the limit, this algorithm has one very appealing feature: that only those weights whose basis function is nonzero will be updated. This means that implementing the LMS algorithm requires only $(\rho + 1)$ floating point multiplications, whereas the RLS algorithm required $O(p^2)$.

The choice of δ must reflect the amount of measurement noise expected, the expected model mismatch, the desired rate of convergence, whether the plant is time varying and the size of the expected inputs $(\mathbf{a}^{\mathrm{T}}\mathbf{a})$. In order to determine a maximum value, the efficiency with which B-splines store patterns must be investigated, i.e. the normalized error reduction after a training pattern $[\hat{y}(t), \mathbf{x}(t)]$ has been presented must be determined. So

$$\begin{aligned}\varepsilon_y(t) &= [\hat{y}(t) - \mathbf{a}^{\mathrm{T}}(t)\mathbf{w}(t-1)]\\ \Delta\dot{\varepsilon}_y(t) &= \Delta\,[\hat{y}(t - \mathbf{a}^{\mathrm{T}}(t)\mathbf{w}(t-)]\\ &= -\mathbf{a}^{\mathrm{T}}(t)\,\Delta\mathbf{w}(t-1)\end{aligned} \tag{5.32}$$

where $\Delta\varepsilon_y(t) = \underline{\varepsilon}_y(t) - \varepsilon_y(t)$ and $\underline{\varepsilon}_y(t)$ is the a posteriori output error. Now because the LMS adaptation law is being used,

$$\Delta\varepsilon_y(t) = -\mathbf{a}^{\mathrm{T}}(t)\,\delta\varepsilon_y(t)\mathbf{a}(t) \tag{5.33}$$

and so the following two relationships hold for the output error reduction

$$\underline{\varepsilon}_y(t) = [1 - \delta\mathbf{a}^{\mathrm{T}}(t)\mathbf{a}(t)]\,\varepsilon_y(t)$$

$$\left|\frac{\Delta\varepsilon_y(t)}{\varepsilon_y(t)}\right| = \delta\mathbf{a}^{\mathrm{T}}(t)\mathbf{a}(t) \tag{5.34}$$

and hence, in order to ensure stable learning, it is required that

$$0 < \delta < 2 * \left(\max_t \{\mathbf{a}^{\mathrm{T}}(t)\mathbf{a}(t)\}\right)^{-1} \tag{5.35}$$

Now from (5.34), the variance in the normalized error reduction is dependent upon the (scaled) variance of $[\mathbf{a}^{\mathrm{T}}(t)\mathbf{a}(t)]$. This term is dependent upon the order of the basis splines and the position of the interior and exterior knots. So initially consider the case when the simple interior knots have intervals of equal width and the exterior knots form coincident (multiple) knots at the ends of the intervals. Then

$$\begin{aligned} \text{order } 1 &\Rightarrow \mathbf{a}^{\mathrm{T}}\mathbf{a} \equiv 1 \\ \text{order } 2 &\Rightarrow 0.5^n \leqslant \mathbf{a}^{\mathrm{T}}\mathbf{a} \leqslant 1 \\ \text{order } k(>2) &\Rightarrow 0 < \mathbf{a}^{\mathrm{T}}\mathbf{a} \leqslant 1 \end{aligned} \tag{5.36}$$

The unity upper bound is a sharp bound; however, for basis functions of orders greater than 2 it is only achieved at the corners of the hyper-rectangle defined by the exterior knots. In fact for basis functions of orders >2 and large input spaces, the value of av$(\mathbf{a}^{\mathrm{T}}\mathbf{a})$ may be close to zero, which will mean that using (5.31) may result in a slower than expected convergence rate. The unity upper bound (for B-splines of orders >2) is cased by the coincident exterior knots, with the maximum value of $(\mathbf{a}^{\mathrm{T}}\mathbf{a})$, for the basis functions defined on the interior of the lattice, often being close to zero (especially for high-dimensional spaces). For these cases it is required that

$$0 < \delta < 2 \tag{5.37}$$

or, more usually, δ is selected so that

$$0 < \delta < 1 \tag{5.38}$$

The motivation for choosing coincident exterior knots is that this results in an increased rate of convergence at the boundaries; however, δ must then be chosen so that slower convergence results across the vast majority of the input space. Choosing equi-spaced interior *and* exterior knots allows a value of δ to be chosen which is greater than 1. This results in improved convergence across the interior of the state-space.

The actual value of δ will be determined by a compromise between choosing a large value for fast convergence and modelling time varying plants and a value close to zero in order to reject measurement and modelling noise.

Equation (5.34) provides the motivation for the development of the NLMS adaptation rule, where the factor $[\mathbf{a}^T(t)\mathbf{a}(t)]^{-1}$ is introduced into the right-hand side of (5.31) in order to make the normalized error reduction *independent* of the size of $[\mathbf{a}^T(t)\mathbf{a}(t)]$:

$$\mathbf{w}(t) = \mathbf{w}(t-1) + \delta[\hat{y}(t) - \mathbf{a}^T(t)\mathbf{w}(t-1)]\, \frac{\mathbf{a}(t)}{\mathbf{a}^T(t)\mathbf{a}(t)} \tag{5.39}$$

Note that $K(t)$ in (5.30) is being approximated by $\delta\mathbf{a}(t)/[\mathbf{a}^T(t)\mathbf{a}(t)]$. This simple scaling improves the rate of convergence for a wider variance of the input's magnitude and may be preferred to (5.31) when the input space dimension is large and the basis functions are of order $\geqslant 2$. The introduction of the scaling factor also means that (5.39) becomes an error correction rule rather than a gradient descent rule as it reduces the output error at $\mathbf{x}(t)$ by a factor of $(1-\delta)$. The weight updating factor is now self-scaling and all that is required for convergence is that it lies in the interval $(0, 2)$. Finally it should be noted that in the limit (5.39) does not minimize the mean squared error but rather a biased version of it. This can be seen by rewriting (5.39) as (Widrow and Lehr, 1990):

$$\mathbf{w}(t) = \mathbf{w}(t-1) + \delta\left[\frac{\hat{y}(t)}{|\mathbf{a}(t)|} - \frac{\mathbf{a}^T(t)}{|\mathbf{a}(t)|}\,\mathbf{w}(t-1)\right]\frac{\mathbf{a}(t)}{|\mathbf{a}(t)|} \tag{5.40}$$

where $|\mathbf{a}(t)|$ is defined to be $[\mathbf{a}^T(t)\mathbf{a}(t)]^{0.5}$

This is of the form of (5.31) for a training set given by $\{\hat{y}(t)/|\mathbf{a}(t)|, \mathbf{a}(t)/|\mathbf{a}(t)|\}_{t=1}^{T}$ and hence, in the limit, the weight vector will converge with the MSE for this training set, which is, in general, a biased version of the original training set. This bias will be greater when the variance of $|\mathbf{a}(t)|$ is large; however, it would be precisely for this reason that the NLMS training rule is preferred over the LMS training rule. So for a particular network the use of (5.39) instead of (5.31) must be justified by examining the variance of $|\mathbf{a}(t)|$.

Numerous studies have been performed comparing the rate of convergence of the LMS/NLMS algorithms with the RLS algorithm. Whilst the RLS generates an 'optimal' weight vector at each training step, the training data may have been heavily corrupted by noise and the problem structure may be such that there is no significant advantage in using this more complex rule. An illustrative example is given in Section 5.6.

5.5.3 Stochastic Approximation LMS and NLMS

For the LMS and the NLMS adaptation rules, the weight updating factor δ is assumed constant. In this section two simple modifications are proposed: (a) assign an individual learning rate to each basis function, and (b) reduce δ_i through time as the confidence in a particular weight increases. These modifications are made in order to retain the fast initial convergence rate whilst, in the long term, filtering out measurement and modelling noise.

Therefore let the ith basis function have an updating factor $\delta_i(t)$ associated with it and let it also be a function of time. Then necessary conditions on $\delta_i(t)$ for the weight vector to converge are given by

$$\delta_i(t) > 0$$

$$\sum_{t=1}^{\infty} \delta_i(t) = \infty \tag{5.41}$$

$$\sum_{t=1}^{\infty} \delta_i^2(t) < \infty$$

where the third condition implies that $\delta_i(t) \Rightarrow 0$. One such function which satisfies these constraints is given by

$$\delta_i(t) = \frac{c_1}{(c_2 + t_i)} \tag{5.42}$$

where c_1 and c_2 are positive and nonnegative constants respectively and t_i is the number of times that the ith basis function has been nonzero.

Then (5.31) and (5.39) become

$$w_i(t) = w_i(t-1) + \delta_i(t)[\hat{y}(t) - \mathbf{a}^{\mathrm{T}}(t)\mathbf{w}(t-1)]\, a_i(t) \tag{5.43}$$

$$w_i(t) = w_i(t-1) + \delta_i(t)[\hat{y}(t) - \mathbf{a}^{\mathrm{T}}(t)\mathbf{w}(t-1)]\, \frac{a_i(t)}{\mathbf{a}^{\mathrm{T}}(t)\mathbf{a}(t)} \tag{5.44}$$

where $\delta_i(t)$ is of the form given by (5.41). One modification of the learning rules (5.31), (5.39) and (5.44) is to set the weight equal to the desired output if the weight vector has been initialized to NULL. This overcomes the problem of the weight being initialized when the basis function output is very small, which results in slow convergence. This heuristic is justifiable because the basis functions have a local, compact support and because they are normalized.

Also if the plant is subject to sudden parameter changes it may be necessary to reset $\delta_i(t)$ to a nominal value. This is necessary because whilst the constraints given in (5.40) guarantee parameter convergence (subject to the usual conditions), convergence may occur very slowly. Hence if the plant is detected as having changed sufficiently, a reasonable counteraction is to make the plant model more plastic.

5.5.4 The Albus CMAC Updating Rule

In the original Albus CMAC, there exist ρ basis functions each with an output of one, the rest have a zero output. To train the weight vector Albus recommended a rule of the form

$$\mathbf{w}(t) = \mathbf{w}(t-1) + \frac{\delta}{\rho}\,[\hat{y}(t) - \mathbf{a}^{\mathrm{T}}(t)\mathbf{w}(t-1)]\,\mathbf{a}(t) \tag{5.45}$$

where $0 < \delta < 2$. Now because $\mathbf{a}^{\mathrm{T}}(t)\mathbf{a}(t) \equiv \rho$, $\forall t$, this rule is equivalent to both the LMS and NLMS training algorithms as given previously. Indeed this illustrates the case when the NLMS rule does converge to the MSE, i.e. when the input vector $\mathbf{a}(t)$ has a constant magnitude for all the training cases. Then the constant factor can be taken out of the training set and what is left is the original training set. (Note that the two rules are also equivalent for B-splines of order 1 when $\mathbf{a}^{\mathrm{T}}(t)\mathbf{a}(t) \equiv 1$.)

Suppose that the basis functions in the Albus CMAC were normalized (like B-splines), then each would output either $1/\rho$ or 0. In this case (5.45) is simply

$$\mathbf{w}(t) = \mathbf{w}(t-1) + \delta\rho\,[\hat{y}(t) - \mathbf{a}^{\mathrm{T}}(t)\mathbf{w}(t-1)]\,\mathbf{a}(t) \tag{5.46}$$

where $0 < \delta < 2$ and noting that $\rho a_i(t)$ is either 1 or 0. Each weight in the network would then simply be a factor of ρ greater than that in the original Albus CMAC and the output would be identical ($\mathbf{a}^{\mathrm{T}}(t)\mathbf{a}(t) \equiv 1/\rho$). Finally it should be noted that the convergence of the Albus CMAC has been rigorously investigated in Parks and Militzer (1989a,b).

5.5.5 Parameter Convergence over a Restricted Domain

In this section, it will be illustrated how the LMS family of adaptation rules (LMS, NLMS, SALMS, SANLMS) learn new information locally. In particular, when the desired function is modified locally, the weight changes recommended by the LMS algorithm decay exponentially, for the weights far away from this domain. This is a very important result, as it relaxes the requirement that the input be persistently exciting over the global domain, to a condition that the input be persistently exciting over the local domain. This is because the effect of changing weights in a new area of the state-space (relative to the defined lattice) modifies the original weights by a factor which is dependent upon c^d ($c < 1$), where d is the distance (defined on the lattice) between the two areas. The decay factor c is dependent upon the order of the B-splines (shapes of the fuzzy sets), the knot placement strategy and the statistics of the input signal.

So consider the case of a B-spline ANN exactly reproducing the desired function with a weight vector $\mathbf{w}(-1)$ and considering how an error εw_0, introduced at $w_0(-1)$ is propagated throughout the weight vector. For simplicity, univariate B-splines of order 2 (hat functions) with equispaced knots will be investigated. The ith input is assumed to lie in the interval I_i and a LMS updating algorithm is used with $\delta \equiv 1$. Now define

$$w_i(0) = \begin{cases} w_0(-1) + \varepsilon w_0 & \text{if } i = 0 \\ w_i(-1) & \text{otherwise} \end{cases}$$

Hence, by induction, $w_i(i-1)$ remains unchanged. Now consider when the input lies in the ith interval, so that $(i-1)+x_i \in I_i$ $(x_i \in [0, 1))$. Then

$$\begin{aligned} y(i) &= x_i w_{i-1}(i-1) + (1-x_i)w_i(i-1) \\ \hat{y}(i) &= x_i w_{i-1}(i-2) + (1-x_i)w_i(i-1) \\ \varepsilon_y(i) &= \hat{y}(i) - y(i) = -x_i\, \Delta w_{i-1}(i-2) \end{aligned} \tag{5.47}$$

where $\Delta w_j(i) = w_j(i+1) - w_j(i)$. Therefore the change in $w_i(i-1)$ caused by this error is given by

$$\begin{aligned} \Delta w_i(i-1) &= -x_i\, \Delta w_{i-1}(i-2) * (1-x_i) \\ \Delta w_i(i-1) &= -c_i\, \Delta w_{i-1}(i-2) = \prod_{j=1}^{i} (-c_j) * \varepsilon w_0 \end{aligned} \tag{5.48}$$

where the decay factor c_j is defined to be $x_j(1-x_j)$. Now as $c_j \in [0, 1/4]$, the absolute value of $\Delta w_i(i-1)$ decreases exponentially fast as i increases. The decay is slowest when $x_i = \frac{1}{2} \forall i$ and hence $c_j \equiv \frac{1}{4}$. The expected value of c_j (for a uniform input probability density function on the unit interval) is 1/6, and if $x_i = 0$ for any i, then $\Delta w_{i+k}(i+k-1) = 0$, $\forall k \geqslant 0$. A plot of c_i against x_i is shown in Figure 5.13. This analysis has been performed for equispaced knots, however it holds for a more general nonlinear (simple) knot placement strategy.

For quadratic B-splines with the input lying on the knots the recurrence relation generated is identical to (5.48) with the decay factor, c_i, equal to $\frac{1}{4}$ and, in general, any input causes the error to be propagated along the weight vector. It is difficult to provide a more complete analysis as the change in the weight vector (measured w.r.t. the one norm) depends upon the previous two changes, unlike (5.48) which depends only on the last change.

If this property did not hold for basis splines then there would be *no*

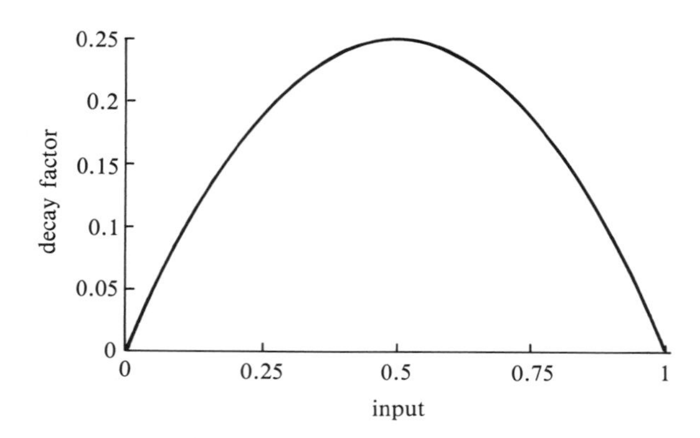

Figure 5.13 Decay factor for order 2 B-splines

justification for using the LMS family of adaptation rules to modify the weight vector. Research is continuing, extending the analysis to higher-order splines and multivariate bases.

5.6 AN EXAMPLE: NONLINEAR TIME SERIES PREDICTION

Consider the following two-input single-output nonlinear time series process given by

$$y(t) = \{0.8 - 0.5 \exp[-y^2(t-1)]\}\, y(t-1) - \{0.3 + 0.9 \exp[-y^2(t-1)]\} \times y(t-2) + 0.1 \sin[\pi y(t-1)] + \eta(t) \quad (5.49)$$

where $\eta(t)$ is a Gaussian white noise sequence whose mean is zero and variance 0.01. Defining an input vector $\mathbf{x}(t) = [y(t-2), y(t-1)]$ and the resulting output to be $y(t)$, then a set of two-input single-output training samples can be collected representing the dynamics of (5.49).

The origin in (5.49) is an unstable equilibrium which is enclosed by a stable attracting limit cycle. This limit cycle is unique as, for large inputs, (5.49) represents a contraction mapping. The noiseless output surface over the domain $[-1.5, 1.5] \times [-1.5, 1.5]$ is shown in Figure 5.14 and an iterated mapping consisting of 1000 data points from the initial condition $(0.1, 0.1)$ is shown in Figure 5.15.

In order to model this time series two data sets were constructed, each consisting of 1000 data points. The first consisted of iterating (5.49) from an initial condition $(0, 0)$. The noisy measurements ensured that the iterated sequence was driven away

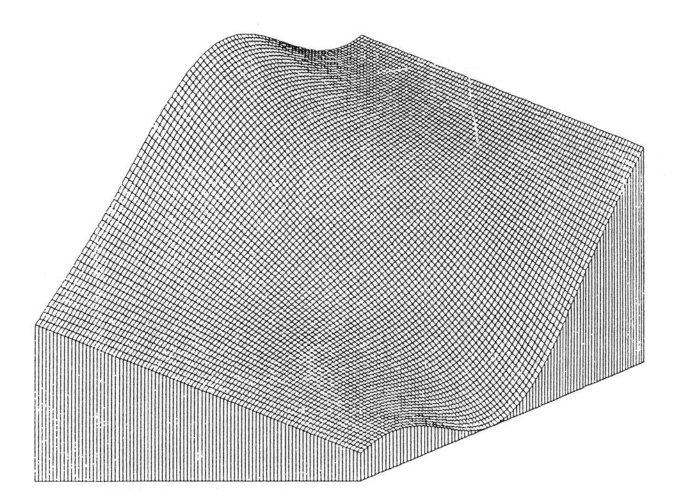

Figure 5.14 A noiseless output surface

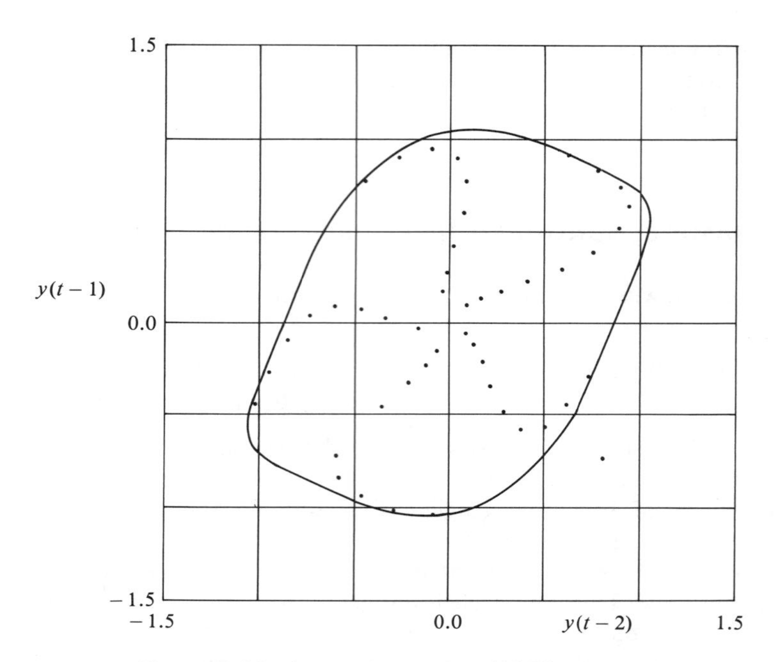

Figure 5.15 Iterated mapping (1000 points)

from the unstable equilibrium. There was a very sparse population of training samples about the origin and around the boundary, whereas about the limit cycle there exists a large number of noisy measurements. The second training set consisted of the union of iterating (5.49), with zero noise, from the eight initial conditions $(0.1, 0.1)$, $(-0.1, 0.1)$, $(0.1 - 0.1)$, $(-0.1, -0.1)$, $(1.3, 1.3)$, $(1.3, -1.3)$, $(-1.3, 1.3)$ and $(-1.3, -1.3)$. The length of each iteration was 125 and so the training set again consisted of 1000 training examples. This resulted in a more even distribution of the input vectors; however, the areas around the boundary were still sparsely populated.

The B-splines used to model this time series were defined upon the compact input space $[-1.5, 1.5] \times [-1.5, 1.5]$, and all the exterior knots were set equal to the appropriate endpoints. On both axes, the interior knots are members of the set $\{-1.0, -0.5, 0, 0.5, 1.0\}$. This provided a linear distribution of the interior knots across each axis. A SALMS updating rule was used with the adaptation parameters c_1 and c_2 being estimated to be both 30 for the noisy data set and 100 for the exact data set. The latter was arrived at because the receptive field of multivariate B-splines of order 2 cover approximately 10% of the total input space, which contains 1000 data points. This choice ensures that, on average, the learning rate halves with each pass through the training set, filtering out model mismatch and so converging on an MSE solution. The parameters associated with the noisy training data were chosen to be one-third of those associated with the noise-free case, an estimate which will (obviously) vary with the size of the noise. The dynamical behaviour of the models was tested by starting the time series from the initial conditions (0.1, 0.1) and (1.0, 1.0) and seeing how well the iterated data points approximated (Figure 5.15).

5.6.1 Piecewise Linear B-splines

For this B-spline ANN, piecewise linear basis functions (order 2) are defined across each input, resulting in a model which has 49 adjustable weights. The noisy data set was then cyclically presented to the network and the model's dynamics after one and five cyclic passes are shown in Figures 5.16 and 5.17 respectively. The piecewise linear nature of the model is obvious from Figure 5.16; however, the network has learnt the basic underlying dynamics of (5.49) with the origin being weakly unstable and the limit cycle approximating that obtained in Figure 5.15. After five cycles through the data the time series model output has been smoothed out where sufficient data exists, and locally generalizes where sparse, noisy data exist. The limit cycle on the iterated map is similar to the previous one, and now the familiar five-armed spiral away from the origin has formed, despite the lack of noisy training data near the origin.

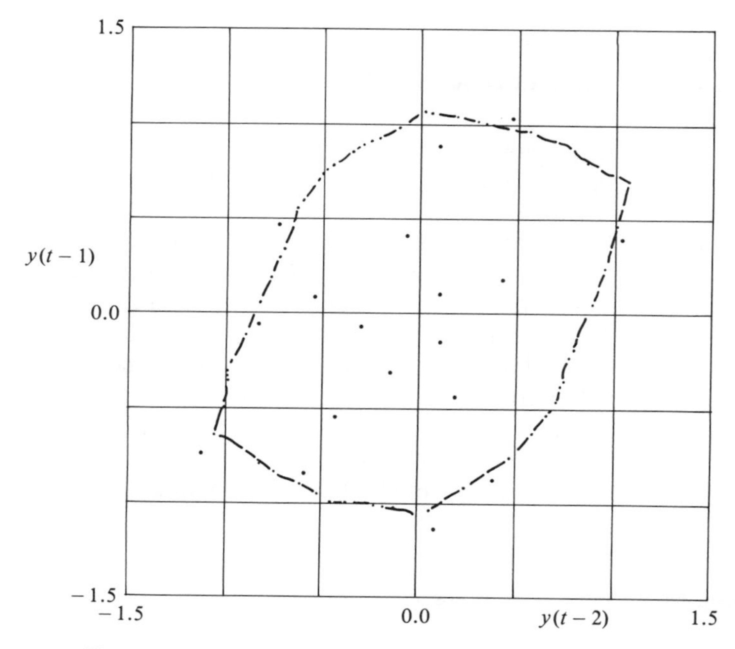

Figure 5.16 Model dynamics after one cycle pass

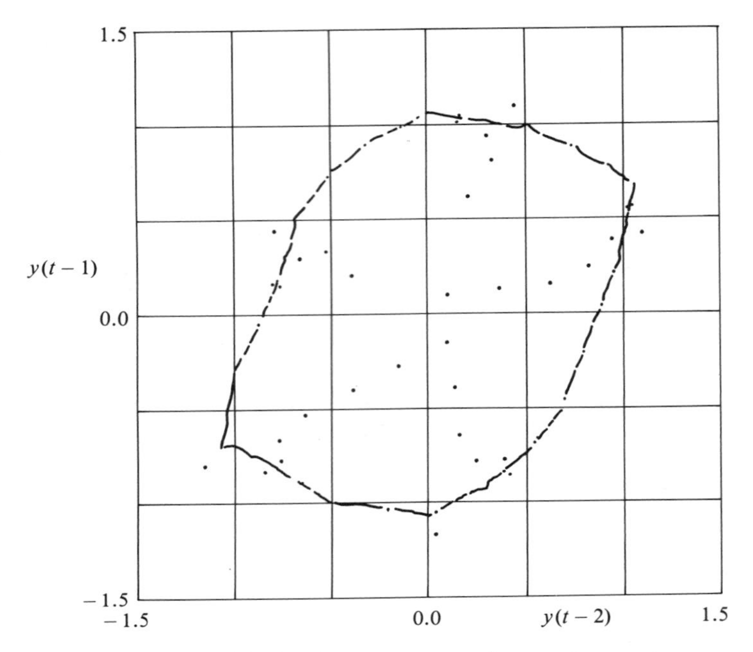

Figure 5.17 Model dynamics after five cycle passes

5.6.2 Piecewise Linear and Quadratic B-splines

This time a B-spline network was used which had piecewise linear B-splines representing $y(t-2)$ and piecewise quadratic B-splines representing $y(t-1)$, which resulted in a model having 56 weights. The exact data set was cyclically presented to the network and the output surface and dynamics are shown in Figures 5.18 and 5.19, respectively, after 20 cyclic presentations. The characteristics of the true plant can be seen over the state-space where enough training data existed. In the centre the model can clearly be seen to be linear w.r.t. $y(t-2)$, and nonlinear w.r.t. $y(t-1)$. The shape of the limit cycle and the placement of the points on the five spirals are a very good approximation to Figure 5.15.

This means that the mismatches in the shape of the limit cycles in Figure 5.16 and Figure 5.17 are due to the distribution of the noisy data samples and the model mismatch. However, a satisfactory model has been obtained.

5.6.3 Comparison with Radial Basis Functions

Radial basis function ANNs have also been applied to this problem (Chen, 1992),

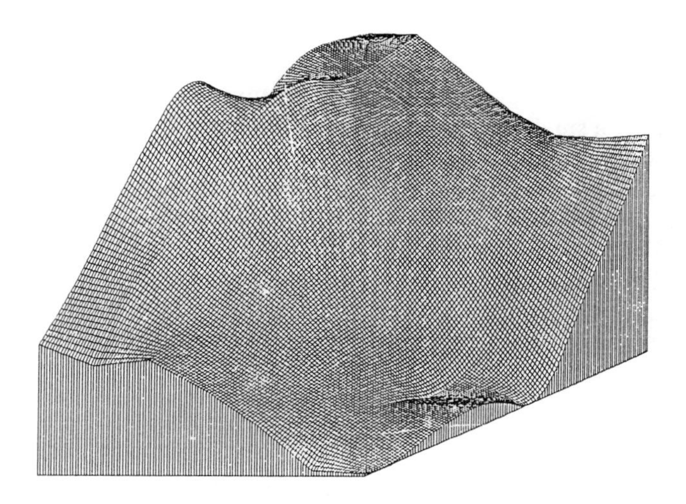

Figure 5.18 Output surface

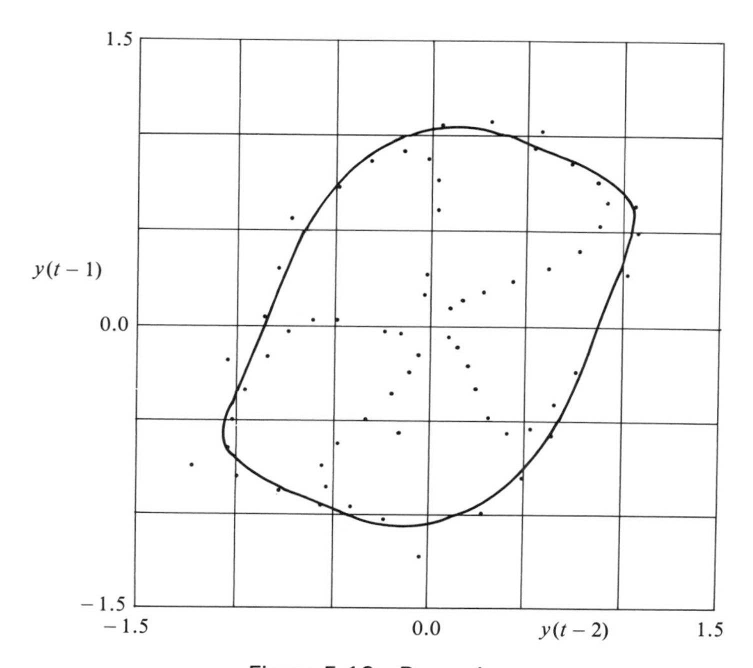

Figure 5.19 Dynamics

where the nonlinearity was chosen to be the thin-plate-spline function. This function has the property that $f(x) \Rightarrow \infty$ as $x \Rightarrow \infty$, and so its support is infinite, unlike the B-spline functions which have compact support. Hence, for on-line modelling, all the weights would be changed in response to input/output data, unlike B-spline ANNs where only a small subset is adapted. Training the weights was performed by pseudo-inverting the coefficient matrix and so the weight vector was optimal in the MSE sense, whereas cyclically training the B-spline ANNs only ensures convergence to an MSE as time tends to infinity (however, satisfactory convergence is generally obtained after a few cyclic presentations). The function centres were selected according to the distribution of the input data and also as an 'optimal' subset of the training data. Surprisingly, the former performed the better, generating a much smoother limit cycle, and determining the centres was much quicker due to the linear form of the training rule used. Neither limit cycle had the derivative discontinuities of Figures 5.16 and 5.17 because of the piecewise nature of the basis functions; however, a detailed comparison is impossible owing to the distribution of the noisy training data.

A polynomial NARMAX model (Chen and Billings, 1992) was also generated, for which excellent results were obtained. However, it was necessary to use a subset of the polynomial terms up to order 7. For on-line modelling this could create a highly oscillatory output surface as the weight vector was adapted in order to incorporate new information.

5.7 CONCLUSIONS

This chapter has introduced a new class of ANNs which have many advantages when applied in the field of intelligent control. The temporal stability of the learning rule coupled with the transparent relationships which are extracted make the networks suitable for both off-line and on-line design. The algorithm is highly suitable for a parallel implementation, which should allow it to be applied in higher dimensional spaces. Fuzzy rule bases have been widely applied in control and this network can be viewed as extracting the useful concepts from this technique (fuzzy sets with compact support being defined on each variable), whilst generating a smooth model surface and guaranteeing convergence subject to the usual conditions. Details of a software tool for viewing and editing a B-spline ANN can be found in Brown (1992).

The model (controller) structure assumed is minimal. These are approximation techniques and so all the designer assumes is that the output can be satisfactorily modelled by a piecewise polynomial of order k (note that B-splines of different orders can be defined upon different input variables as in Section 5.6.2). Using piecewise polynomials to approximate data is widely used in graphical applications and in Section 5.5 a set of training rules have been derived for use when the model is time varying but the underlying structure remains constant.

APPENDIX

This example illustrates the convergence to different optimal weight vectors when minimizing cost functions J_u and J_y, in the case of model mismatch.

So consider a single input, unbiased network which consists of a single weight and a sigmoid which nonlinearly transforms the weighted input. Then there exists an optimal weight. The network is trained in batch mode using the two patterns P_1 and P_2 given by (5.A1) and (5.A2).

$P_1 : (\hat{y}(1), -1)$

$$\begin{aligned} \hat{y}(1) &= \mathrm{f}(x(1)) = \frac{1}{1 + \mathrm{e}^{-(1*-1)}} \\ &= 0.269 \ (\hat{w} = 1) \end{aligned} \tag{5.A1}$$

$P_2 : (\hat{y}(2), 2)$

$$\begin{aligned} \hat{y}(2) &= \mathrm{f}(x(2)) = \frac{1}{1 + \mathrm{e}^{-(2*2)}} \\ &= 0.982 \ (\hat{w} = 2) \end{aligned} \tag{5.A2}$$

The two desired patterns are sigmoids with different optimal weights. Now initially choose $w(0) = 1.5$ and set the learning rate $\delta = 0.2$. Then

$$y(1) = 0.182 \quad \mathrm{f}'(x(1)) = 0.149 \quad [\mathrm{f}'(x(1))]^{-1} = 6.72$$

$$y(2) = 0.953 \quad \mathrm{f}'(x(2)) = 0.0448 \quad [\mathrm{f}'(x(2))]^{-1} = 22.3$$

Now, from (5.12), performing gradient descent on J_y gives an update rule of the form:

$$\Delta w(0) = \frac{\delta}{2} * \sum_{t=1}^{2} \varepsilon_{y(t)}(1)\mathrm{f}'(x(t))x(t) \tag{5.A3}$$

which evaluates to:

$$\begin{aligned} \Delta w(0) &= 0.1 * (0.87 * 0.149 * (-1) + 0.029 * 0.0448 * 2) \\ &= 0.1 * (-0.0129 + 0.00260) \\ &= -0.00103 \end{aligned}$$

hence

$$w(1) < w(0) \text{ and } \hat{w} < w(0) \tag{5.A4}$$

By computer simulation, to a tolerance of 10^{-7}, this algorithm converged to the weight $\hat{\mathbf{w}} = 1.25186$ after 2088 iterations and the final output errors were:

$$\varepsilon_{y(1)} = 0.0466, \varepsilon_{y(2)} = 0.0576$$

Also from (5.14) performing gradient descent on J_u gives an update rule of the form:

$$\Delta w(0) = \frac{\delta}{2} * \sum_{t=1}^{2} \varepsilon_{y(t)}(1)\,[f'(x(t))]^{-1} x(t) \tag{5.A5}$$

which evaluates to:

$$\begin{aligned}\Delta w(0) &= 0.1 * (0.087 * 6.72 * (-1) + 0.029 * 22.3 * 2)\\ &= 0.1 * (-0.584 + 1.293)\\ &= 0.0709\end{aligned}$$

hence

$$w(1) > w(0) \text{ and } \hat{w} > w(0) \tag{5.A6}$$

Again by computer simulation, to a tolerance of 10^{-7}, this algorithm converged to the weight $\hat{\mathbf{w}} = 1.71001$ after 33 iterations. The final output errors were:

$$\varepsilon_{y(1)} = 0.116, \ \varepsilon_{y(2)} = 0.0137$$

Comparing (5.A4) and (5.A6) it can be seen that the optimal weight for the cost functions J_y and J_u are different, the rates of convergence to these values are substantially different and the relative values of the final output errors are weighted differently.

As a final comparison, the first example was repeated with the learning rate δ set to its 'optimal' value of 34 (obtained from simulations). The algorithm then converged after four iterations. Note how close this optimal learning rate is to the expected value of the square of the inverse Jacobian, which is 30. The second example was also repeated with the learning rate set to the 'optimal' value of 0.51 and convergence again occurred after four iterations.

REFERENCES

AN, P. C. E., MILLER, W. T. and PARKS, P. C. (1991) 'Design improvements in associative memories for cerebellar model articulation controllers', in *Proceedings of the International Conference on Artificial Neural Networks, Helsinki*, Vol. 2, pp. 1207–10, Amsterdam: North Holland.

BROWN, M. (1992) 'bsEditWeight: An interactive software tool for editing and viewing a B-spline network, PANORAMA report Release 3.1', Dept Aeronautics and Astronautics, University of Southampton, UK.

BROWN, M. and HARRIS, C. J. (1991) A nonlinear adaptive controller: a comparison between fuzzy logic control and neurocontrol', *IMA J. Math. Control Inform.*, **8**, 239–65.

CADZOW, J. A. (1990) 'Signal processing via least squares error modelling', *IEEE ASSP Magazine*, October, 12–31.

CHEN, S. and BILLINGS, S. A. (1992) 'Neural networks for nonlinear dynamic system modelling and identification', accepted by *Int. J. Control*, special issue on intelligent control, Vol. 56, no. 2, 319–46.

CHEN, V. C. (1989) 'Learning control with neural networks', PhD Thesis, Case Western Reserve University, USA.

COX, M. G. (1990) *Algorithms for Spline Curves and Surfaces*, NPL Report DITC 166/90.

COX, M. G., HARRIS, P. M. and JONES, H. M. (1990) 'A knot placement strategy for least squares spline fitting based on the use of local polynomial approximations', in *Algorithms for Approximation II*, (eds J. C. Mason and M. G. Cox) London: Chapman and Hall, pp. 37–45.

HARRIS, C. J., MOORE, C. G. and BROWN, M. (1992) *Intelligent Control: Aspects of Fuzzy Logic and Neural Networks*, London: World Scientific.

JOHNSON, C. R. (1988) *Lectures on Adaptive Parameter Estimation*, Englewood Cliffs, NJ: Prentice Hall.

JORDAN, M. J. (1989) 'Generic constraints on unspecified target trajectories', in *Proceedings of the International Joint Conference on Neural Networks, Washington*, Vol. 1, pp. 217–25.

MILLER, W. T. (1987) 'Sensor-based control of robotic manipulators using a general learning algorithm', *IEEE J. Robotics Automation*, **3/2**, 157–65.

MOORE, C. G. (1991) 'Indirect adaptive fuzzy controllers', PhD Thesis, Dept of Aeronautics and Astronautics, University of Southampton, UK.

NARENDRA, K. S. and PARATHASARATHY, K. (1990) 'Identification and control of dynamical systems using neural networks', *IEEE Trans. Neural Networks*, **1**, 4–27.

PAO, Y. H. (1989). *Adaptive Pattern Recognition and Neural Networks*, Reading, MA: Addison-Wesley.

PARKS, P. C. and MILITZER, J. (1989a) 'Convergence properties of associative memory storage for learning control systems', *Automation and Remote Control*, **50**(2), Part 2, 254–86.

PARKS, P. C. and MILITZER, J. (1989b) 'Improved convergence properties for associative memory storage', in *Proceedings of the IFAC Symposium on Adaptive Systems in Control and Signal Processing*, Vol. 2, Oxford: Pergamon Press, pp. 565–72.

PSALTIS, D., SIDERIS, A. and YAMAMURA, A. (1987) 'Neural controllers', in *IEEE First International Conference on Neural Networks*, Vol. 4, pp. 551–8.

RIGLER, A. K., IRVINE, J. M. and VOGL, T. P. (1991) 'Rescaling of variables in backpropagation learning', *INNS Neural Networks*, **4/2**, 225–9.

RUMELHART, D. E., McCLELLAND, J. L. and the PDP Research Group (1986) *Parallel Distributed Processing*, Vol. 1, Cambridge, MA: MIT Press.

SAERENS, M. and SOQUET, A. (1989) 'A neural controller', in *IEE Conference on Artificial Neural Networks, London*, pp. 211–15.

SUTTON, R. and JESS, I. M. (1991) 'A design study of a self-organising fuzzy autopilot for ship control', *Proc. Inst. Mech. Eng.*, **205**, 35–48.

WIDROW, B. and LEHR, M. A. (1990) '30 years of adaptive neural networks: perceptron, madaline and backpropagation', *Proc. IEEE*, **78/9**, 1415–41.

6

Parallel Processing for Self-organizing Control Systems

D. A. Linkens

6.1 INTRODUCTION

In systems engineering there has been much interest in recent years in the concepts of self-adaptation and self-tuning. Many practical applications of self-adaptive control have been investigated covering disciplines as wide apart as industrial processes, aerospace and biomedicine. The emphasis has been mainly on strongly algorithmic approaches to given limited structures with well-defined, and usually linear, models. In contrast, the work described in this chapter follows an alternative approach which considers the possibility of systems which have a self-organizing capability. Such systems do not have a rigidly fixed structure chosen a priori, but rather allow for on-line modifications to structures as well as to parameters. A further feature is that the self-organizing concepts may include heuristics for decision making, and hence mixed arithmetic and symbolic computational aspects must be considered. Clearly, this means that such techniques are closely related to artificial intelligence (AI) principles and developments.

Three AI-related self-organizing structures are considered, these being relevant to the areas of system identification and control, and pattern recognition. Self-organizing systems tend to be complex and computing-intensive, and hence should be candidates for possible implementation via transputer parallelism. The group method of data handling (GMDH) technique is a multilayer self-organizing algorithm based on a nonlinear mathematical model of data. Although known for some time, its use has been limited because of the very heavy computing power required for realization. It is, however, ideally suited to parallelism, and has been successfully coded in Occam and tested as a diagnostic tool on medical data and for use with weather prediction. A comparison between the GMDH and a multilayer perceptron neural network applied to these latter data has been made. The second structure described is the self-organizing controller (SOC) due to Barron (1968). Again, this is a complex system and has been coded in Occam. The resulting SOC system had been applied to a medical application involving anaesthesia. The third structure is fuzzy logic control, again with extension to self-organizing concepts.

This has been developed in both single-variable and multivariable formulations and applied to anaesthesia and laboratory rigs. In the latter case fast sampling (20 ms) is required, and transputer based implementations have been necessary.

Not surprisingly, all three self-organizing methods described in the following sections are computationally very intensive. Thus, transputer power and parallelism is an attractive proposition for such systems, which require fast processing if they are to be used for on-line control purposes. The methods are very different in style, with the GMDH being heavily algorithmic and intended for the production of a detailed mathematical model of a process. In contrast, the other two methods are intended to provide control strategies which give adequate performance even in the absence of a detailed mathematical model of the plant being controlled. In fuzzy logic control, the merging of qualitative and quantitative concepts is desirable. To perform this work, linking an alien language to Occam was necessary. The knowledge base for the rules is best coded in an AI language such as Lisp. The available Lisp was written in C language and operates in an interpretive mode. Parallel C was used to compile the C code and then linked to the Occam program containing the final control algorithm and also the process simulation.

6.2 GROUP METHOD OF DATA HANDLING

The design of a successful control system depends on the ability to predict the process response during given operating conditions. This information can be either extracted from the differential equation describing the dynamics of the system or can be extrapolated from its measured input—output response map. A comprehensive mathematical description is generally not available, so an approach based on system identification is often used.

A deterministic approach to complex modelling and control often fails because the dynamics of the subcomponents and their interconnection are not easily understood. The information available is not enough to construct differential equations for the system, so a different approach based on predictive polynomials can be used. The prediction polynomial is a regression equation which connects future values of all input and output variables. Regression analysis can be used to evaluate the coefficients of the polynomial by the criterion of mean square error. The polynomial description of a system determines its performance and invariance, and can be used for synthesis of a control system strategy.

The polynomial description can be determined in two ways:

1. From differential equations, by replacing the time derivatives by finite differences.
2. By performing regression analysis on the sampled input and output observation of the system.

The group method of data handling (GMDH), based on the principle of heuristic self-organization, belongs to the second group. GMDH is intended for the solution of diverse interpolation problems of engineering cybernetics, such as identification of the static and dynamic characteristics of plants, pattern recognition, prediction of random processes and events, optimal control and storage of information etc. It was first presented by Ivankhnenko (1968), and since then a great deal of attention has been given to the method both in and outside the Russian Federation. A survey of its development and use can be found in Farlow (1980). It has found use in ecological modelling (Ivankhnenko *et al.* 1971), real-time estimation and control (Ikeda *et al.*, 1976), materials modelling (Kokot and Patareu, 1980), wheat productivity (Khomevnenko and Kolomiets, 1980) and long range planning (Pokrass and Golubeva, 1980). Its utilization has, however, been impeded by the heavy computational burden imposed in spite of the simplification from the Kolmogorov–Gabor polynomial. It is considered to be a powerful approach to the identification of nonlinear systems because it avoids the increasing computational load for determining the order and parameters in the identified models.

To clarify the method, suppose that the input consists of N observables $X_1, X_2, X_3, \ldots, X_N$. Also suppose that the output may be considered as the estimate of some property of the input process. In general, Y will be some nonlinear function of X as follows:

$$Y(t) = \mathrm{f}(X_1, X_2, \ldots, X_N) \tag{6.1}$$

The problem is to determine the unknown structure $\mathrm{f}(X_1, X_2, \ldots, X_N)$ from the available past data. For practical use the prediction algorithm should have simplicity, need a small amount of computation time, be well suited for real-time operation and be applicable to a small amount of available data. Let us assume that $\mathrm{f}(X_1, X_2, \ldots, X_N)$ is represented by a polynomial of a certain order with respect to X_i. The Kolmogorov–Gabor polynomial for a stationary stochastic process provides a conceptual basis for (6.1):

$$Y = \alpha_0 + \sum_{i=1}^{N} \alpha_i X_i + \sum_{i=1}^{N} \sum_{j=1}^{N} \alpha_{ij} + \sum_{i=1}^{N} \sum_{j=1}^{N} \sum_{k=1}^{N} \alpha_{ijk} X_i X_j X_k \tag{6.2}$$

The Kolmogorov–Gabor polynomial requires a large amount of data together with computation of high-dimensional matrices to determine the large number of coefficients in (6.2). The GMDH provides an alternative approach to deal with matrices of large dimension, and makes it possible to solve complex problems when the data sequence is relatively small.

The GMDH can be realized by many algorithms which differ with respect to the structure of the complete description. The most commonly used is the multi-layered perceptron-type structure shown in Figure 6.1. It uses partial polynomials of second order and self-sampling thresholds.

The basic steps in constructing a GMDH description of a process are as follows:

1. The original data are divided into training and testing sequences, the separation rule being a heuristic one. Usually the training and the testing

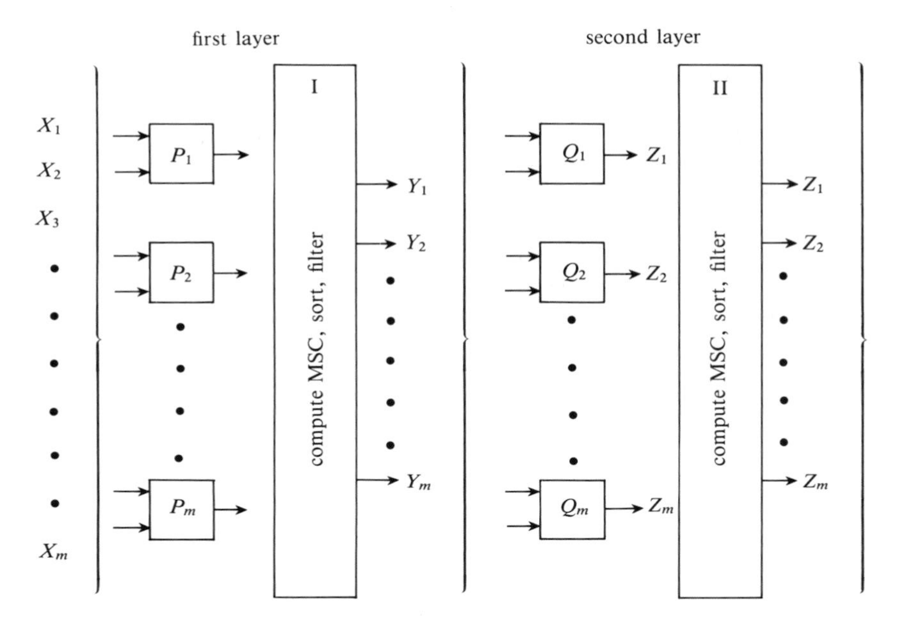

Figure 6.1 GMDH schematic

sequences are taken alternately or on the basis of the magnitude of the variance from the mean value.

2. Quadratic polynomials are formed for all possible combinations of X_i variables taking two at a time. For example, in the case of three variables X_1, X_2, X_3, the following set of polynomials is formed:

$$\begin{aligned} Y_1 &= P_1(X_1, X_2) \\ &= a_{10} + a_{11}X_1 + a_{12}X_2 + a_{13}X_1X_1 + a_{14}X_2X_2 + a_{15}X_1X_2 \\ Y_2 &= P_1(X_1, X_3) \\ &= a_{20} + a_{21}X_1 + a_{22}X_3 + a_{23}X_1X_1 + a_{24}X_3X_3 + a_{25}X_1X_3 \\ Y_3 &= P_1(X_2, X_3) \\ &= a_{30} + a_{31}X_2 + a_{32}X_3 + a_{33}X_2X_2 + a_{34}X_3X_3 + a_{35}X_2X_3 \end{aligned}$$

3. For each polynomial a system of normal Gaussian equations is constructed using all the data points in the training set. By solving these equations, the values of the intermediate variables Y_i are determined.

4. The models are used to predict the system response in the training set data region. The predictions are passed through some form of selection criteria, the most widely used being the mean square error (MSE):

$$\text{MSE} = \frac{1}{NC} \sum_{i=1}^{NC} [Y(t) - \hat{Y}(t)]^2$$

where $\hat{Y}(t)$ denotes the predicted value, and NC is the number of data points in the testing set.

5. The models Y_1, Y_2, Y_3, are ordered with respect to the smallest MSE. The models with MSE less than a specified threshold are allowed to pass to the next level of GMDH. The number of functions selected at this level is arbitrary.
6. At the next level the independent variables for the new training and testing sets are found by mapping the original training and testing sets through the single layer which has been formed.
7. New polynomials are formed according to step 2, and for each layer, steps 2 to 6 are repeated. As each new layer is formed, the smallest MSE is stored, and plotted as a function of layer number. The procedure is terminated when the smallest overall MSE is reached at any level. The global minimum is the point of optimum complexity for this choice of network heuristics. The Ivankhnenko polynomial is formed at this point by choosing the best element in the layer of optimum complexity.

6.2.1 Selection Criteria

Since two different selection methods will rarely select the same group of functions, selection of variables is of prime importance in GMDH. The criteria should be able to take forward the best model in a manner so as not to lose any significant information. The MSE has a significant weakness since it selects models which sometimes duplicate the information at the expense of discarding useful information. Some other possible criteria of interest are as follows:

1. Regularity of criteria.
2. Minimum bias criteria.
3. Balance of variable criteria.
4. Residual criteria.

6.2.2 GMDH and Parallel Processing

Different system identification methods with specific ranges of applicability have been invented, but a common feature of all such methods is the fact that they are realized in the form of sequential computer programs which significantly increases the computer running time. GMDH offers wide possibilities of parallelism for the information processing operations required in the construction of its models.

In constructing the multilayer GMDH, all independent variables in one layer are combined, two at a time, to predict the local polynomial. These combinations can be considered as separate independent blocks performing the same task on different inputs. GMDH is ideally suited for a SIMD machine, but this approach requires design of specific hardware for the particular job. Therefore the transputer which is a MIMD machine, was used and is capable of emulating the SIMD architecture without great loss of speed or efficiency. In a hardware realization it is important to isolate blocks, and to organize communication between these blocks in such a way that it is independent of the specific realization of these blocks. The transputer is a high-performance single chip computer whose architecture facilitates the implementation of these blocks in the construction of a parallel processing system.

The GMDH algorithm could be made more efficient if an array of transputers is used, where each block is calculated by a different transputer, thus reducing the computer running time very significantly. In order to code the GMDH algorithm in Occam, the main algorithm is divided into a number of small processes. Each process is independent and communicates with other processes by channels. If we have three independent variables in the first layer, three combinations result in estimating three local polynomials as shown in Figure 6.2, which are then processed further by the given selection criteria. In the second layer, another set of local polynomials is determined in a similar manner. In order to make the algorithm more efficient, an array of three transputers has been used as shown in Figure 6.3. The process to calculate the Ivankhnenko polynomial, from the combination of two independent variables, is transmitted from the host to the three slave transputers and each one is provided with a different input. The three transputers perform their work simultaneously and calculate the required polynomials. These calculated values are transmitted back to the host and further calculations regarding the MSE are performed, prior to moving to the next layer of the GMDH algorithm. The flow diagram of the GMDH algorithm for four transputers is shown in Figure 6.4.

A benchmark data set obtained from the literature (Draper and Smith, 1966) was used to validate the technique. It consists of three independent variables, (X_1) radiation in relative gram calories, (X_2) average soil moisture tension and (X_3) air temperature in degree C. The dependent variable is milligrams of vitamin B2 in

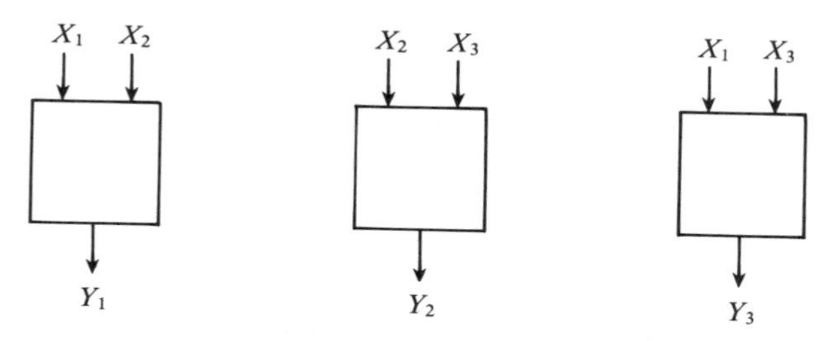

Figure 6.2 GMDH and parallelism

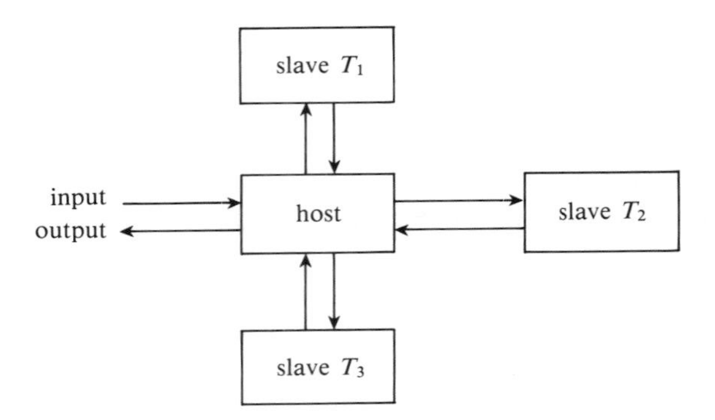

Figure 6.3 Transputer arrangement for three variables

turnip green. These data were first used to predict the Ivankhnenko polynomial on one transputer and then on four transputers. It was noted that the time required by four transputers was less than half the time required by one transputer. This difference in time increases with an increase in the number of independent variables. Parallelism can be extended to any number of transputers depending upon the independent variables present in the data. For example, four independent variables require six partial polynomials to be predicted and need seven transputers; similarly for five independent variables, eleven transputers are required. Other forms of parallelism are described in Hasnain (1989) which can be exploited within the GMDH algorithm.

The technique has also been applied to the area of decision-making in medicine. This involves a decision rule applied to a patient's measurement vector to decide between possible diseases in a differential diagnosis leading to possible treatments. Two case studies have been undertaken using published data, one being concerned with the detection of the cardiac abnormality in patients with diabetes. The other study considered the effects of human growth hormones on the level of blood glucose, also in diabetic patients. In both cases the number of points in the data sets was small, but the GMDH method was able to give good model fitting. It was also successful when used in a predictive mode; the full details of all these aspects are given in Hasnain and Linkens (1988).

The GMDH has a multilayer structure and is self-organizing in style, and therefore has some similarities with neural networks. Recently, a comparison has been made between the performance of GMDH and the multilayer perceptron neural network with a back-propagation algorithm, using the medical data sets referred to above. Both techniques have also been applied to a climatological problem involving the need for frost protection via predictive methods. In addition, a method of learning via machine induction has been applied to the same data, but gave inferior results. Neural networks and GMDH gave similar performance, with GMDH being much faster computationally. The predictive performance was encouraging for both GMDH and neural networks (Hasnain, 1989).

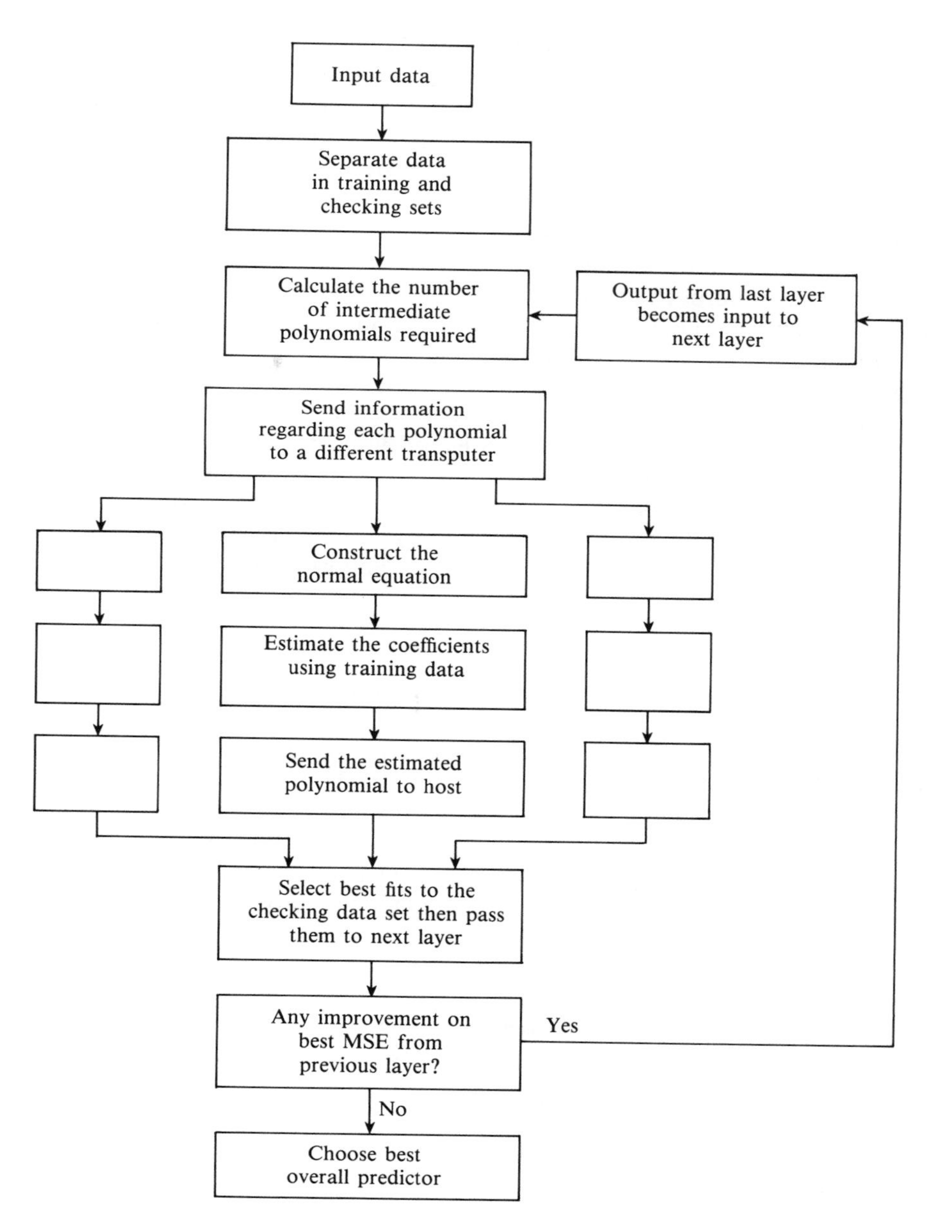

Figure 6.4 Flow diagram of GMDH algorithm on transputers

6.3 SELF-ORGANIZING CONTROL

When the parameters and characteristics of a controlled plant are variable and depend on the magnitude of disturbances or other operating conditions in ways which are unknown or difficult to predict, the overall control system must be designed to have very low sensitivity to changes in operating conditions. A control system can be called 'adaptive' if it achieves invariant control response throughout the operational envelope of the controlled vehicle. In common usage, however, the term 'adaptive' is used for those systems which adjust themselves on-line to improve their behaviour in terms of computed control performance indices.

The bionics approach to adaptive control has emphasized development of self-organizing control (SOC) structures. The SOC can readapt many times within the closed-loop response period of the plant by modification of its internal signal pathways. The adaptive controller obtains information while the plant is operating and uses this acquired knowledge to improve system performance. This technique places minimum dependence on a priori modelling, requires no pretraining and possesses unique capabilities for control of plants having multiple response variables with strong interactions among the variables, especially the actuators.

The self-organizing controller, invented by Barron (1968), is an evolutionary form of adaptive controller in which guided random search results are used to achieve flexibility and speed of adaptation.

6.3.1 Elementary SOC and Principle of Operation

The key to SOC lies in the modification of internal signal paths by on-line changes in controller probability state variables (PSVs). The SOC employing the PSV principle is shown in Figure 6.5 in its most elementary form. The controller portion consists of two elements, a performance assessment (PA) module and a PSV module. In essence, a PA module evaluates system performance on the basis of the error signal (e) and informs the PSV module through the binary reward/

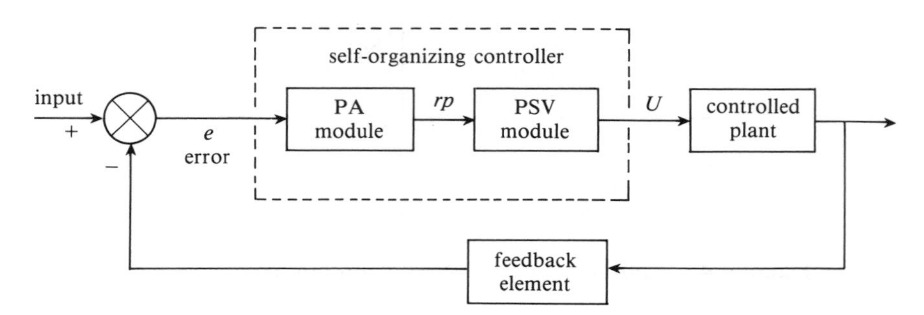

Figure 6.5 Elementary SOC

punishment (*rp*) signal of the results of the evaluation. The *rp* signal may be thought of as a 'good/bad' comment of the PA module on the last incremental change (Δu) made by the PSV module in its output signal (U). The PSV module performs a self-evaluation by comparing the most recent ΔU with *rp*. As a result of this self-evaluation it either incrementally increases or decreases (as appropriate) the probability that the next ΔU will have the same sense as the last one.

Through this iterative process of system performance measurement and self-evaluation, the SOC controls the plant in a manner which is suited to the plant dynamics characteristics. Although SOC systems cannot have an explicit 'teacher' and must rely on self-assessment of performance, empirical modelling can usually be guided by an explicit, stored database with gauging of performance by means of a goodness-of-fit criterion.

6.3.2 High-speed SOC for Multiple Goal/Multiple Actuator Control

Plants having multiple goals and multiple actuators often exhibit an interdependence among goals. Thus, the operation of one of the actuators for control of one of the goals will result in a significant change in the remaining plant variables, thereby requiring that the system be capable of control despite the interdependence of the several goals and actuators in the plant. High-speed SOC provides simultaneous, multiple goal/multiple actuator control in which the instantaneous influence of each actuator or error signal is identified and a self-organizing controller compensates for changes in the polarities of actuator, direct and cross-coupled effects (Barron, 1967).

A schematic diagram of the entire system using SOC is shown in Figure 6.6 where multiple command inputs $X_{c_1}, \dots, X_{c_k}$ from an external source, such as the stick of an aircraft, are all simultaneously fed to a summing point with measured

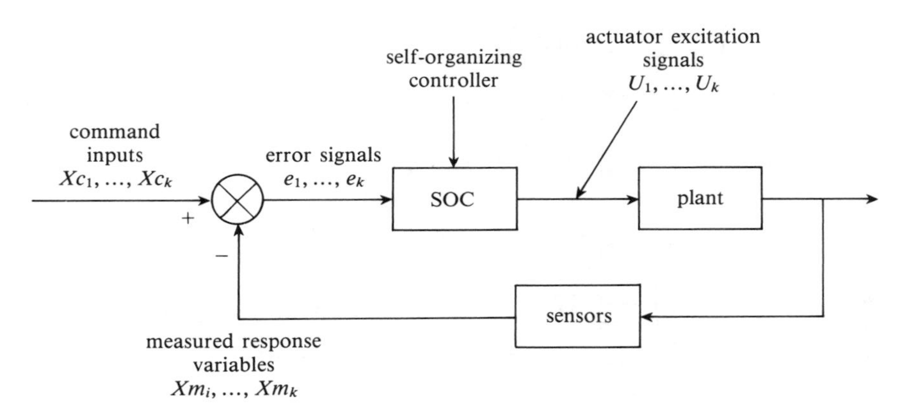

Figure 6.6 Multi-sensor SOC

response variables $X_{m_i}, \ldots, X_{m_k}$ which are provided by the sensors, there normally being one sensor for testing the response of each of the variables or goals of the plant. The summing point provides a plurality of error signals $e_1, \ldots, e_k$ to a SOC system which provides the actuator excitation signals $U_1, \ldots, U_k$ based upon the evaluation of the system performance.

A SOC for a multiple actuator system is shown in Figure 6.7. The controller includes a combination of performance assessment units $PA_1, \ldots, PA_k$, there being one PA unit for each goal of the system and all working in parallel. All PA units synchronize to evaluate the performance relating to the variable with which each is associated. The output from each PA unit is fed into a combination of logic circuits $AL_{ii}, \ldots, AL_{ji}, \ldots, AL_{ki}$, each actuation logic circuit being composed of a coupling unit, a clock and a PSV unit. For each PA unit the number of actuation logic units is equal to the number of actuators. All these actuation logic units work in parallel and the output signals $U_i, \ldots, U_k$ are fed to the goal weighting logic (GWL) circuits $GWL_i, \ldots, GWL_k$, there being one GWL unit for each actuator. The GWL circuits provide actuator excitation signals $U_i, \ldots, U_k$ in response to signals applied, and thereby alter plant operation.

The complex arrangement of SOC with the large amount of logic presents a problem for sequential computing and simulation. The parallel arrangement inside the SOC, especially for multiple goal/multiple actuator control can be realized

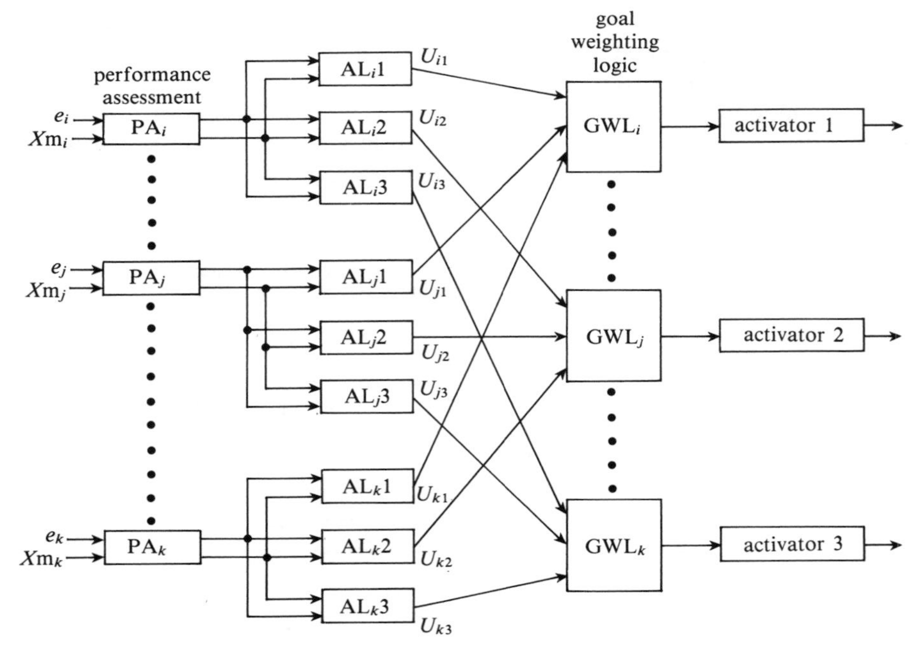

Figure 6.7 Multiple actuator ALU

naturally via transputers, thereby enhancing the power and speed of control. The SOC is not suitable for a SIMD machine as the process to be controlled needs a separate processor for its simulation if efficiency is kept in mind. The algorithms were first developed and validated using Fortran on a mainframe computer. The process used for investigation was that of drug administration for muscle relaxation in operating theatres, for which a well-developed nonlinear model is available.

The pharmacology of muscle relaxant drugs comprises two important parts, pharmacokinetics and pharmacodynamics. Pharmacokinetics concerns the absorption, distribution and excretion of drugs, whereas pharmacodynamics is a term employed in drug pharmacology to describe the relationship between drug concentration and the therapeutic effects of these drugs.

For the drug Pancuronium the body is considered to consist of two compartments between which reversible transfer of drugs can occur. The drug is injected into compartment 1 which represents the plasma volume and the perfused tissues such as heart, lung, liver from which it is excreted into urine. The drug is exchanged between the first compartment and the connecting peripheral compartment. The plasma concentration of drug as a function of time is given by the biexponential equation

$$c(t) = A_1 \exp(-\alpha_1 t) + A_2 \exp(-\alpha_2 t) \tag{6.3}$$

where A_1 and A_2 are complex functions, $A_1 \exp(-\alpha_1 t)$ reflecting the distribution phase and $A_2 \exp(-\alpha_2 t)$ reflecting the elimination phase.

It is known that the interaction between the injected drug and the components of the body induce a certain therapeutic effect. The steady-state plasma concentration of the drug is proportional to the amount of drug at the site of action. The plasma concentration of drug against effect relationship can be described mathematically by a Hill equation (Whiting and Kelman, 1980) of the form

$$E = \frac{E_{\max} C^{\alpha}}{C^2 + C_{50}^{\alpha}} \tag{6.4}$$

where E represents the drug effect, $E_{\max}$ is the maximum drug effect, C is the drug concentration and C_{50} is the drug concentration corresponding to a 50% effect.

Using previous pharmacological identification studies (Linkens *et al.*, 1982) the drug Pancuronium can be modelled as a two-time-constant linear dynamic system with a dead-time in series with a nonlinear part comprising a dead space and a limiting device or a Hill equation as illustrated in Figure 6.8. The parameters quoted are those appropriate for dogs. For the drug Atracurium, a three compartment model is considered and the transfer function is third order as shown in Figure 6.9 (Linkens and Mahfouf, 1989). The parameters quoted are those appropriate for humans.

Figure 6.10 shows the response of the total SOC system to a series of step commands, which is considered to be an adequate response in terms of speed and overshoot. The algorithms were coded in Occam, as were the model equations, and the equivalent results are also shown in Figure 6.10, again representing good

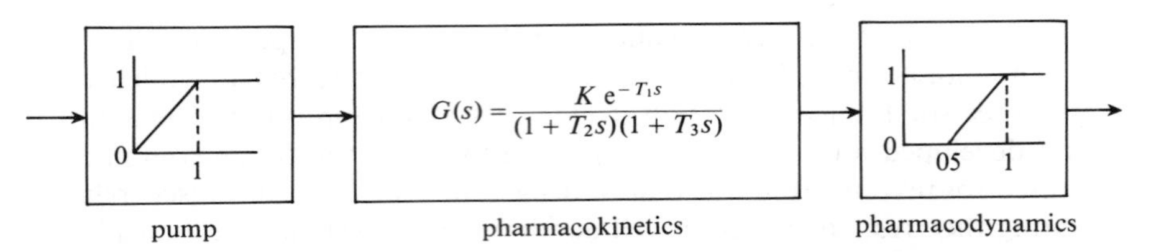

Figure 6.8 Pancuronium bromide drug muscle relaxant model

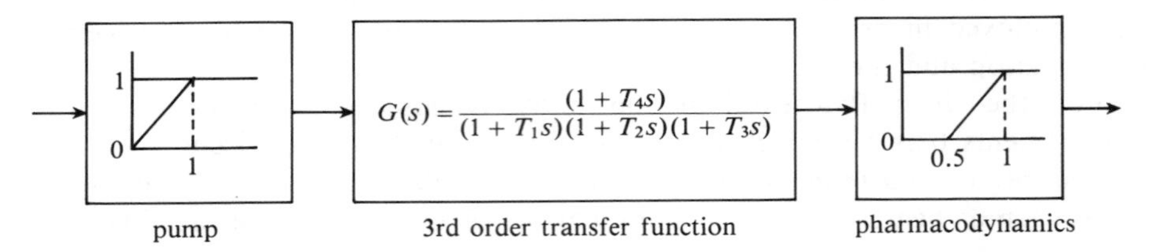

Figure 6.9 Representation of muscle relaxant system related to Atracurium

performance. The method is currently being extended for multivariable anaesthesia providing simultaneous control of unconsciousness and muscle relaxation, and is a suitable candidate for parallel implementation.

6.3.3 Adaptive Learning Control

The adaptive learning controller (ALC) is another form of SOC, being a promising learning network that demonstrates problem solving capability. ALC consists of two neuron-like adaptive elements: associative search element (ASE) and adaptive critic element (ACE) (Barto *et al.*, 1983). The most attractive features of the system are as follows:

1. Knowledge of system dynamics, i.e. a mathematical model of the system, is not necessary in order to develop a control law. The controller learns to develop by association of input and output signals.
2. The system to be controlled can be time-varying and/or nonlinear.
3. A wide class of measures of performance can be optimized.
4. A non-uniform sampling rate can be used.
5. The algorithms are naturally adaptive. They can be used to control the system directly or to optimize the performance of an existing control system.

The ASE operates by generating an output pattern, by receiving an evaluation from its environment in the form of scalar payoff or reinforcement, updating the contents of its memory, and then repeating this generate-and-test procedure. As this kind of learning proceeds, each input causes the retrieval of better choices for the pattern

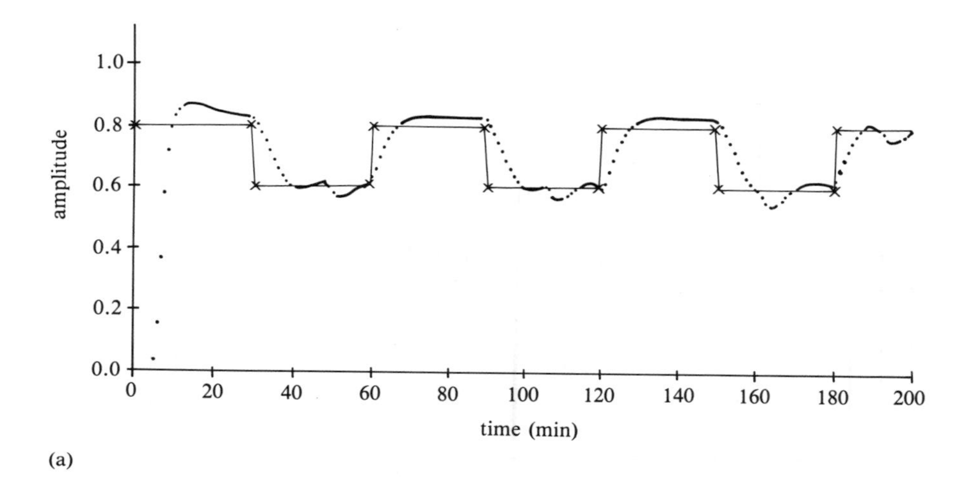

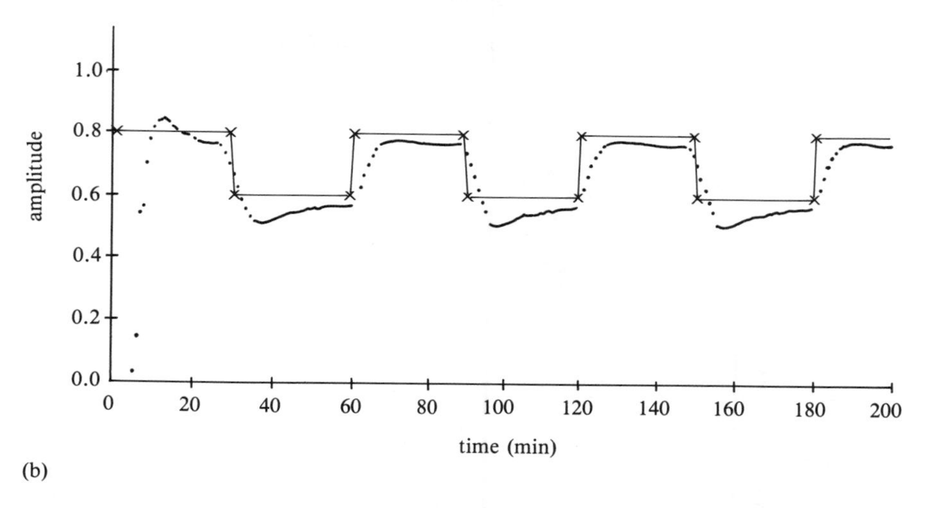

Figure 6.10 (a) SOC on mainframe; (b) SOC on transputer: × input, ● output

to be associated to that of the input. In order to discover what response leads to improvement in performance, ASE employs a trial and error, or generate-and-test search process. As shown in Figure 6.11, the state vector of the system is sampled and fed into a decoder, which transforms each state vector into an n-component binary vector, whose components are all zero except for a single one in the position corresponding to the state of system at that instant. This vector is provided as an input to the ASE, the adaptive element receives the signal through the reinforcement pathway and this information is used in the ASE.

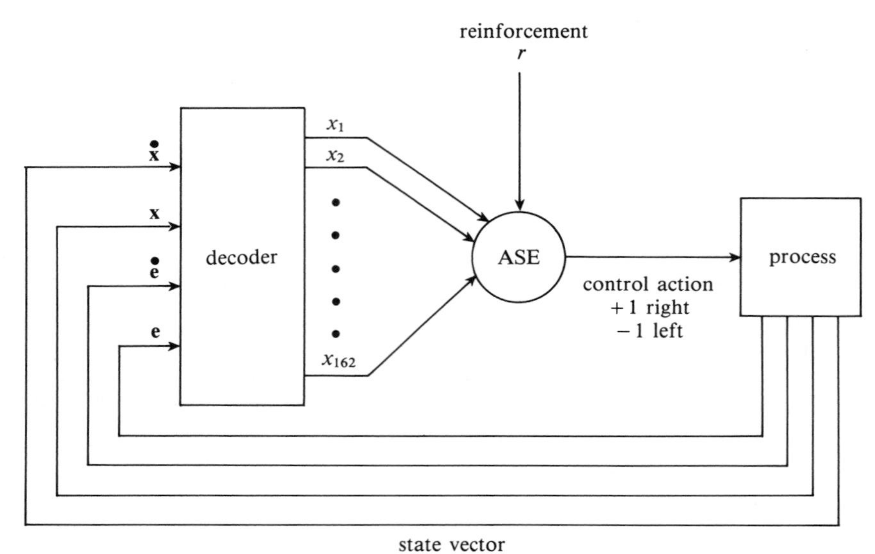

Figure 6.11 Block diagram for ASE

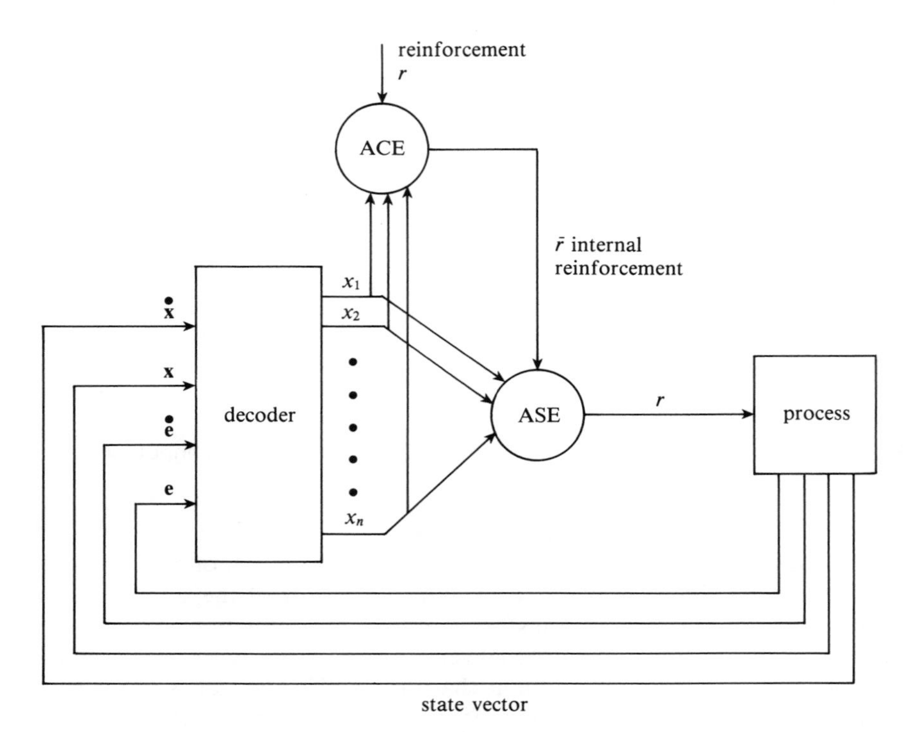

Figure 6.12 Block diagram for ALC

The learning action needs to be more distinctive to ensure convergence that leads to least punishment in cases in which only punishment is available. An ACE is introduced to overcome this problem. Among other functions, the ACE constructs predictions of reinforcement so that if punishment is less than its expected level then it acts as a reward. As shown in Figure 6.12 the ACE receives the externally supplied signal, which it uses on the basis of the current vector, to compute an improved reinforced signal that it sends to the ACE. The central idea behind the ACE algorithm is that predictions are formed that predict not only reinforcement but also future predictions of reinforcement.

Considering a system such as an inverted pendulum, the four-state vectors are divided into 162-component binary vectors. As depicted in Figure 6.12 the ALC is quite complex and the computing time involved is enormous. In order to use the ALC for a control application, it must be fast enough to deal with the changing variables. Parallel computing seems a good solution to overcome the problems of logic operations and handling the number of component binary vectors which result at every sampling interval.

6.4 FUZZY LOGIC CONTROL

It is well known that humans can often provide good control of processes which are ill-defined mathematically or physically. They are capable of manipulating processes which are probably highly nonlinear and possibly with varying dynamic structure. To do this, they learn by experience, and translate this into an internal control strategy which is often hard to articulate. The strategy is often couched in imprecise qualitative terms.

To support the translation of the more vague, nonnumeric statements that might be made about such a control strategy a semiquantitative calculus is required. It was against this background that Zadeh introduced and developed fuzzy set theory (1965) and approximate reasoning (1973). The later paper introduced the concept of fuzzy linguistic variables and the fuzzy set, which seemed to provide a means of expressing linguistic rules in such a way that they can be combined into a coherent control strategy.

Since its introduction by Zadeh, fuzzy logic has been used successfully in a number of control applications. The first application of fuzzy set theory to the control of dynamic processes was reported by Mamdani and Asilian (1975). They were concerned with the control of a small laboratory scale model of a steam engine and boiler combination. The control problem was to regulate the engine speed and the boiler pressure. Despite the nonlinearity, noise and the strong coupling in the plant, they managed to get acceptable control using a fuzzy logic controller. Kickert and Van Nauta Lemke (1976) applied fuzzy logic to design a controller for a laboratory scale warm water plant. The water tank was divided into several compartments. The aim was to control the temperature of the water in one of the compartments by altering the flow rate through a heat exchanger contained in the tank. A secondary control task was to ensure fast response to step changes in the outlet water temperature set point. The first experiment on an industrial plant with a fuzzy logic controller was undertaken by Rutherford and Carter (1976). They developed a controller for the permeability at the Cleveland sinter plant, and showed that the fuzzy logic controller worked slightly better than the PI controller.

Tong (1976) applied fuzzy logic to a pressurized tank containing liquid. The problem was to regulate the total pressure and the level of liquid in the tank by altering the rates of flow of the liquid into the tank and the pressurizing air. Good control was achieved despite the nonlinearity and strong coupling. However, it was no better than performance obtained by a controller designed using conventional techniques. Tong *et al.* (1980) also examined the behaviour of an experimental fuzzy logic control algorithm on an activated sludge waste water treatment process. They concluded that a fuzzy algorithm based on practical experience can be made to work on this difficult process. The success obtained by Mamdani and Assilian in the control of a steam engine led them to study temperature control of a stirred vessel which formed the batch reactor process – a nonlinear time varying gain and time delay process (Mamdani and Assilian, 1975). The results obtained showed that processes can be controlled effectively using heuristic rules based on fuzzy control, but to achieve good control the fuzzy rules must be correctly formulated to take account of time delays when they occur. In the same decade the techniques of fuzzy logic were applied by independent groups over a wide variety of processes. Ostergaard (1976) applied it successfully on a heat exchanger; Van Amerongen *et al.* (1977) applied fuzzy sets to model the steering behaviour of ships. Larsen (1979) reported work to implement fuzzy logic control on a coal-fired wet process umax-cooler kiln. Sheridan and Skjoth (1983) attempted to replace kiln operators at the Durkee plant of the Oregon Portland cement company using fuzzy algorithms. In this and related work on cement kiln control it was noted that human operators could successfully control their operation. Attempts to obtain adequate mathematical models of the kiln process have been generally unsuccessful. Thus, conventional control techniques have not been widely acceptable for cement kiln control. In contrast, fuzzy logic control has been applied successfully, reducing product variability, including shift-to-shift characteristic differences between operators (Haspel, 1989).

The major shortcomings of previous work is the slow response time of the fuzzy logic controller, caused by the large calculations essential for the algorithm. This makes it unsuitable for dynamic operations. In this work an attempt has been made to speed up the calculation time by splitting the fuzzy logic controller into different processes and running them in parallel on one or a combination of transputers. A novel technique to store the linguistic rules in Lisp (an AI language) for use by the fuzzy logic controller on a transputer network has also been developed.

The self-organizing fuzzy logic controller (SOFLC) algorithm initially presented by Procyk (1977) for a single input/output process provides an adaptive rule-learning capability to complement a fuzzy logic control strategy. Yamazaki and

Mamdani (1984) examined the problems of poor settling time and occasional instability associated with SOFLC, and proposed an improved version which overcomes these problems, by applying it on single input/output and multi-input multi-output processes. The main problem which restricted the application of SOFLC over a wide area was the unclarity of the rule learning and storing procedure. It was later modified by Daley (1984) and applied to a process of greater complexity and higher mathematical dimension, i.e. attitude control of a space craft. Larkin (1985) has applied SOFLC to aircraft control and Harris and Moore (1989) have applied it to the control of autonomous guided vehicles. A description is now given of an attempt to control the nonlinear muscle relaxant model of Pancuronium with SOFLC. The simulations were first performed on a sequential machine and then parallelism was introduced in the SOFLC to extend the implementation to a multiprocessor system. The robustness of SOFLC is examined with consideration of parameter changes in the process model and introducing white noise to the process.

6.4.1 Sequential Fuzzy Logic Control

The simple fuzzy logic controller described here is designed to regulate the output of a process around a given set-point. The output at regular intervals is sampled and sent to the controller. The controller in Figure 6.13, shows the configuration in relation to a single-input single-output process. In general there are two inputs and

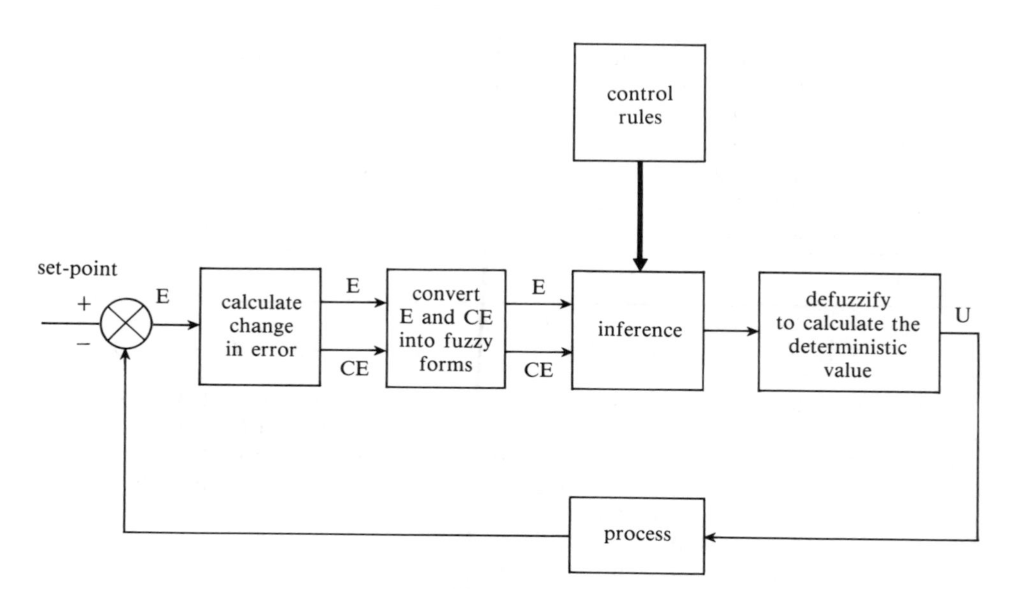

Figure 6.13 Rule-based fuzzy logic controller

one output to the controller. One of the inputs is the process error, E, the other input, CE, is the change in error, obtained by subtracting the error at the last sampling instant from the present one. Movements towards the set-point are positive changes in error, and movements away are negative changes in error. The control action, U, is the change in input to be applied to the process.

The linguistic synthesis of the fuzzy controller relates the state variables, E and CE, to the action variable, U, by means of the linguistic protocol. This linguistic

Table 6.1 The fuzzy set definitions

	Error													
	−6	*−5*	*−4*	*−3*	*−2*	*−1*	*−0*	*+0*	*+1*	*+2*	*+3*	*+4*	*+5*	*+6*
PB	0	0	0	0	0	0	0	0	0	0	0.1	0.4	0.8	1.0
PM	0	0	0	0	0	0	0	0	0	0.2	0.7	1.0	0.7	0.2
PS	0	0	0	0	0	0	0	0.3	0.8	1.0	0.5	0.1	0	0
P0	0	0	0	0	0	0	0	1.0	0.6	0.1	0	0	0	0
N0	0	0	0	0	0.1	0.6	1.0	0	0	0	0	0	0	0
NS	0	0	0.1	0.5	1.0	0.8	0.3	0	0	0	0	0	0	0
NM	0.2	0.7	1.0	0.7	0.2	0	0	0	0	0	0	0	0	0
NB	1.0	0.8	0.4	0.1	0	0	0	0	0	0	0	0	0	0

	Change in error												
	−6	*−5*	*−4*	*−3*	*−2*	*−1*	*−0*	*+1*	*+2*	*+3*	*+4*	*+5*	*+6*
PB	0	0	0	0	0	0	0	0	0	0.1	0.4	0.8	1.0
PM	0	0	0	0	0	0	0	0	0.2	0.7	1.0	0.7	0.2
PS	0	0	0	0	0	0	0	0.9	1.0	0.7	0.2	0	0
Z0	0	0	0	0	0	0.5	1.0	0.5	0	0	0	0	0
NS	0	0	0.2	0.7	1.0	0.9	0	0	0	0	0	0	0
NM	0.2	0.7	1.0	0.7	0.2	0	0	0	0	0	0	0	0
NB	1.0	0.8	0.4	0.1	0	0	0	0	0	0	0	0	0

	Output														
	−7	*−6*	*−5*	*−4*	*−3*	*−2*	*−1*	*−0*	*+1*	*+2*	*+3*	*+4*	*+5*	*+6*	*+7*
PB	0	0	0	0	0	0	0	0	0	0	0	0.1	0.4	0.8	1.0
PM	0	0	0	0	0	0	0	0	0	0.2	0.7	1.0	0.7	0.2	0
PS	0	0	0	0	0	0	0	0.4	1.0	0.8	0.4	0.1	0	0	0
Z0	0	0	0	0	0	0	0.2	1.0	0.2	0	0	0	0	0	0
NS	0	0	0	0.1	0.4	0.8	1.0	0.4	0	0	0	0	0	0	0
NM	0	0.2	0.7	1.0	0.7	0.2	0	0	0	0	0	0	0	0	0
NB	1.0	0.8	0.4	0.1	0	0	0	0	0	0	0	0	0	0	0

protocol consists of a set of fuzzy rules which define individual control situations to form a fuzzy conditional sentence or algorithm. A section of such an algorithm might be

```
if error is...
then if change in error is...
then output is...
else
if error is...
then change in error is...
then output is...
```

Each rule is a fuzzy conditional statement connecting the output to the input.

In the fuzzy controller the universe of discourse is discrete, thus membership vectors rather than membership functions are considered. The variables are therefore defined by fuzzy subsets in the following manner:

error	$E = (e, \mu_R(e))$	$e \in E$
change in error	$CE = (c, \mu_{CE}(c))$	$c \in CE$
change in input	$U = (u, \mu_U(u))$	$u \in U$

Error and its rate of change are first calculated and then converted into fuzzy variables after being scaled. Zadeh's rule of inference is used to infer the fuzzy output. The deterministic input is then calculated by defuzzifying the input using the mean of maxima method.

The fuzzy sets are formed on a discrete support universe of 14 elements of error; 13 elements of error rate and 15 elements of the controller output. Appropriate membership functions are assigned to each element of the support set. The resulting sets representing the linguistic terms are listed in Table 6.1.

The linguistic elements used are the same as those used in most applications and these items have the following meaning:

PB	positive big
PM	positive medium
PS	positive small
P0	positive zero
Z0	zero
N0	negative zero
NS	negative small
NM	negative medium
NB	negative big

The terms N0 and P0 are introduced to obtain finer control about the equilibrium state, N0 being defined as values slightly below zero and P0 terms slightly above zero.

6.4.2 Parallel Fuzzy Logic Control

The fuzzy control of a process is quite complex as indicated by Figure 6.14. Fuzzification, defuzzification and scanning of the control rules to find the appropriate one makes the system very slow. Parallel processing presents a possible solution. In order to experiment with a transputer-based system, an experimental process (muscle-relaxant drug model), the fuzzy control logic and the control rules were all coded in Occam and run on a single transputer. The results were encouraging, with a considerable gain in speed. Trials showed that since the control rules are linguistic and Occam is not particularly suitable for handling linguistic problems some alternative language should be used.

There are many problem-solving systems that are based on matching simple rules to given problems. They are often called rule-based expert systems, and sometimes they are called situation-action systems or production-rule systems. LISP is a powerful language to implement rule-based systems with great potential to handle linguistic rules and commands. The available transputer Lisp system was originally written in C and capable of running on a single transputer. In order to implement it as an interactive system which can advise the control algorithm on other transputers it has to be linked with Occam.

Inside the TDS (transputer development system), C is considered an alien language. The facilities provided for linking it with an Occam program results in slow communication to the knowledge base files required by the Lisp program. The 3L parallel C compiler provides the necessary software needed to link C programs with Occam outside the TDS. The Occam program was compiled using a stand-alone compiler and then linked to the C compiled programs. A configuration file which explains the physical description of the links and processor descriptions connects the C and Occam programs and produces an object file which can be executed on a transputer network.

It is worth mentioning at this point that the major limitation of the fuzzy algorithm was the off-line calculation of the rule map. Any change in the control rules were to be observed by the operator first, and then followed by the change in the fuzzy linguistic algorithm and calculation of the rule map, which is subsequently applied to the process. The whole procedure is repeatedly performed until an acceptable level of control is achieved. This leads us to the concept of a self-organizing fuzzy logic controller.

6.4.3 Self-organizing Fuzzy Logic Control

The self-organizing fuzzy logic controller (SOFLC) is an extension of the simple fuzzy logic controller as shown in Figure 6.14. The SOFLC must be able to assess its performance in order to improve its control strategy. The basic functions of SOFLC can be summarized as follows:

1. To issue appropriate control action whilst evaluating the performance.
2. To modify the control action based on this evaluation.

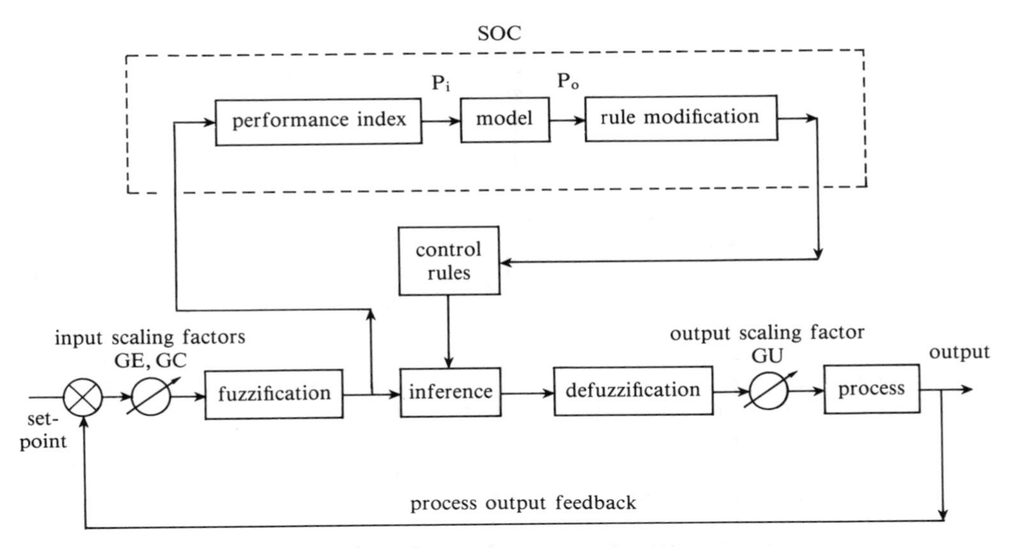

Figure 6.14 Self-organizing fuzzy logic controller

The performance of the controller in relation to the process output is measured by the deviation of the control response from the desired response. The response of an output can be monitored by its error and change in error. The performance index takes the form of a decision maker which assigns a credit or reward based on the knowledge of error and change in error. This credit value is obtained by the fuzzy algorithm which linguistically defines the desired performance. The performance index matrix used for muscle relaxant control using Pancuronium is shown in Table 6.2.

If the input states are assumed to be fuzzy singletons, i.e. fuzzy sets with membership function equal to zero everywhere except at the measured value where

Table 6.2 SOC performance index matrix

				Change in error			
Error	*NB*	*NM*	*NS*	*Z*	*PS*	*PM*	*PB*
NB	NB	NB	NB	NM	NM	NS	Z0
NM	NB	NB	NM	NM	NS	Z0	NS
NS	NB	NB	NS	NS	Z0	PS	PM
N0	NB	NM	NS	Z0	Z0	PM	PB
P0	NB	NM	Z0	Z0	PS	PM	PB
PS	NM	NS	Z0	PS	PS	PB	PB
PM	NS	Z0	PS	PM	PM	PB	PB
PB	Z0	PS	PM	PM	PB	PB	PB

Table 6.3 Performance index look-up table

	Change in error												
Error	*−6*	*−5*	*−4*	*−3*	*−2*	*−1*	*−0*	*+1*	*+2*	*+3*	*+4*	*+5*	*+6*
−6	−7.0	−6.5	−7.0	−5.0	−7.0	−7.0	−4.0	−4.0	−4.0	−2.5	−1.0	0.0	0.0
−5	−6.5	−6.5	−6.5	−5.0	−6.5	−6.5	−4.0	−4.0	−4.0	−2.5	−1.5	0.0	0.0
−4	−7.0	−6.5	−7.0	−5.0	−4.0	−4.0	−4.0	−1.0	−1.0	−2.5	0.0	1.5	1.0
−3	−6.5	−6.5	−4.0	−4.0	−4.0	−3.0	−2.5	−2.5	0.0	0.0	1.5	2.5	2.5
−2	−7.0	−6.5	−7.0	−4.0	−1.0	−1.0	−1.0	0.0	0.0	0.0	1.0	4.0	4.0
−1	−6.5	−6.5	−6.5	−4.0	−1.5	−1.5	−1.5	0.0	0.0	2.5	1.5	4.0	4.0
−0	−7.0	−6.5	−4.0	−3.5	−1.0	−1.0	0.0	0.0	0.0	2.5	4.0	6.5	7.0
+0	−7.0	−6.5	−4.0	−2.5	0.0	0.0	0.0	1.0	1.0	3.5	4.0	6.5	7.0
+1	−4.0	−4.0	−1.5	−2.5	0.0	0.0	1.5	1.5	1.5	4.0	6.5	6.5	6.5
+2	−4.0	−4.0	−1.0	0.0	0.0	0.0	1.0	1.0	1.0	4.0	7.0	6.5	7.0
+3	−2.5	−2.5	−1.5	0.0	0.0	2.5	2.5	3.0	4.0	4.0	4.0	6.5	6.5
+4	−1.0	−1.5	0.0	2.5	1.0	1.0	4.0	4.0	4.0	5.0	7.0	6.5	7.0
+5	0.0	0.0	1.5	2.5	4.0	4.0	4.0	6.5	6.5	5.0	6.5	6.5	6.5
+6	0.0	0.0	1.0	2.5	4.0	4.0	4.0	7.0	7.0	5.0	7.0	6.5	7.0

it equals unity, the linguistic performance rules can be transformed into a look-up table (Table 6.3) using the standard techniques of fuzzy calculus.

The rule modification procedure can be described as follows (Daley and Gill, 1986):

1. Let the input to the controller be $e(kT)$ and $ce(kT)$ and output of the controller be $u(kT)$.
2. At some instant kT the process input reward is $p_i(kT)$.
3. If the process input rT samples earlier contributed most to the present state, then the controller output caused by the measurements $e(kT-rT)$ and $ce(kT-rT)$ should have been $u(kT-rT)+p_i(rT)$.

The rule modification can be approached by forming fuzzy sets around the above single values:

$$\mathrm{E}(kT-rT)=F_z(e(kT-rT)) \tag{6.5}$$

$$\mathrm{CE}(kT-rT)=F_z(ce(kT-rT)) \tag{6.6}$$

$$\mathrm{U}(kT-rT)=F_z(u(kT-rT)+p_i(rT) \tag{6.7}$$

where F_z is the fuzzification procedure.

The controller is modified by replacing the rules that most contributed to the process input rT samples earlier with the rule

$$\mathrm{E}(kT-rT)\rightarrow \mathrm{CE}(kT-rT)\rightarrow \mathrm{U}(kT-rT) \tag{6.8}$$

If the controller is empty then an initial rule must be created. Further rules will be generated when required by the modification procedure. The initial rule will be formed by the fuzzification of the initial condition e_i, ce_i and p_i, and controller outputs u_i. This is done by forming a fuzzy kernel to provide a spread of values about the single support element, thus creating the following fuzzy sets:

$$\mathrm{E} = (e - x), \mu_k(e - x) \tag{6.9}$$

$$\mathrm{CE} = (ce - x), \mu_k(ce - x) \tag{6.10}$$

$$\mathrm{U} = (u + p_i - x), \mu_k(u + p_i - x) \tag{6.11}$$

where

$$\begin{matrix} \mu_k(e - x) \\ \mu_k(ce - x) \\ \\ \mu_k(u + p_i - x) \end{matrix} = \begin{cases} 1.0 & x = 0 \\ 0.7 & x = \pm 1 \\ 0.2 & x = \pm 2 \\ 0.0 & 3 \leqslant x \leqslant -3 \end{cases}$$

The single elements and the input and output of the controller are stored for use by the modification procedure. To reduce computation time, Daley (1984) suggested storage of each element of the rules in 3×5 sublocations of the rule matrix containing the following:

1. A single value giving the number of nonzero membership values in the set.
2. The desired support value of the set.
3. The membership value of the set.

Thus if the present instant is kT and the modification is made to the controller output rT samples earlier, the rule to be included results from the fuzzification of

$$e(kT - rT), ce(kT - rT), u(kT - rT) + p_i(kT) \tag{6.12}$$

The output U is obtained from the rules by using a modification procedure as follows. A control rule written as:

```
IF 'E' THEN IF 'CE' THEN 'U'
```

is a fuzzy relation

$$\mathrm{R} = \mathrm{E} \times \mathrm{CE} \times \mathrm{U} \tag{6.13}$$

then

$$\mu(e, ce, u) = \min\{\mu_{\mathrm{E}}(e), \mu_{\mathrm{CE}}(ce), \mu_{\mathrm{U}}(u)\} \tag{6.14}$$

If the measured fuzzy sets at some instant are $\hat{\mathrm{E}}$ and $\hat{\mathrm{C}}\mathrm{E}$ then an implied output set $\hat{\mathrm{U}}$ can be obtained by the above rule using the process of fuzzy composition

$$\hat{\mathrm{U}} = \{\hat{\mathrm{E}} \circ [\hat{\mathrm{C}}\mathrm{E} \circ (\mathrm{E} \times \mathrm{CE} \times \mathrm{U})]\} \tag{6.15}$$

which has a membership function given by

$$\mu_0(u) = \max_e \min\{[\max_{ce} \min\{\mu_R(e, ce, u), \mu_{\mathrm{CE}}(ce)\}], \mu_e(\mathrm{E})\} \tag{6.16}$$

For several control rules the output set U_o is characterized by the membership function

$$\mu_{U_o}(u) = \max_{\text{rules}}\{\mu_O(u)\} \tag{6.17}$$

and the deterministic control action is obtained from U_o by the 'mean of maxima' procedure.

Selection of process reference model

The process reference model reflects the degree of input–output coupling. For a single variable process, the control rules modifications are given by the following:

$$P_i(nT) = KP_o(nT) \tag{6.18}$$

where P_i is the performance index output, P_o is the required output changes, and K is the gain (usually = 1). For a multivariable process:

$$P_i(nT) = M^{-1}\ P_o(nT) \tag{6.19}$$

where

$$M = \begin{bmatrix} m_{11} & m_{12} \\ & \\ m_{21} & m_{22} \end{bmatrix}, \quad P_i(nT) = \begin{bmatrix} P_1(nT) \\ \vdots \\ P_q(nT) \end{bmatrix}, \quad P_o(nT) = \begin{bmatrix} P_1(nT) \\ \vdots \\ P_j(nT) \end{bmatrix}$$

M is an incremental model of the process. The elements of the model must be scaled and normalized to their maximum values (± 1). Daley and Gill (1986) derived the process reference model from the state-space model of the flexible and rigid model of the process. Procyk (1977) referred the choice of the model to the steady-state gain of the process. The model should be scaled from the real values to normalized (± 7) values using the scaling factors for each input:

$$\text{G}\hat{\text{Y}} = \begin{bmatrix} \text{GY}_1 & 0 \\ 0 & \text{GY}_2 \end{bmatrix}, \quad \text{G}\hat{\text{T}} = \begin{bmatrix} \text{GT}_1 & 0 \\ 0 & \text{GT}_2 \end{bmatrix} \tag{6.20}$$

Thus

$$P_i(nT) = (\text{G}\hat{\text{T}})^{-1}(\text{M})^{-1}(\text{G}\hat{\text{Y}})^{-1}P_o \tag{6.21}$$

and P_i should not exceed the maximum or the minimum range (± 7), the condition being satisfied if:

$$(\text{S}_1)^{-1} = \frac{m_{11}}{\text{GY}_1\text{GT}_1} + \frac{m_{12}}{\text{GY}_1\text{GT}_2} \leqslant 1 \tag{6.22}$$

$$(\text{S}_2)^{-1} = \frac{m_{21}}{\text{GY}_2\text{GT}_1} + \frac{m_{22}}{\text{GY}_2\text{GT}_2} \leqslant 1 \tag{6.23}$$

This yields

$$\hat{M} = \text{scale}(G\hat{T})^{-1}(M)^{-1}(G\hat{Y})^{-1} \tag{6.24}$$

where

$$\text{scale} = \min[S_1, S_2] \tag{6.25}$$

The state buffer

The state buffer is a first-in first-out register which records the values of the error, change in error and the controller output. The input to the register will be at the output after a time equal to the delay in reward.

Scaling factors

The selection of the scaling factors is not entirely subjective since their magnitude is a compromise between the sensitivity during the rise time and the required steady-state accuracy.

1. The error scaling factor (GE) is chosen according to the error percentage (EP) value given by the equation:

$$\text{Full error} = \text{error percentage} \times \text{set point} \tag{6.26}$$

$$GE = \frac{6.0}{\text{full error}} \tag{6.27}$$

   The value of EP decides the error scaling factor sensitivity. High values of EP mean small scaling factors and low values mean large scaling factors.
2. The change in error scaling factor (GC) is chosen according to the maximum real change-in-error for each variable to be scaled to the maximum discrete support value, thus:

$$GC = \frac{6.0}{\text{max. real error change}} \tag{6.28}$$

3. The output scaling factor is chosen to accommodate the controller output to the process input according to:

$$GT = \frac{\text{max. process input}}{\text{max. controller output}} \tag{6.29}$$

Delay in reward

The delay in reward parameter (DLS) plays a vital role in the rule's modification. It specifies which input in the past contributes to the present performance. Procyk (1977) suggested that this parameter should be equal to the dead-time of the process.

6.4.4 Simulation of Simple Fuzzy Logic Control

The fuzzy linguistic algorithm for the muscle relaxant process of Pancuronium is written by studying the general behaviour of the process under different conditions to provide fast convergence and high steady-state accuracy (Table 6.4).

The rules are written in the form of a matrix and interpreted as:

```
If 'Error is NO' and 'CE is NS' then controller output is
'PS'
```

The nonlinear muscle relaxant process model for Pancuronium was used to test the fuzzy logic controller. The theory of fuzzy logic calls for a large amount of computation, i.e. multiplication of $A \times (B \times C \times D)$ matrices, where A is the number of rules given in the fuzzy linguistic algorithm (state action diagram, Table 6.4). For the Pancuronium process 14 rules were initially selected. B, C and D are matrices representing the fuzzy sets error, change in error and controller output (Table 6.1). This induced a calculation of $14 \times (14 \times 13 \times 15)$ matrices, to calculate the rule map. Two Fortran programs were written, one to calculate the rule map and the second to implement the fuzzy logic controller using the rule map. The process was simulated using PSI (an interactive simulation package (1980)) and connected to the Fortran program using a special communication routine written for PSI called PSICON.MES.

The parameters selected for the muscle relaxant transfer function

$$G(s) = \frac{Ke^{-T_1 s}}{(1 + T_2 s)(1 + T_3 s)} \tag{6.30}$$

were as follows: $T_1 = 1$ min; $T_2 = 2$ min; $T_3 = 20$ min. A sampling interval of 1 min was used with a total experiment length of 200 min for each simulation. The initial fuzzy linguistic rules were written based on the required response of the process with a step input. In order to get acceptable results these rules were modified by close observation in the subsequent runs. The final rule matrix gave an acceptable response to a constant set-point.

Table 6.4 System state action diagram

			Change in error				
Error	*NB*	*NM*	*NS*	*Z*	*PS*	*PM*	*PB*
NB							
NM							
NS			PS				
N0			PS	PS	Z		
P0			PS	PS	PS		
PS			PB	PS			
PM			PB	PM			
PB			PB	PS			PB

The parallel version of fuzzy logic, as described earlier, was implemented on an array of transputers (Figure 6.15). The Lisp code containing the rules (knowledge) was loaded on the root transputer, and the process (which was the muscle relaxant model in this case) along with the fuzzy control algorithms were loaded on the slave transputers. Whenever a decision point is reached by any slave transputer it can gain advice from the Lisp system by communicating over the channels. An example of the Lisp-based fuzzy control applied to the muscle relaxant model is shown in Figure 6.16. The output response to a step command shows an adequate rise-time, but with some steady-state offset which could be detrimental in certain cases.

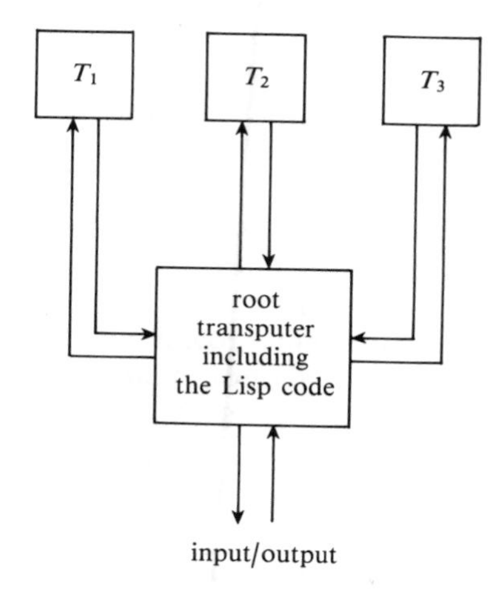

Figure 6.15 Transputer arrangement for fuzzy control

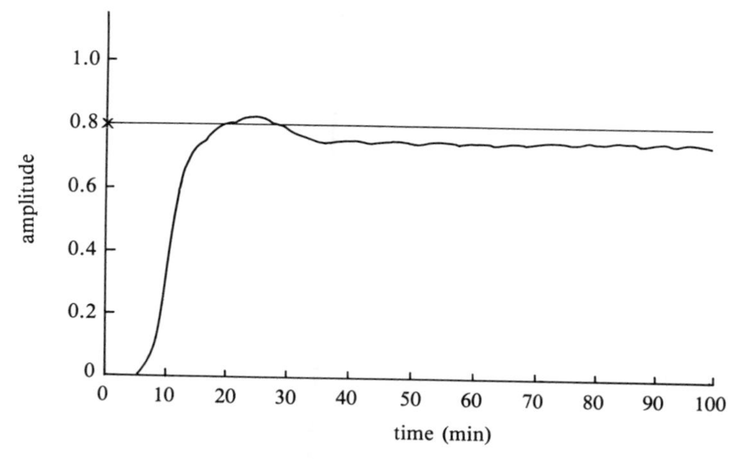

Figure 6.16 Fuzzy control with Lisp on transputer for Parcuronium drug: time constants of 2 and 20 (× input, ● output)

6.4.5 Simulation of Self-organizing Fuzzy Logic Control

The SOFLC agorithm was used to control the single-variable muscle relaxant model for the drug Pancuronium. The simulations were first performed on a Sun Unix-based system coded in Fortran and then repeated on a transputer system with a parallel version of the SOFLC to compare the timing details.

The results are presented for time delay of $T_1 = 1$ min, and time constants of $T_2 = 1$ min and $T_3 = 10$ min in the equation of the Pancuronium anaesthetic model (6.30).

The scaling factors were determined by trial and observation starting from initial scaling factors, selected by the procedure described by Procyk (1977). The simulation started with a completely empty controller and subsequently the entire control protocol was learned. Constant and varied set-points were used (Figures 6.17 and 6.18) and convergence was indicated when the number of rule changes fell to zero.

The parallel implementation of fuzzy logic has been demonstrated earlier in this chapter. The SOFLC and the Pancuronium process were coded in Occam for the parallel implementation. A sample run with parameters of $T_1 = 1, T_2 = 1$ and $T_3 = 10$ in (6.30) of the Pancuronium process was first performed on a T414 transputer (without floating point hardware) and then repeated on the T800 transputer (with floating point hardware). The result of this simulation is presented in Figure 6.19. The timing details in Table 6.5 compare the execution time for the same controlled process on different machines, showing that the parallel implementation of SOFLC is ten times faster than on a sequential machine.

Multivariable SOFLC has been used in the control of a real-time coupled-

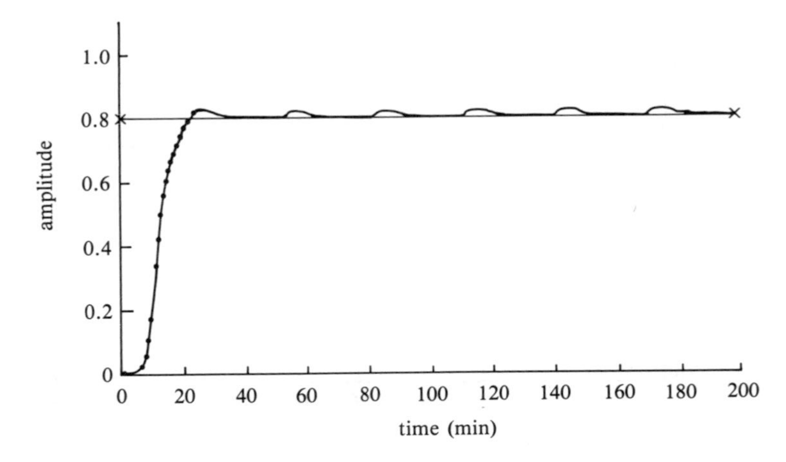

Figure 6.17 SOFLC on Sun Unix system for Parcuronium drug – constant set point: time constants of 1 and 10 (× input, ● output)

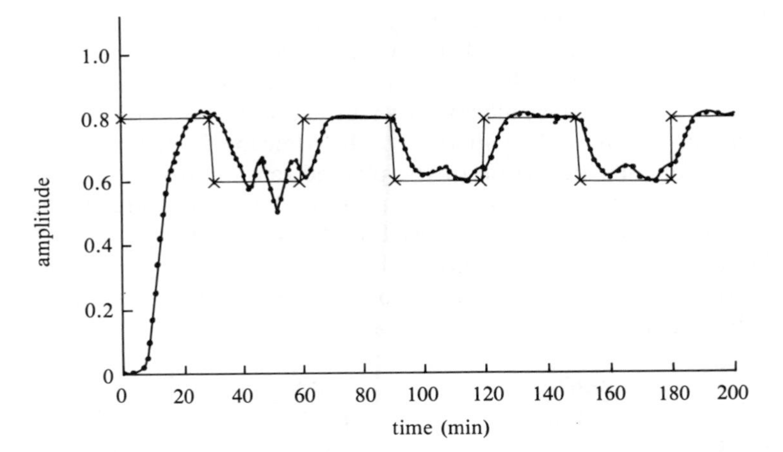

Figure 6.18 SOFLC on Sun Unix system for Parcuronium drug – variable set point: time constants of 1 and 10 (× input, ● output)

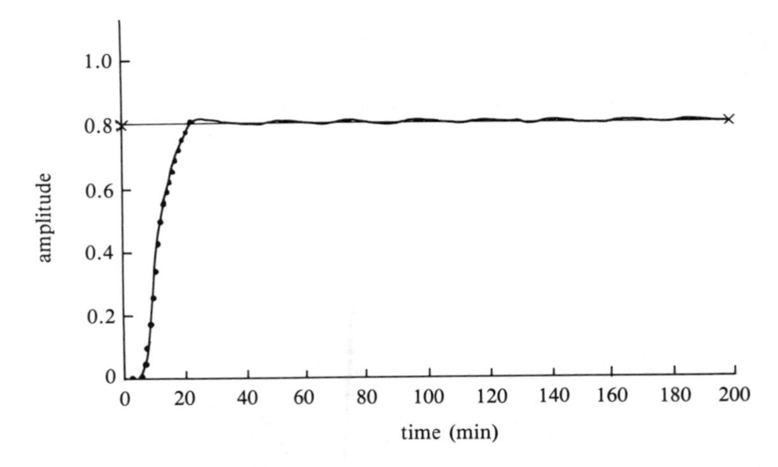

Figure 6.19 SOFLC on transputer for Parcuronium drug: time constants of 1 and 10 (× input, ● output)

Table 6.5 Timing details

Processor	*Fuzzy logic SOC*
Sun Unix system	51.00 s
T414 transputer	11.455 s
T800 transputer	5.119 s

electric drive rig. A recurring industrial problem is control of tension and speed in material handling and transport. The coupled electric drives apparatus is a simple example, consisting of two DC servo motors which drive a jockey pulley via a continuous belt. The jockey pulley velocity measurement is obtained from a tacho generator driven by the pulley spool. The belt tension is measured indirectly by monitoring the angular deflection of the pivoted tension arm to which the jockey pulley is attached (Figure 6.20). The inputs to the motors are in the range of 0–10 volts, while, for the outputs, a speed of 10 rev s^{-1} corresponds to 2 volts, and a deflection of 0.5 degree corresponds to 1 volt. Owing to the operating conditions limitation, the inputs and the outputs to the process are taken in the range of 0.3 volt. The apparatus can be modelled as follows:

$$\begin{bmatrix} Y_1 \\ Y_2 \end{bmatrix} = \begin{bmatrix} \dfrac{0.67}{1+0.6s} & \dfrac{0.8}{1+0.6s} \\ \dfrac{482.44}{s^2+32.82s+1015.66} & \dfrac{-5.41.69}{s^2+38.27s+1355.24} \end{bmatrix} \begin{bmatrix} V_a \\ V_b \end{bmatrix} \tag{6.31}$$

where V_A = input to drive A, V_B = input to drive B, Y_1 = pulley speed and Y_2 = jockey arm deflection.

The process has a dead-time of 120 ms and because of the high noise associated with the signals, an active filter was used on each output signal. The sampling time was chosen to be 20 ms. This means the control algorithm must be executed faster

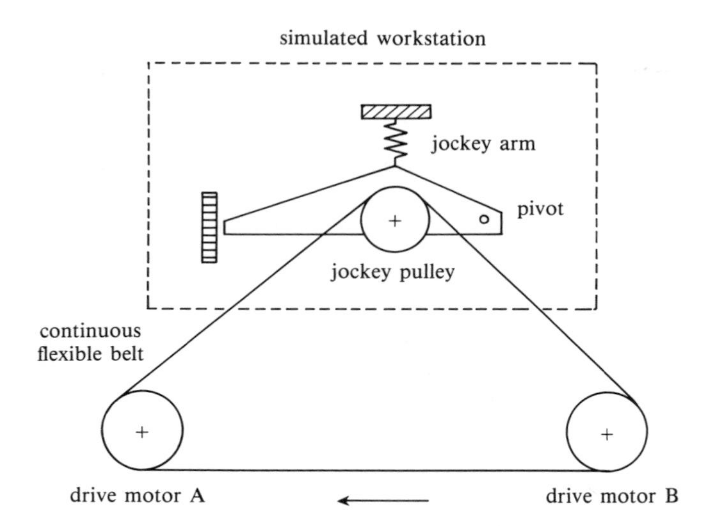

Figure 6.20 Coupled-electric drives apparatus

than it is possible with a PC, and therefore parallel processing was used in order to speed up the execution. The control algorithm was designed to perform multitasking. The first procedure is to sample the output of the coupled-electric drives apparatus through the analog to digital (A/D) converters, then calculate the control signals by executing the fuzzy control algorithm, then write to the screen the process status and read from the keyboard the operator task, and finally send the control signals to the process through the digital to analog (D/A) converters. According to the required sampling time, all these procedures must be done within 20 ms. Therefore, it was decided to use a transputer system network in order to do the processing in parallel. The algorithm is designed to comprise two processes running together in parallel. The first process is to deal with the A/D converter and the D/A converter, while the second process contains three processes running together in parallel and communicating with each other as well. The first one is the keyboard handler and the second one is the screen handler which writes the process status to the screen every second, while the main process is the self-organizing fuzzy logic controller algorithm. The algorithm is written in the Occam-2 programming language and organized to be run on one transputer board (T800). The time taken by the program to execute all the procedures did not exceed 20 ms with the controller having less than 110 rules. Figure 6.21 shows the block diagram of the algorithm.

Extensive experiments have been performed to investigate robustness, effect of scaling factors and knowledge base size. The delay in reward parameter was chosen to be 6. Small scaling factors were experimented with and gave less rules and small steady-state error, Figure 6.22. Large scaling factors were tested and gave more rules and better accuracy, Figure 6.23. Improving the controller performance could be achieved by extending the algorithm to include switching of the scaling factors used for Figure 6.24. A problem arose in determining for which variable the switch should occur. Therefore three algorithms were tested. In the first algorithm the switches occurred depending on the speed output. In the second algorithm the switch occurred depending on the tension output. Finally in the third one each switch occurred depending on its related variable. The scaling factors were:

before the switch

$$GE_S = 3, \quad GC_S = 15, \quad GT = 0.5$$
$$GE_1 = 3, \quad GC_T = 3, \quad GT = 0.5$$

after the switch

$$GE_S = 9, \quad GC_S = 30, \quad GT = 0.4$$
$$GE_T = 9, \quad GC_T = 5, \quad GT = 0.4$$

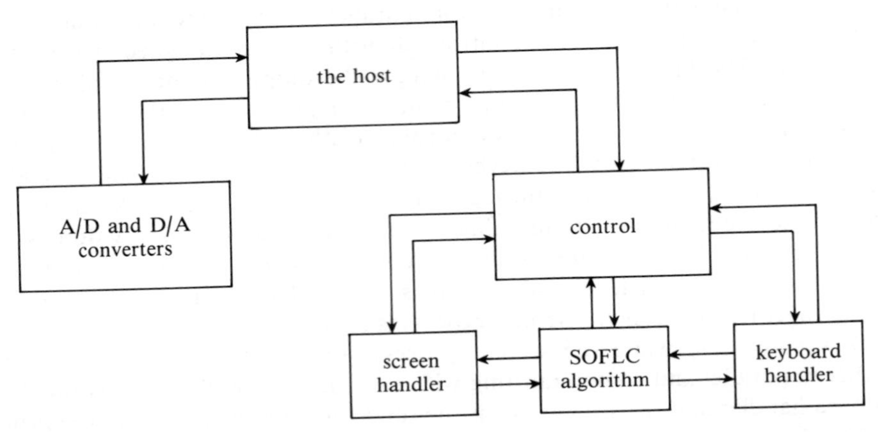

Figure 6.21 Block diagram of the control algorithm on transputers

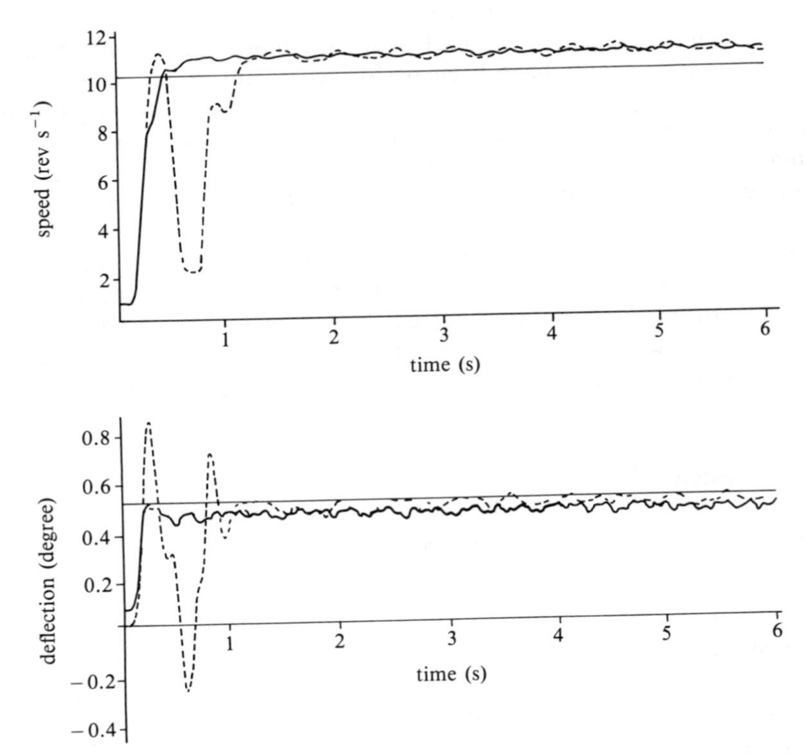

Figure 6.22 Response of SOC (small scaling factors): --- 1st run, — 3rd run

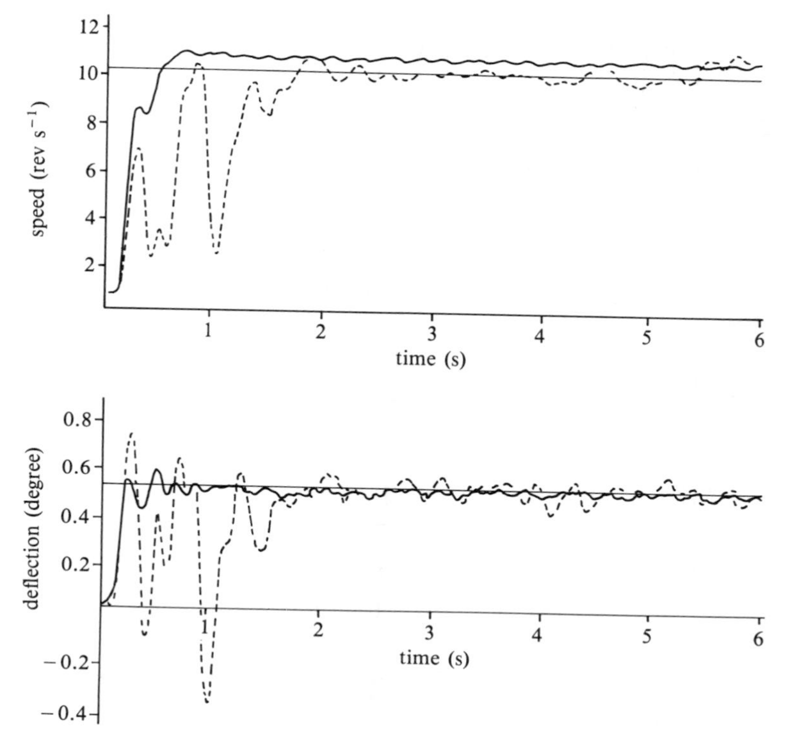

Figure 6.23 Response of SOC (large scaling factors): --- 1st run, — 3rd run

A multirun procedure was adopted in order to improve the controller performance by starting each run with initial rules from the previous run. The related controller parameters for the third run are shown in Table 6.6.

Table 6.6 Controller parameters

Fig.	*Set-points*		*Algorithm*	*Rules*	
	Speed	*Deflection*		*Speed*	*Deflection*
6.24a	10	0.5	1	70	29
6.24b	10	0.5	2	57	35
6.24c	10	0.5	3	70	43

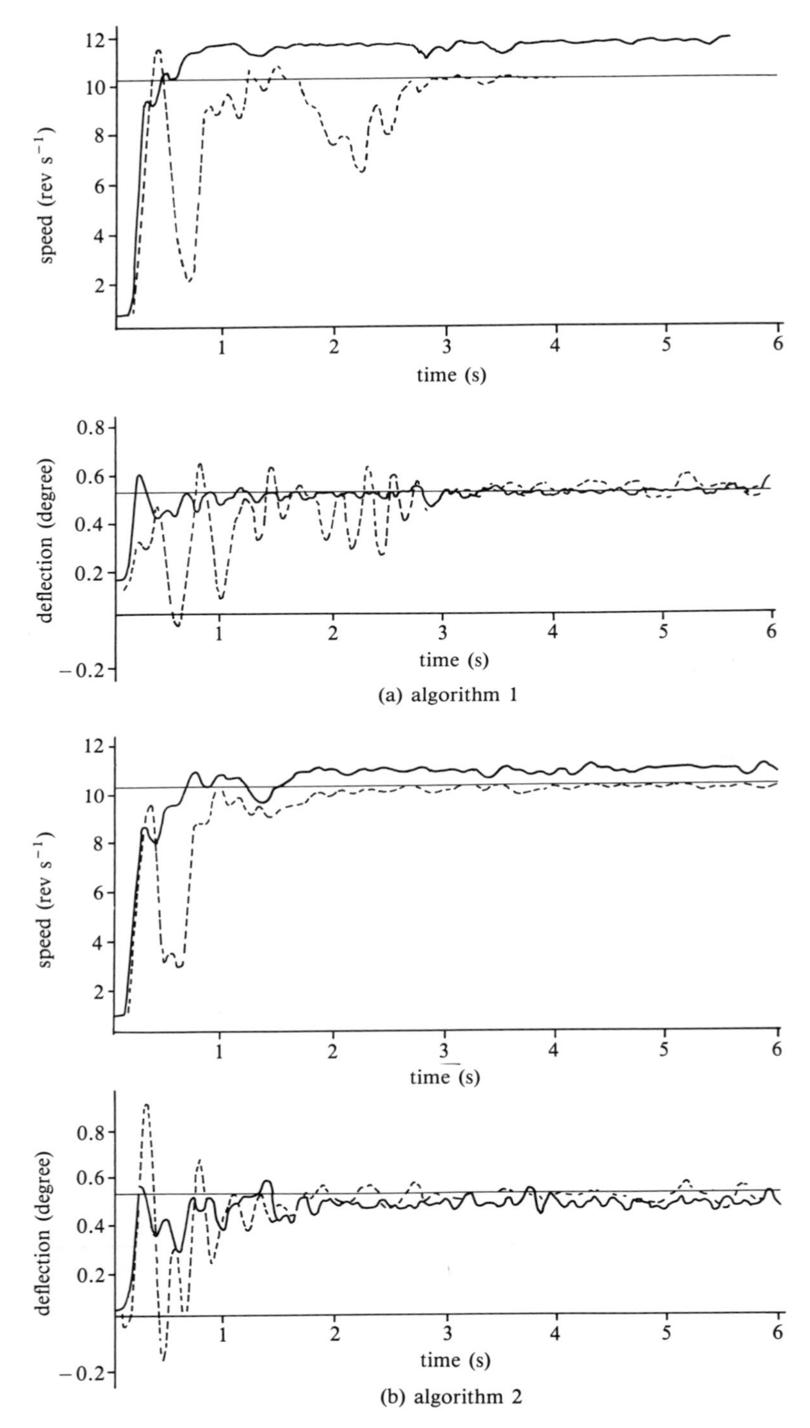

(a) algorithm 1

(b) algorithm 2

Figure 6.24 Response of extended SOC: --- 1st run, — 3rd run

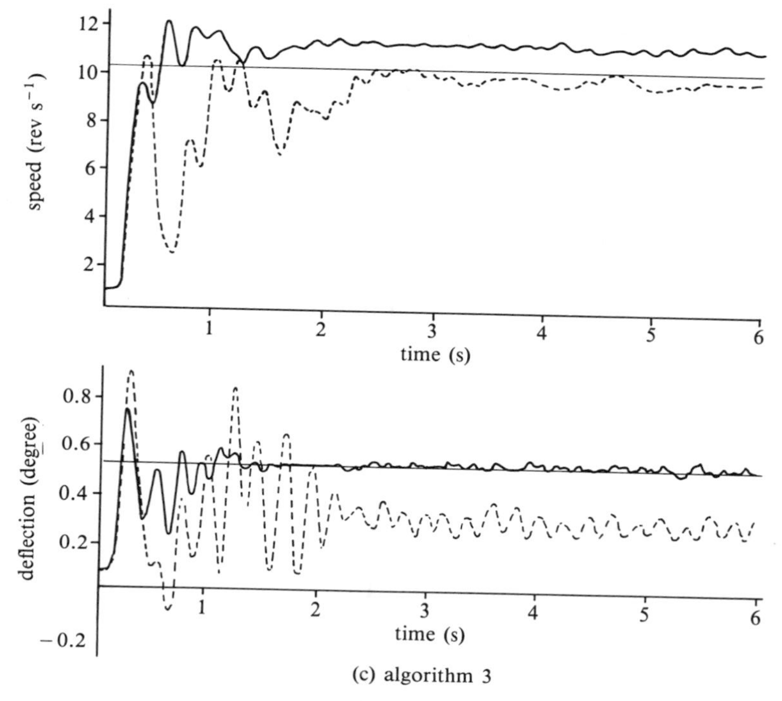

(c) algorithm 3

Figure 6.24 *(continued)*

6.5 CONCLUSIONS

Self-organizing methods for control systems offer a wide range of methodologies extending from complex nonlinear dynamic identification to probing self-adaptive controllers. The three examples described in this chapter all have high computational loads, which pose a challenge for real-time implementation. The concept of parallelism offers, therefore, an attractive proposition for such self-organizing systems, particularly for fast devices such as the coupled-electric drives rig.

The GMDH identification method has natural parallelism and can be coded simply to achieve large benefits in terms of speed of execution. It has similarities with neural networks in terms of its layered structure, and experiments indicated a much faster execution time for GMDH both for training and testing. The SOC method also has natural medium granularity for parallel processing in terms of its computational elements for multiple actuator control. For SOFLC, the hierarchical structure of the system lends itself to parallelism, as does functional decomposition for the multivariable situation.

The question of merging AI-related symbolic computations with numeric calculations offers the challenge of integrating different languages via a multi transputer platform. This has only been explored here in a rudimentary fashion using a C-coded Lisp interpreter. With the advent of other compilers for transputer based systems, including AI languages such as Prolog, the possibility of multiple language environments or toolkits becomes an interesting avenue of investigation.

Thus, exploitation of increasing computer power becomes an exciting possibility via parallel processing, together with the ever-increasing motivation to provide integrated systems engineering solutions to complex control problems. These must address the human–computer interaction aspects, as well as deal with software engineering concepts which inevitably increase the computing power and memory necessary to accommodate these features. Advanced control methodologies can thus benefit from increased computing power both for individual algorithmic evaluations and integration of systems techniques and functions.

ACKNOWLEDGEMENTS

The research endeavours which form the basis of this chapter are largely due to two postgraduate students in the Department of Automatic Control and Systems Engineering who have performed their research work under the author's supervision. S. B. Hasnain and M. F. Abbod have worked enthusiastically to master the concepts which are described and contributed many hours in producing the results which are presented.

REFERENCES

BARRON, R. L. (1967) 'Self-organizing control systems for providing multiple-goal, multiple actuator control', US patent No. 3519, 998, Sept. 29.

BARRON, R. L. (1968) 'Self-organizing control', *Control Eng.*, January.

BARTO, A. G., SUTTON, R. S. and ANDERSON, C. W. (1983) 'Neuron like adaptive elements that can solve difficult learning control problems', *IEEE Trans. Systems, Man. Cybernetics*, **SMC-13**(5), 834–6.

DALEY, S. (1984) 'Analysis of fuzzy logic controller', PhD Thesis, University of Leeds, UK.

DALEY, S. and GILL K. F. (1986) 'A design study of a self-organising fuzzy logic controller, *Proc. Inst. Mech. Eng.*, **200**, 59–69.

DRAPER, N. R. and SMITH, H. (1966) *Applied regression analysis*, New York: Wiley.

FARLOW, S. J. (1980) *Self-organizing Methods in Modelling GMDH Type Algorithms*, New York: Marcell Dekker.

HARRIS, C. J. and MOORE, C. G. (1989) 'Intelligent Identification and Control for Autonomous Guided Vehicles using Adaptive Fuzzy Based Algorithms', Research Report, Dept of Aeronautics and Astronautics, University of Southampton, UK.

HASNAIN, S. B. (1989) 'Self-organising control systems and their transputer implementation', PhD Thesis, Dept of Automatic Control and Systems Engineering, University of Sheffield, UK.

HASNAIN, S. B. and LINKENS, D. A. (1988) 'The use of transputer parallelism for the group method of data handling (GMDH) self-organizing identification algorithm', in *Parallel Processing in Control: The Transputer and Other Architectures*, London: Peter Peregrinus.

HASPEL, D. (1989) 'The use of an expert system for cement manufacture' in *Proceedings of ISA '89 Advanced Control Conference, Birmingham*, pp. 8.1–9.

IKEDA, S., OCHIAI, M. and SAWARAGI, Y. (1976) 'Sequential GMDH algorithm and its application to river flow prediction', *IEEE Trans. Systems Man. Cybernetics*, **SMC-6**, 473–9.

IVANKHNENKO, A. G. (1968) 'The group method of data handling: a rival of stochastic approximation', *Soviet Automatic Control*, **13**(3), 27–45.

IVANKHNENKO, A. G., KOPPA, Y. V., TODVA, M. M. and PETRACHE, G. (1971) 'Mathematical simulation of complex ecological systems', *Soviet Automatic Control*, **4**, 15–26.

KHOMEVNENKO, M. G. and KOLOMIETS, N. G. (1980) 'Self-organization of a system of single partial models for predicting the wheat harvest', *Soviet Automatic Control*, **13**, 22–9.

KICKERT, W. J. M. and VAN NAUTA LEMKE, H. R. (1976) 'Application of fuzzy logic controller in a warm water plant', *Automatica*, **12**, 301–8.

KOKOT, V. S. and PATAREU, S. G. (1980) 'Processing of experimentally measured densities of metals using GMDH', *Soviet Automatic Control*, **13**, 67–70.

LARKIN, L. I. (1985) 'A fuzzy controller for aircraft flight control', in *Industrial Applications of Fuzzy Control* (ed. L. I. Larken), Amsterdam: Elsevier, North Holland.

LARSEN, P. M. (1979) 'Industrial applications of fuzzy logic control' *Int. J. Man–Machine Stud.*, **12**, 67–70.

LINKENS, D. A. and MAHFOUF, M. (1989) 'Generalised predictive control of muscle relaxant anaesthesia', in *IFAC Workshop Decision Support for Patient Management, City University, London*, pp. 251–61.

LINKENS, D. A., ASBURY, A. J., RIMMER, S. J. and MENAD, M. (1982), 'Identification and control of muscle-relaxant anaesthesia', *IEE Proc. D*, **129**, 251–61.

MAMDANI, E. H. and ASSILIAN, S. (1975) 'An experiment in linguistic synthesis with a fuzzy logic controller', *Int. J. Man—Machine Stud.*, **7**, 1–13.

OSTERGAARD, J. J. (1976) *Fuzzy Logic Control of a Heat Exchanger Process*', Internal Report 7601, Technical University, Denmark.

POKRASS, V. L. and GOLUBEVA, L. V. (1980), 'Self-organization of a mathematical model for long range planning of the cost of coal mining', *Soviet Automatic Control*, **13**.

PROCYK, T. J. (1977) 'Self-organising controller for dynamic processes', PhD Thesis, Queen Mary College, London, UK.

PSI (1980) 'Interactive simulation package', User's Manual.

RUTHERFORD, D. A. and CARTER, G. A. (1976) 'A heuristic adaptive controller for a sinter plant', in *Proceedings of the 2nd IFAC Symposium on Automation in Mining, Mineral and Metal Processing, Johannesburg*, pp. 315–24.

SHERIDAN, S. E. and SKJOTH, P. (1983) 'Automatic kiln control at Oregon Portland cement company's Durkee plant utilising fuzzy logic', in *Proceedings of the 25th IEEE Cement Industry Technical Conference, San Antonio, Texas.*

TONG, R. M. (1976) *Some problems with the design and implementation of fuzzy logic controllers*, Internal Report CUED/F-CAMS/TR 127, University of Cambridge, UK.

TONG, R. M., BECK, M. B. and LATTEN, A. (1980) 'Fuzzy control of activated sludge waste water treatment process', *Automatica*, **16**, 659–701.

VAN AMERONGEN, J., VAN NAUTA LEMKE, H. R. and VAN DER VEEN J. C. T. (1977) 'An autopilot for ships designed with fuzzy sets', in *Digital Computer Applications to Process Control* (ed. H. R. Van Nauta Lemke), Amsterdam: North-Holland.

WHITING, B. and KELMAN, A. W. (1980) 'The modelling of drug response', *Clinical Sci.*, **59**, 311–17.

YAMAZAKI, T. and MAMDANI, E. H. (1984) 'On the performance of a rule-based self-organising controller', in *Proceedings of the IEEE Conference on Application of Adaptive and Multivariable Control, Hull*, pp. 50–5.

ZADEH, L. A. (1965) 'Fuzzy sets', *Information and Control*, **8**, 28–44.

ZADEH, L. A. (1973) 'Outline of a new approach to the analysis of complex systems and decision processes', *IEEE Trans*, **SMC-3**, 338–53.

Part III

Transputer Networks

Instead of silicon fabrication, VLSI-oriented architectures, such as those detailed in the previous sections, can also be mapped onto, and simulated by, a parallel network configured by the user from commercially available microprocessors. This has obvious cost benefits but also retains benefits arising from the original architecture. These include modularity, spatial and temporal locality and boundary input/output (I/O) interfacing from the systolic array.

The application of processor networks to control engineering can be traced back to the 1970s when decentralized and distributed machines began to be used in implementations. Currently the design and application of transputer based networks is a very active subject across a very broad spectrum of areas. This section contains a representative cross-section of control oriented applications.

In effect, a transputer is a microprocessor with a reduced instruction set computer (RISC) architecture, which supports parallelism at the hardware level and can be used as a 'building block' for concurrent networking. Basically, all members of the transputer family have a main processor, local memory, external memory interface and four intertransputer communication 'links' integrated on a single silicon chip, together with external interrupt and internal timers for real-time applications. The most important feature of the transputer is, from the parallel processing point of view, the on-chip inclusion of four communication links, which permit 'easy' construction of concurrent networks and 'fast' intertransputer communications. An in-depth treatment of the transputer system can be found in, for example, the references cited in Chapters 9 and 10.

Occam is the 'companion language' of the transputer system. This is a high-level language but is the lowest-level coding dedicated to transputers. It can directly map and match a concurrent procedure/module in software to one transputer in the hardware network. One source of an in-depth treatment of the features of Occam is again the references cited in Chapters 9 and 10. Note also that other languages such as Parallel C, Pascal and Fortran can be run on a transputer network under certain compilers. Given that Occam is an 'easy to read' code, it, rather than other languages or pseudo-codes, is used to demonstrate the algorithms developed in the chapters of this section.

Applications to the control of robotic systems is the subject of Chapter 7 which consists of three main sections. These are (a) problem definition for the UMI RTX robotic manipulator, (b) system specification and high-level design issues using the DeMarco methodology, and (c) implementation issues. A case study is used to highlight the very strong benefits of using transputers in this very topical applications area and the ease with which they can be interfaced with other integrated circuit devices.

Target tracking, an area which should benefit significantly from appropriate application of parallel processing, is the subject of Chapter 8. The particular aspect of this very wide ranging area considered is the problem of tracking a large number of objects using information, corrupted by noise, from one or more sensors. Further, the work detailed here on track partitioning, distribution and clustering configuration should also be of interest in developing similar type parallel implementations of other schemes.

Chapter 9 is based on the so-called heterogeneous system approach to parallel control and simulation problems. The motivation for this is other work which has revealed shortcomings in the ability of the transputer to cope with the demands of real-time control software. In particular, there is evidence to suggest that the architecture granularity (a measure of computation power to interprocessor communications overhead) is not appropriately matched. This, in turn, suggests that fusing the finer granularity of parallel digital signal processing (DSP) chips, such as the IMS A100, with the transputer's ability to handle irregular computations and manage parallel operations should prove highly effective.

Use of parallel processing enables the computations to be organized in a distributed sense. Hence it is possible to provide a fault tolerance capability, i.e. an operational failure results in performance degradation rather than complete operational failure. This is one of the major benefits of 'parallel processing for control' and is the subject of Chapter 10. Both software (Occam based) and hardware (transputer systems) methods are developed, including some configured to operate in real time.

7

Design and Implementation of a Transputer Based Robot Control System

M. I. Barlow, S. E. Burge, A. P. Roskilly

7.1 INTRODUCTION

The Inmos transputer family is a set of very large scale integration (VLSI) components ideally suited to real-time control applications. Unlike conventional sequential processors, the inherent ability of the transputer to perform tasks concurrently gives the system designer extra freedom to achieve performance requirements inexpensively and without recourse to complex architectures. This is particularly important when developing an entire system as opposed to just the implementation of a control algorithm. Indeed, as system requirements become more demanding, several levels of control are frequently found necessary. In these situations the transputer is ideally suited. It is possible, however, despite the power of the transputer, to develop an inefficient and cumbersome system, unless an appropriate top down and structured design methodology is adopted.

This chapter presents a design and implementation case study of a transputer based control system for a RTX robotic manipulator. By describing the life cycle of this case study from user requirement through to implementation, it is intended to demonstrate how an appropriate systems design methodology can ease the design and development of relatively complex real-time digital control systems. Naturally, during the course of the design and development of the transputer based system several problems were encountered. Some of these problems are highlighted and their solutions discussed.

The chapter is divided into three main sections. The first covers the problem definition for the case study with a brief description of the RTX robot and an abridged version of the user requirements. The second section describes the system specification and high-level design for the robot control system. The simple, but extremely powerful, systems analysis (and design technique) of DeMarco (1980) is presented and used as an appropriate methodology to obtain an efficient and effective transputer based system.

The DeMarco analysis presented in the second section results in a set of technical requirements for both hardware and software. How these were satisfied for the RTX case study is covered in Section 7.3. This section also presents some of the problems that were encountered during the development of the RTX control system and their solution.

7.2 PROBLEM DEFINITION

This section describes the basic user requirements for the RTX robot control system. An abridged version is presented here and a full description can be found in Roskilly (1991). It is important, however, to give some description of the existing RTX robot, and accordingly the first subsection is devoted to this.

7.2.1 The UMI RTX Robotic Manipulator

The UMI RTX robotic manipulator is a six degree of freedom enhanced SCARA configured robot. It has one translational joint which allows movement of the other manipulator linkages in the vertical plane. The remaining linkages are all of the revolute joint type. There are also two additional driven 'axes', the gripper and one spare axis for use in controlling some external device if required.

The manipulator joints are driven by 24 V d.c. permanent magnet motors through a transmission system. Apart from the one translational joint and gripper, the transmission systems are relatively high geared. Furthermore, the majority of the transmission systems incorporate belt drives which introduce a certain amount of unwanted flexibility when driving the linkages and relatively unsophisticated gearboxes which add both backlash and stick-slip friction to the system.

Directly mounted on each motor armature is an incremental optical encoder to allow the position of each link to be determined in relation to some initial position. Because the encoders are mounted on the motor armatures they are relatively inexpensive and of low resolution, but consequently are susceptible to errors due to backlash in the transmission system.

The existing control hardware comprises one board which comprises two IP (intelligent peripheral) subsystems. One subsystem controls the gripper, spare and two wrist motors, while the second controls the z-axis, shoulder and elbow wrist yaw. Each subsystem has an Intel 8031 microcontroller embedded with a proportional plus integral plus derivative control algorithm for each motor. The control-laws operate at an iteration rate of 62.5 Hz, and the pulse-width-modulated (PWM) outputs are passed to Sygnetics L293E motor drive chips.

In order to command the existing system, a host PC with appropriate software is connected directly through an RS232 interface. There is no local memory on the IP boards for the storing of robot sequences, and it is therefore dependent upon the host PC for these resources.

7.2.2 User Requirements

Below is an abridged version of the user identified requirements for the RTX control system. A full set of user requirements can be found in Roskilly (1991).

1. The robot system should be capable of receiving commands from a PC and a six-axis interceptor.
2. The robot system should be capable of receiving commands stored in memory as a result of 1 and the end-effector being 'led by the nose'.
3. Access to the controller parameters must be available at any time to facilitate development.
4. There should be a repeatable initialization routine so that the 'zero' datum can always be achieved from any manipulator position and orientation.
5. Each individual joint can be commanded, as can the position and orientation of the end-effector.
6. Motor speeds can be set as a percentage of the maximum motor speed.
7. The operator can specify if the manipulator joints should be locked or free at the end of a manoeuvre.
8. The robot system should include checks to identify software or hardware errors.
9. User input errors should be registered by an audible response; this includes out of envelope manoeuvres.
10. The robot system should be such that 'play-back' of a command sequence can be performed without the need of a dedicated PC.
11. The robot system should allow full diagnostics via a monitor to the user.
12. The robot system shall be 'safe'.

7.3 CONTROL SYSTEM SPECIFICATION/DESIGN METHODOLOGY

One of the crucial features of transputer based systems is the intimate relationship between hardware and software processes. Indeed, when generating high-level system designs, it is not possible to separate the software and hardware; moreover, this should not occur. With the design of any embedded system that incorporates both hardware and software, the decision as to what functions should be implemented in hardware or software should be left as long as possible. Early decisions can lead to sub-optimum designs. What is necessary is a technique for describing the required system that does not compromise exploitation of concurrency attributes of the transputer, but is sufficiently abstract to model both hardware and software functionality. It is also beneficial if such a technique is structured to permit a top-down approach, and simple to use but rich enough to cover most practical applications.

The need for such a tool has been apparent for a considerable time and several methodologies/techniques have been developed. All of these purport to have the attributes required; however, for reasons that will become apparent, the DeMarco (1980) technique is used for the robot control system.

The next subsection gives a brief overview of the DeMarco technique which is followed by an analysis and specification of the robot control system. It is then shown that this specification forms an ideal high-level design for determining the portioning between the hardware and software elements. Furthermore, it is also shown that the structured approach of the DeMarco technique maps almost directly to transputer implementation requirements.

7.3.1 The DeMarco Technique

Although the techniques developed by DeMarco (1980) for data processing systems have been extended by Ward and Mellor (1985/86) to real-time systems, the methodology presented herein is based on DeMarco's original approach, which in part has been used by other workers (Kerridge, 1987). The reason for this is that while Ward and Mellor's extensions are useful in real-time applications they have lost a great deal of the simplicity and flexibility of the original approach which can be used to good effect with a transputer based system. The technique uses three basic tools:

1. The data flow diagram (DFD), which is a network representation of the system. DeMarco noted that virtually any system can be described in terms of processes which communicate with each other. Note that this is conceptually similar to the arguments proposed by Inmos (1988) for the transputer. In order to describe these systems DeMarco proposed four graphical conventions:

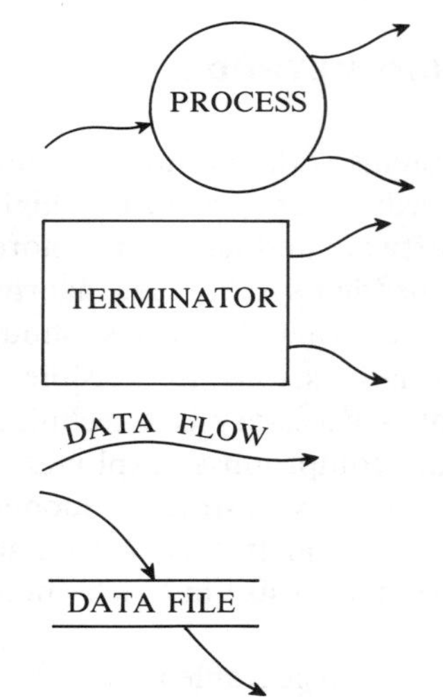

PROCESS which transforms incoming data flows to outgoing data flows;

TERMINATOR which is a process external to the systems under consideration;

DATA FLOW which is a 'vector' indicating a well-defined flow of 'information';

DATA FILE which is a temporary store of data.

2. The data dictionary (DD) which defines the data flows and data files on the DFDs. DeMarco developed a simple but quite rigorous method of defining these based on the relational operators, IS EQUIVALENT TO; AND; EITHER OR; ITERATIONS OF; and OPTIONAL.
3. The process specification (PS) which describes the internal working of the component processes and how the process converts incoming data flows to outgoing data flows. Again DeMarco developed a set of conventions to help in the writing of process specifications called Structured English. This 'language' follows simple but formal rules of layout, vocabulary, and syntax which is restricted to:

   simple procedural sentences
   closed end decision constructs
   closed end repetition constructs

It is quite clear, however, that for anything other than the simplest of systems a single DFD would be far too extensive to be comprehended easily. Therefore a top-down approach is used to produce a set of 'levelled' DFDs. The start of this decomposition process is a DFD which shows the whole system at a level of detail which can be represented on one diagram. This top level diagram is known as the context diagram, because it has an important role in that it defines the boundary of the system under consideration. To continue the decomposition, the 'processes' of the context diagram are considered as subsystems which are decomposed on separate child diagrams or lower-level DFDs. This activity of successive decomposing is continued to the primitive level where processes cannot usefully be further decomposed. If all the primitive processes are then defined with a PS and the various dataflows and files defined the result is a clear, concise and unambiguous set of documentation which models the system.

The simple but rich set of conventions also allows for abstraction in that it is possible to model processes as elements of hardware, data flows as physical quantities such as voltages, etc. Thus the distinction between hardware and software elements in a system need not be made until necessary.

This section will present a method based on the structured top-down methods of DeMarco. It will be shown that the DFDs for a real-time system can be used to identify 'processes' and 'data structures/flows'. The 'data structures/flows' can subsequently be defined as a set of protocols in the DD. This step is considered vital since it is these protocols that will perform interprocess communication.

7.3.2 DeMarco Analysis of Robot System Requirements

The application of the DeMarco technique to the user requirements given in Section 7.2.2 has resulted in the set of DFDs shown in Figures 7.1–7.4. Although strictly speaking the DeMarco technique is an analysis and specification tool, it is inevitable that some design decisions are made during the analysis. Indeed, it had already been agreed to use a six-axis inceptor device, and to utilize the IP boards as much as possible for data acquisition and driving of the motors. It was also clear that two levels of control were necessary. The lowest level of control would be concerned with the control of the motors which is reflected in Figure 7.4 as PROCESSes 2.3 to 2.10. The higher level of control would translate console and inceptor commands into demands from each of the motors. This philosophy is reflected in Figure 7.2 as PROCESS 1 for the high-level control and PROCESS 2 for the low-level control.

In accordance with the top-down approach recommended by DeMarco, Figures 7.3 and 7.4 show the respective decomposition of PROCESSes 1 and 2. Considering Figure 7.3 first, the decomposition is clearly logical with some form of input device driver which converts, as necessary, the console and inceptor signals into a suitable

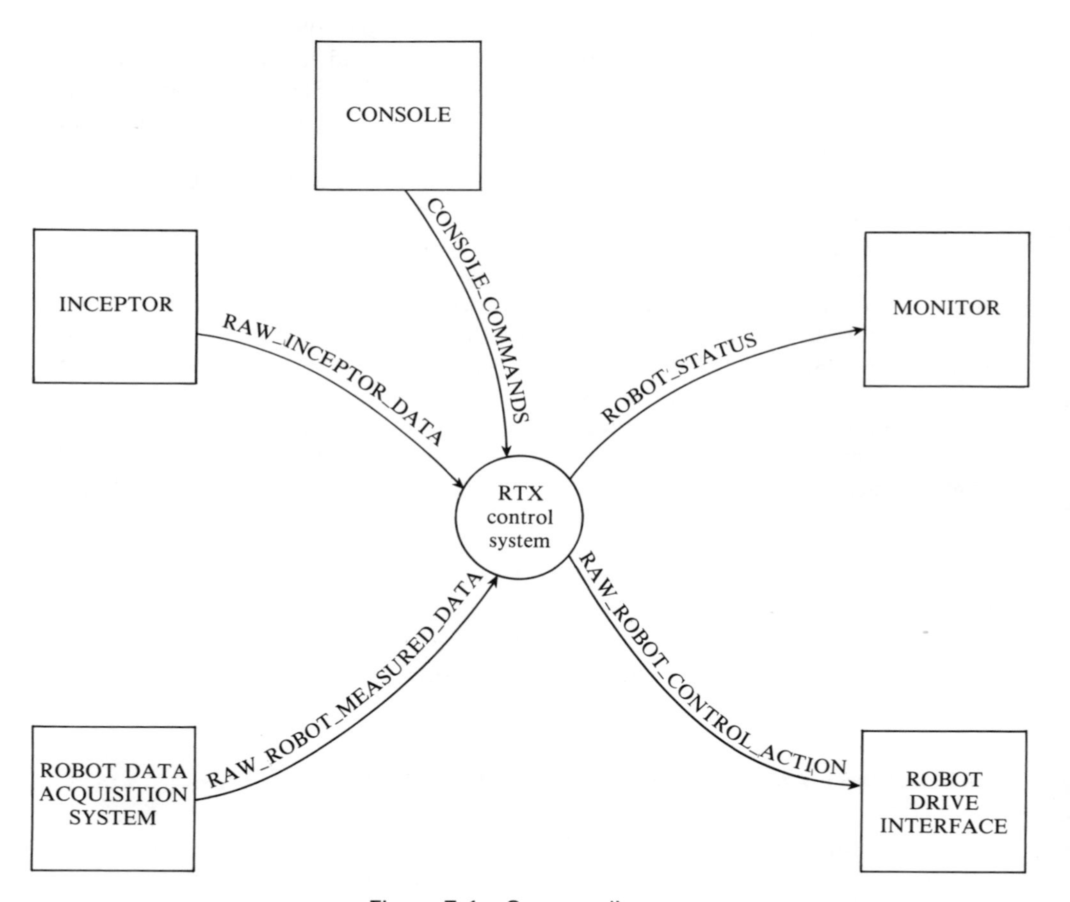

Figure 7.1 Context diagram

form for further processing. The path generator (PROCESS 1.2) operates on the converted input signals to give Cartesian data for the PROCESS 1.3 which performs the inverse geometry calculations to give the motor demands. There is a high-level feedback from the motors which is decoded by PROCESS 1.4 to give status information as well as provide data for the path generator (PROCESS 1.2) as to whether the motors are operating at their limits. This particular aspect was considered to be extremely important to ensure that the system would operate as efficiently as possible, and more detail will be given in Sections 7.4.4 and 7.4.5. Finally, in order to facilitate the development a data logging PROCESS 1.6 was also envisaged.

Considering Figure 7.4 the demands generated by PROCESS 1 are directed by PROCESS 2.1 to their respective motor controller. The controllers would act on these commands and feedback from the optical encoders to generate appropriate outputs. The control outputs would then be sent to some multiplexing process (2.2) which would in turn pass the commands to the IP INTERFACE. The multiplexing process (2.2) would also demultiplex the optical encoder data received via the IP INTERFACE from the IP boards.

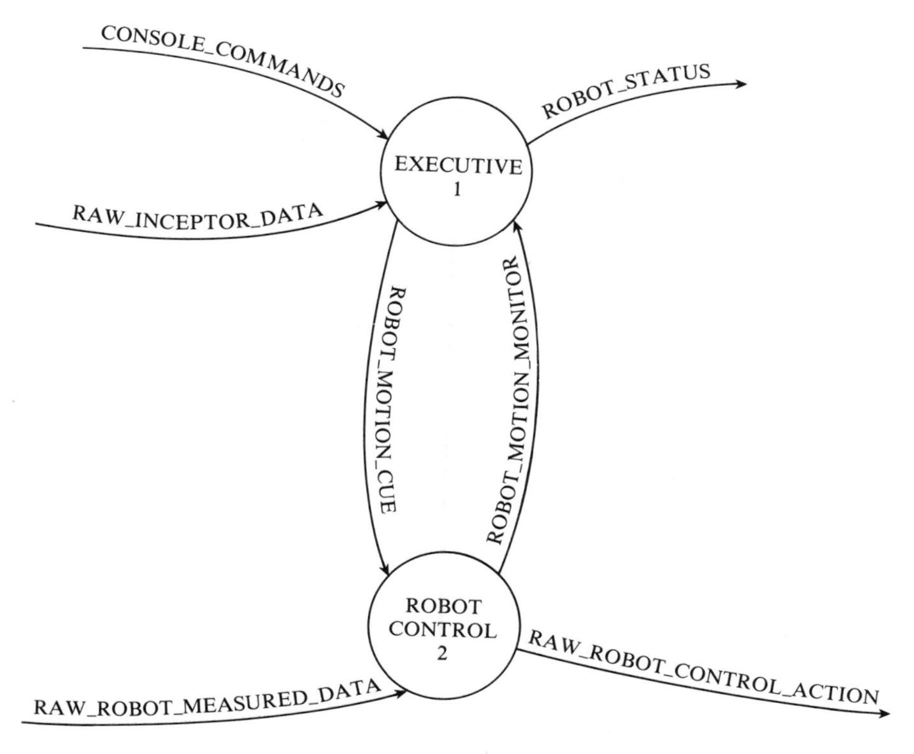

Figure 7.2 Level 1 Diagram 0

To complete the analysis and specification it was necessary to define all the data flows in a DD and define all the primitive level processes as PSs. While it must be emphasized that a complete specification only exists provided that there is a consistent set of DFDs, DD and PSs, only examples of the PSs and DD will be given here. An example of an entry in the DD is:

```
ROBOT_MOTION_CUE = [ EXPIRE
                   | HALT_ALL
                   | ZERO_ENCODERS
                   | ALTER + JOINT + TYPE + COEFF
                   | MOVE + JOINT + TARGET
                   | HOLD + JOINT + TARGET
                   | HOME                         ]
```

In the above example, a short-hand notation is used for the relational operators where ' = ' means IS EQUIVALENT TO, '[ ]' means OR and ' + ' means AND.

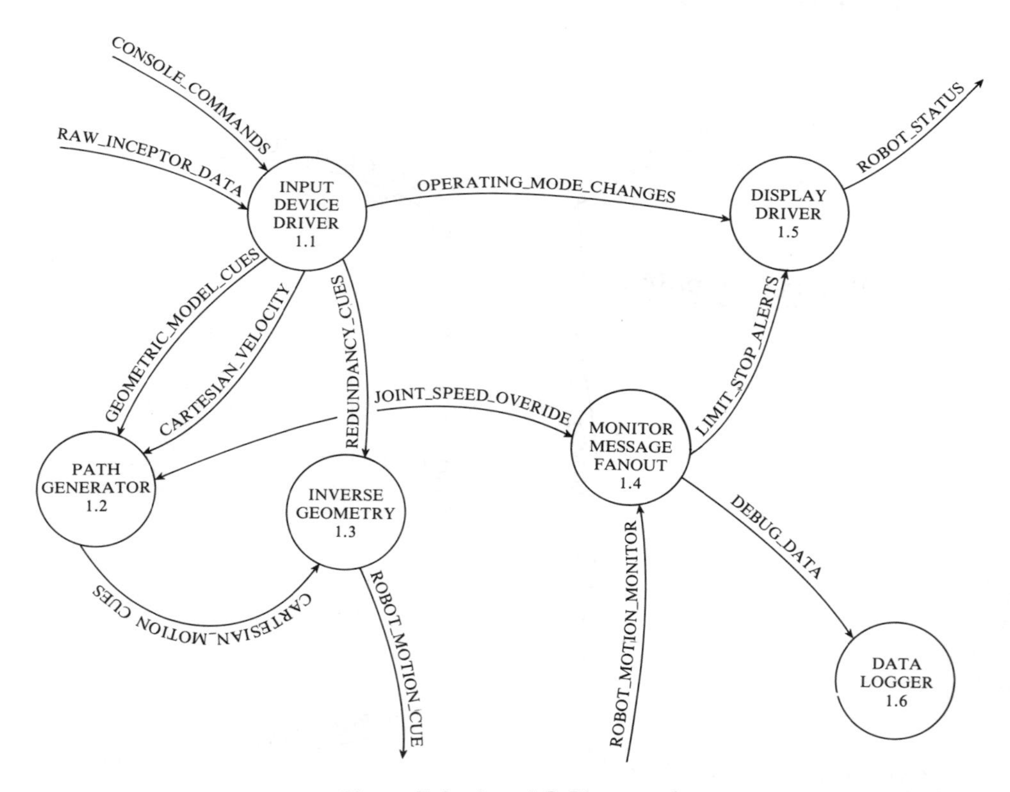

Figure 7.3 Level 2 Diagram 1

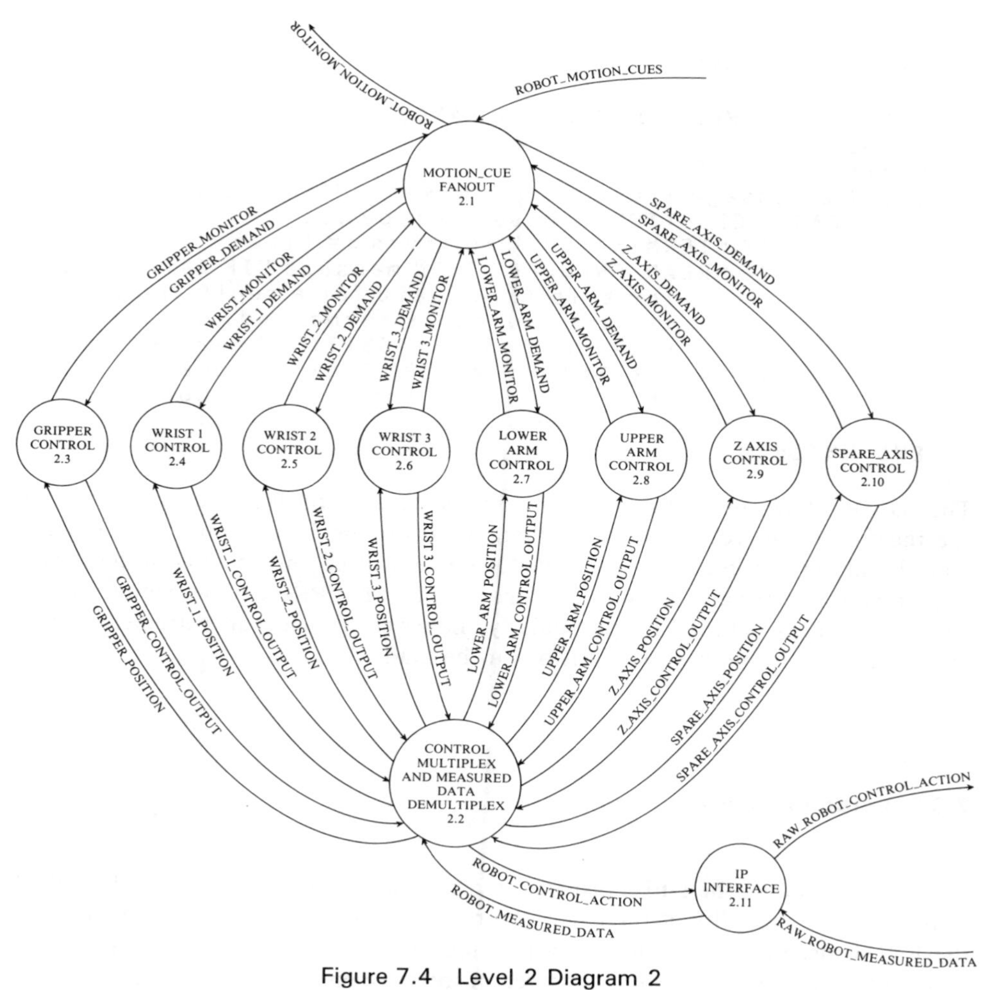

Figure 7.4 Level 2 Diagram 2

This example also shows that the DD serves as a mechanism for decomposing a data item into its constituent elements. Accordingly, therefore, to ensure the DD is complete, each of the data elements on the right-hand side of the above example must themselves become entries in the dictionary. For example:

```
EXPIRE = Operator request to stop all motors and stop the
         robot system.
```

This particular entry represents a primitive data item since the right-hand side (written in lower case) gives an exact specification for the data entry.

An example of PS is:

```
PROCESS 1.4 MONITOR MESSAGE FANOUT
lives:=8
WHILE lives > 0
input a message from any "_CONTROL_OUTPUT" channel
      CASE SELECTION on message type ...
           IF "TERMINATE" THEN lives := lives - 1
           IF "READ_ENCODER" THEN message "IP INTERFACE"
                  for this info; wait for a reply and send
                  it back, suitably formatted, on the
                  relevant "_AXIS_POSITION" channel
           IF "WRITE_MOTOR_POWER" THEN message "IP
                  INTERFACE" with the relevant motor no.
                  and power value
END_WHILE
```

The DFDs, DD and PSs together give a clear, concise and unambiguous specification of the robot control system. At this point the decomposition is quite logical and has not indicated which PROCESSes will be implemented in hardware or software, nor has any decision been made as to what PROCESSes would run on what transputers. The DFDs particularly, however, provide an extremely useful model in order to allow these decisions to be made.

7.3.3 Partitioning of Processes

As indicated above, the DFDs provide an extremely useful diagrammatic model as to how the functionality of the robot control system may be implemented. Indeed, the logical decomposition of the system makes this task trivial. It is clear that PROCESS 1 can be entirely implemented in software on a transputer system. There is no obvious advantage in implementing any of PROCESS 1's child processes in hardware or on any other device. PROCESS 2, however , will be a combination of software and hardware. With reference to DIAGRAM 2 (Figure 7.4), PROCESS 2.1 to 2.10 can be implemented on a transputer system, while clearly PROCESS 2.11 will require some hardware and associated software in order to interface with the RTX's IP board.

While these decisions appear trivial and obvious, it is because of the thorough understanding of the problem gained by using the DeMarco technique. However, the benefits of using this approach do not stop here. The PS and DD entries associated with PROCESS 2.11 serve as a design specification for the IP board interface. Furthermore, the remaining PSs and DD entries serve as the design specifications for the software.

7.4 SYSTEM DESIGN AND IMPLEMENTATION

The previous section has demonstrated the use of systematic analysis and specification tools that has permitted a thorough understanding of the user requirements, and has enabled certain crucial design decisions to be made. This section will build upon these sound foundations by examining the software design and implementation for a target system of transputers. It will be shown that the DeMarco specification is extremely useful in mapping the user requirements almost directly into Occam code. Furthermore, the specification is also shown to have its uses in assigning Occam processes to transputers.

The IP interface design and implementation will also be described, indicating the inherent advantages (and disadvantages) of the transputer over conventional microprocessors in such exercises.

While the actual control algorithms used to control the motors are independent of the type of target processor (the use of transputers did offer some advantage), a subsection is devoted to this aspect. This is primarily because early development work indicated that motor control action limits severely restricted the performance of conventional proportional plus integral plus derivative control laws. To overcome these performance limitations an alternative control algorithm (Roskilly *et al.*, 1990) was adopted.

Finally, during the course of implementation and development several 'problems' arose which were successfully solved. Although some of these were specific to the robot control system, many have general implications on any real-time transputer based control system and these will be discussed.

7.4.1 Software Implementation

The result of the DeMarco analysis and process portioning was a software specification for the robot control system defining exactly what the software had to achieve. It was decided almost immediately that a proprietary real-time kernel was not appropriate for several reasons:

1. Because it was necessary to solve the inverse kinematics problem in real-time, the kernel had to be as efficient as possible. Any proprietary kernel would necessarily incur some overhead because of its general nature.
2. Development time and cost would be expensive for several reasons: the learning curve with such a kernel is high and in order to be able to use it efficiently, debugging would be relatively difficult and the cost relatively high.
3. Many of the tasks performed by a proprietary kernel already exist in hardware on the transputer.
4. Whilst a bare transputer does not offer true deterministic deadline

scheduling over as many levels of priority as a real-time kernel would, by using the (relatively inexpensive) 16-bit T212 or T222 chips with external memory minimized or omitted, an adequate level of determinism is possible using the given two priority levels. This approach is aided by the compactness of compiled Occam; indeed a single T222 easily accommodated PROCESS 2 'ROBOT CONTROL' in its 4 kb internal memory.

5. Real-time kernels are intended for situations where processing is a scarce resource and memory a cheap one; in this application the interest was in investigating approaches to the converse situation. Depending on the definition of 'cheap', this can be said to be true where transputers are concerned and may certainly be expected to become more common in the future.

It was also recognized that the DFDs, DD and PSs are an ideal tool for Occam software design and implementation. The key is that the primitive processes identified on the lower-level DFDs can map directly to Occam processes. The implications of this are quite profound. Firstly, if the DeMarco processes are considered as Occam processes, then the data flows are communication protocols. Therefore the DD entries serve as individual specifications for each protocol definition. An example of this is where the DD entry is:

```
ROBOT_MOTION_CUE =[ EXPIRE                           ]
                  [ HALT_ALL                         ]
                  [ ZERO_ENCODERS                    ]
                  [ ALTER + JOINT + TYPE + COEFF     ]
                  [ MOVE + JOINT + TARGET            ]
                  [ HOLD + JOINT + TARGET            ]
                  [ HOME                             ]
```

And the corresponding protocol definition is

```
PROTOCOL ROBOT.MOTION.CUE
CASE
 expire
 halt.all
 zero.encoders
 alter;INT;INT;REAL32 -- joint, coefficient No., value
 move.to;INT;INT     -- joint, target position
 hold.at;INT;INT    -- joint, target position
 home
:
```

This example clearly demonstrates the relationship between the entry and its Occam implementation. While the above approach is recommended, it follows that the experienced system analyst/designer could write the primitive data dictionary entries directly in Occam protocol statements.

Having defined the communication protocols each process can be treated as a separate module and its internal design and coding developed from the PS. Knowing the communication protocols before designing and implementing the processes is crucial, since although the PS gives the outline of transformation, this must be performed in accordance with precisely defined input and output data. An example of this approach can be illustrated by considering the 'Structured English' PS given in Section 7.3.2 for PROCESS 1.4-MONITOR MESSAGE FANOUT and the corresponding Occam code:

```
INT lives,count,power:
SEQ

 lives := 8                                -- initialize
 WHILE lives > 0                            -- loop

  ALT index =0 FOR 8                           -- wait for any
                                                            input
   from.controller[index]? CASE     -- and select

    terminate                              -- appropriate
     lives := lives - 1               -- action

    read.encoder
     SEQ
      to.ip.interface ! ip.read;index
      from.ip.interface ? count
      to.controller[index] ! encoder.is;count

    write.motor;power
     to.ip.interface ! ip.write;index;power

 SEQ index = 0 FOR 8                           --finally,
  to.ip.interface ! ip.write;index;0 -- stop all motors
```

Note how orderly termination is supported by waiting for all eight communicating processes to signal 'TERMINATE' before this process itself dies. If this were not so, one of its interlocutors might be left waiting forever to exchange messages with it.

Again, it is recommended that the PSs are written in Structured English, but an experienced analyst/designer could write them in Occam (other languages, such as parallel C, could also be used). It is likely that they would be an abridged version, since, for example, the precise details of the inverse geometry calculations may not be known and thus several steps may be covered by a non-executable comment. There is another good reason for ultimately adopting this approach, in that rapid prototyping could be performed using the abridged Occam 'process specifications'. Such a prototype is unlikely to have anywhere near the full functionality of the final system, but the fact that it may be possible to execute it and thus allow it to function as a 'stub' routine, may cut down the overall development time.

7.4.2 Hardware Design and Implementation

The DeMarco analysis of the robot control system resulted in the definition of the primitive PROCESS 2.11 named the IP INTERFACE (see Figure 7.4). As discussed in Section 7.3.2 it had been decided early on in the project to utilize the existing IP subsystems as much as possible for data acquisition and for driving the motors. In consequence the PS for PROCESS 2.11 was very broad:

```
PROCESS 2.11 IP INTERFACE
     receive ROBOT_CONTROL_ACTION from PROCESS 2.2
     translate ROBOT_CONTROL_ACTION into
                  RAW_ROBOT_CONTROL_ACTION
     send RAW_ROBOT_CONTROL_ACTION
     receive RAW_ROBOT_MEASURED_DATA
     translate RAW_ROBOT_MEASURED_DATA into
                  ROBOT_MEASURED DATA
     send ROBOT_MEASURED_DATA to PROCESS 2.2
```

In order to satisfy this broad PS a number of options were considered viable:

1. Interface a transputer to the 8031 processors which would allow the use of the existing interface between the data acquisition system and motor drive chips.
2. Interface a transputer directly to the existing data acquisition system and motor drive chips.
3. Develop a slave processor which would interface the data acquisition system and drive chips to a transputer.

It was recognized that the first two options would be destructive in that significant physical modifications to the board containing the IP subsystems would have to be made. Moreover, these modifications would be irreversible. For this reason alone the first two options were quickly rejected. This decision was made despite the fact that the second option would be the most effective and efficient since the interfacing would require fewer components and therefore be more cost-effective and potentially more reliable. It would also have had minimal communications delay and potentially the shortest development time.

An examination of the existing arrangement indicated that the two 8031 microprocessors have multiplexed address/data buses connected to two I/O devices; a byte-wide write-only latch concerned with motor power control and a resetable byte-wide read location which flags transitions of the corresponding encoders. In each case the byte is divided into bit-pairs, one per motor. The power control bits may be set to indicate (a) power on/off and (b) sense, i.e. positive or negative torque. The encoder bits indicate (a) a transition occurring since the last reset and (b) the direction of rotation at transition. Consequently, it was decided that the slave processor would interface to these buses.

To interface the slave processor to the host transputer network a link-adaptor was chosen. The selection of a link rather than a bus interface provides a much more flexible system. Any microprocessor may be interfaced easily to connect all manner of peripherals, but if a bus system was adopted then the number of bits, the wait state generation, address decoding, etc. will all have to be modified depending on the selected microprocessor. Since these parameters will necessarily differ between microprocessors in terms of the most efficient choices, the software will also differ between implementations. Having to make these significant hardware and software changes under these circumstances not only creates more work but also introduces areas where mistakes can easily be made. With the link implementation the worst case would only involve the creation of new 'write to link' and 'read from link' procedures. Links also help with the physical modularity of a circuit board since their connections are completely standard. The TRAM format is an example of this and can usefully be used for peripherals.

A further advantage of using the chosen interface approach was that the host transputer network, used to run the control system software with communication via the link adaptor to the slave processor, can request data transfers via the link and immediately carry on with other parallel tasks while the data are being obtained. The process that requires these data will be automatically rescheduled when the transfer is complete and in this way the total throughput may be improved (provided sufficient work is found for the transputer to carry on with). It is important when using such schemes to *initiate* communications at the highest priority and to *process* the results at a lower one. This ensures that the link engines are always given useful work as soon as it becomes available. This can be illustrated by the following trivial example:

```
-- to calculate " b*sin(a)" where b must be input
-- inefficient code
REAL32 result,a,b:
SEQ
 source.link ? b
 result := b*sin(a)

-- to calculate " b*sin(a)" where b must be input
-- efficient code
REAL32 result,temp,a,b:
SEQ
 PRI PAR
  source.link ? b
  temp := sin(a)
result := temp*b
```

Additionally, by being decoupled from the slave processor's bus timing, a link-based peripheral can be designed with far greater latitude for the inclusion of local intelligence and storage.

Another major advantage of the transputer/link combination is that it allows rapid prototyping to be carried out. Once the protocol has been designed to support the link interface, software development can proceed without the hardware, by simulating it in Occam. This may be achieved by a 'stub' routine which supplies the required response to messages which can be generated from a file of test data. Throughout the construction stages of the hardware, a 'filter' may be used to carry out tests incrementally. The 'filter' would supply default responses for features not yet complete while passing a certain subset of messages on to the hardware in some unseen manner. Obviously this testing technique may be conducted on conventional systems with parameters and procedures rather than messages and processes. However, when it comes to a full hardware test, this only requires changing the address of the message channel to point to the physical link within the Occam software making this stage extremely simple.

Finally, object-oriented methods of working are easily supported when link-based peripherals are used. A peripheral's 'users' do not require any details of its internal structure because the peripherals capabilities are defined only in terms of its protocol on the link. Therefore, performance upgrades and new technologies can be applied without changing anything at higher levels. To gain this advantage great care must be taken when designing the protocol so that future development is not constrained. In Occam, a variant protocol may be extended with additional members but with a little forethought and an acceptance of a certain amount of redundant information passing (e.g. sending a 32-bit integer where a byte would suffice), it is possible to produce interfaces that are inherently upgradable.

It was identified that the slave processor would have to perform three tasks: the generation of the pulse width modulated (PWM) signals for the Sygnetics motor drive chips; maintaining encoder counts; and communication with the host transputer network. The generation of the PWM signal is time-critical and therefore it was decided that this task would be interrupt-driven to ensure that the PWM was always available at the correct time. The two remaining tasks would thus effectively run as 'background' tasks. With all these points in mind a child DFD to the IP INTERFACE PROCESS (2.11) was constructed and is shown in Figure 7.5.

The choice of a Z80B slave processor was made on the grounds of previous experience of the processor's assembler language and architecture. It was also decided to incorporate a transputer on the prototype board. This additional processor was included mainly to satisfy user requirement 10. The host transputer network could download PROCESSes 2.1–2.10, thus allowing the RTX robot to operate in 'play-back' mode independently from the host transputer network.

Thus the main components of the prototype control board were a CO12 Link Adapter, a 4 MHz Z80B microprocessor, a T212 16-bit 20 MHz transputer, and the associated logic and memory circuit hardware. The Z80B was provided with a small amount of RAM for its stack and data and a ROM for program code.

On the prototype, this contained an implementation of the algorithm described above in Z80 assembler. In future versions it would appear more sensible to provide more RAM, and a small monitor routine in ROM to allow the Z80B to 'boot from link' in a similar manner to the transputer itself by reading, at reset time, a variable length array of bytes, placing them in memory and then executing them as machine

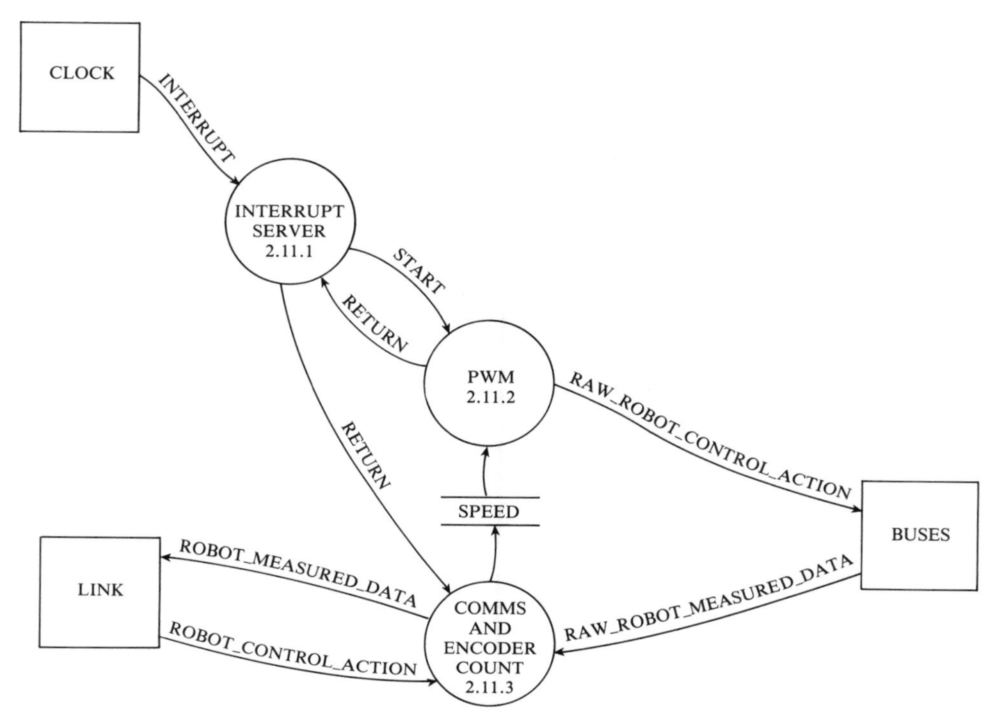

Figure 7.5 Level 3 Diagram 2.11

code. The link-adaptor was provided with manual switching to enable its serial input/output lines to be connected to either an off-board standard Inmos link cable socket or an on-board T212 16-bit 20 MHz transputer with 24 kb RAM. Owing to the high compactness of compiled Occam this was found to be quite sufficient to hold the code and data for processes 2.1 to 2.10 together with, in one experiment, a 1024-place queue of position demands which could be sequenced to provide rudimentary path following.

Considering the tasks performed by the Z80B, Figure 7.6 shows the flow diagram representing the internal structure of PROCESS 2.11.2. This process was given the task of generating the PWM signals required by the Sygnetics motor drive chips. In order to generate these signals for each motor, the Z80B nonmaskable interrupt is clocked from a 3 kHz external source. The interrupt service routine maintains a counter which increments modulo 65 which defines the length of a PWM time-frame. At the start of each frame, using the facilities detailed above, it turns on all motors for which nonzero torque demands are in force, and sets their sense bits according to the sign of the demand. As a time-frame progresses, motors with progressively higher demands are turned off again when the counter equals their demanded level. Motors with maximum torque demands are thus never turned

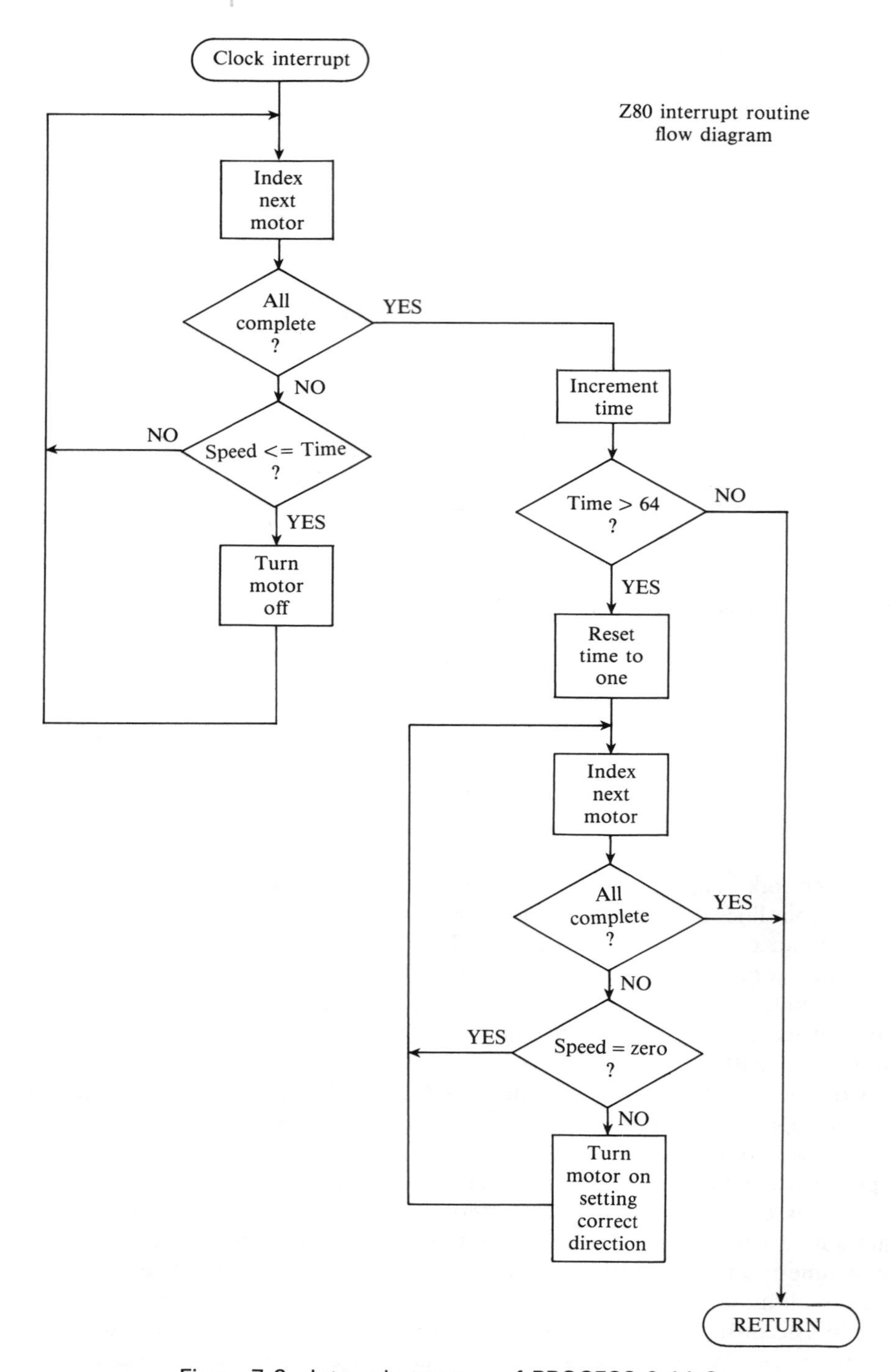

Figure 7.6 Internal structure of PROCESS 2.11.2

off, whilst those with zero demands are never turned on and intermediate levels are represented by an appropriate mark—space ratio square wave. Because the demand levels may be altered as a result of messages received from the transputer in between one interrupt and the next, a 'less than or equal' comparison is actually used between the counter and the demand level to avoid delaying the updating of motors undergoing deceleration by one time-frame.

The two background tasks of communicating with the transputer network and maintaining encoder counts were considered as one process as shown in Figure 7.5. While it might have been possible to separate these tasks as two processes, there was concern that an encoder transition could be missed. Since the RTX hardware does not maintain these data it falls to the IP INTERFACE PROCESS to retain running position counts for generating error signals for the control laws to act on. Figure 7.7 shows the internal structure of PROCESS 2.11.3.

It was decided that the Data flow ROBOT_CONTROL_ACTION would comprise three message types which may be sent to the IP INTERFACE PROCESS:

1. A reset request.
2. A request to set the output torque of a given motor to some value in the range -64 to 64, representing full negative to full positive saturation.
3. A request for the return of a 16-bit integer representing the current cumulative encoder count for a given motor.

The protocol between the transputer network and the Z80B was optimized for the latter, in view of its modest processing power, by packing the messages as follows:

1. *Reset request*, 1 byte: all bits set (# FF hexadecimal).
2. *Write torque setting request*, 2 bytes: byte 0 having bit 7 set, bits 6–3 clear and bits 2–0 encoding the motor number, byte 1 having bit 7 indicating the sense whilst the remaining bits encode the magnitude of the desired torque. (Note that because the maximum torque is 64, it is possible to detect an asynchronous Reset request which may arrive instead of the expected byte 1.)
3. *Read encoder request*, 1 byte: bits 7–3 clear, bits 2–0 encode the motor number. The response to this message is a 16-bit 2s complement value which is sent low-byte-first. (It is interesting to note that this classical communication 'good practice', together with the fact that the link engines deposit subsequent bytes received at increasing memory addresses has led to the transputer's 'little-endian' architecture.)

The interface between the Z80B and the transputer network was provided, as already mentioned, by a CO12 link-adaptor. This device has four byte-wide registers: ReadStatus, ReadData, WriteStatus and WriteData. Although capable of generating interrupts this was used in polled mode. To perform a read or write, the relevant status register was examined until it indicated the device could provide or accept a byte at the corresponding data register.

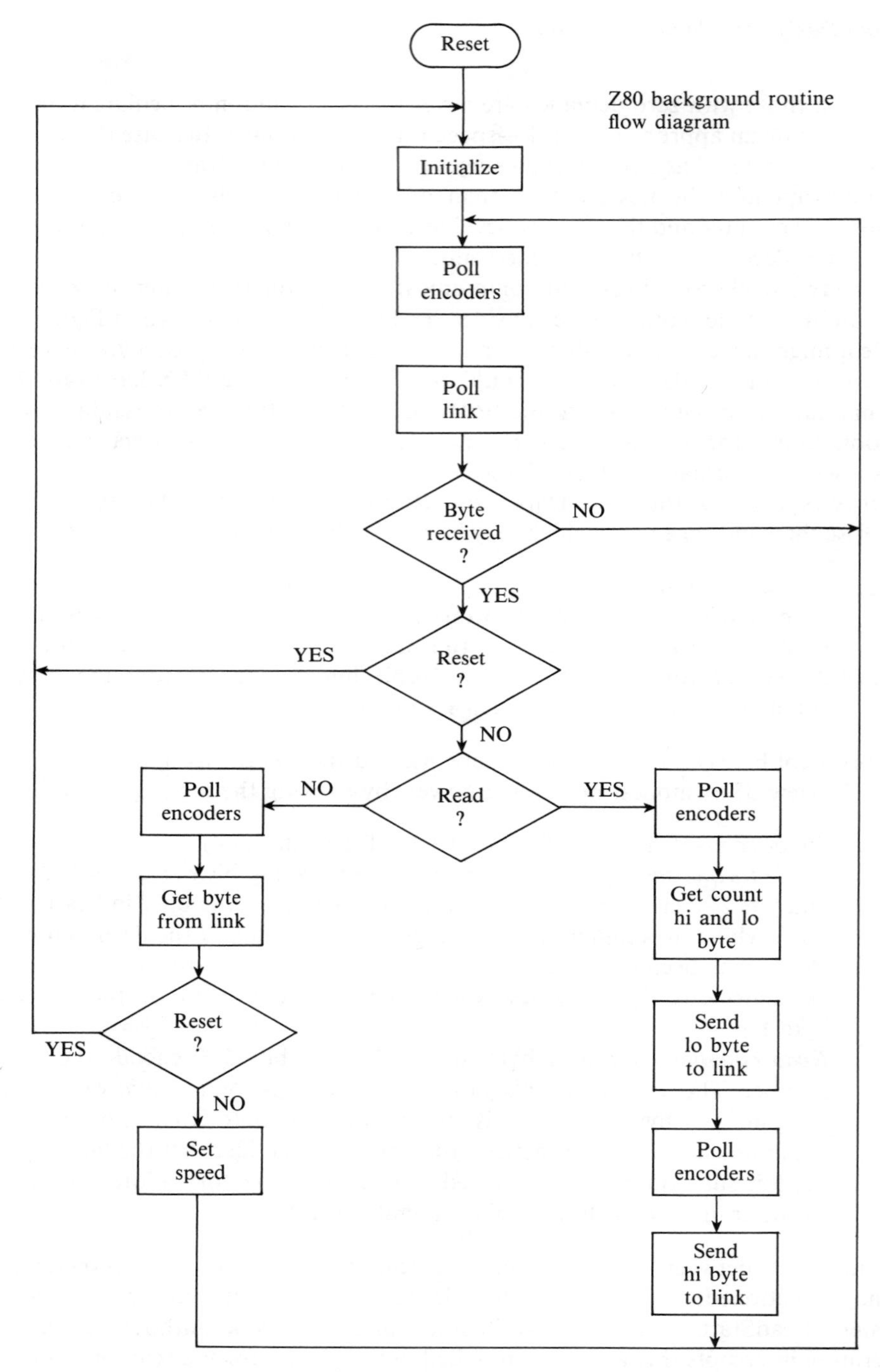

Figure 7.7 Internal structure of PROCESS 2.11.3

Since the transputer is capable of continuing with other parallel tasks once it has delegated a communication to one of its autonomous link engines, whilst missing the occurrence of an encoder transition would lead to an irrecoverable error in the corresponding encoder count, communications were given a higher priority than polling the encoders. Examining the motor manufacturer's quoted maximum rotation speed and the encoder geometry (12 transitions per revolution) showed that the delay introduced by one link-adaptor access could never lead to an encoder transition being missed, even when the most disadvantageous possible combination of interrupt and background events occurred. The algorithm accordingly interpolates calls to poll the encoders between each link access as shown in Figure 7.7.

To avoid problems caused by noise on the 8031 bus lines, which had to be extended to a rather undesirable length in order to reach the Z80 board in the prototype, the encoder latches are polled in a spin-loop until two subsequent values agree. Experiments have shown that this is rarely necessary, but it has been retained in view of the importance of maintaining valid encoder counts if 'drift' in the robot's position is to be prevented. Once reliably read, the latches are reset to prepare them for subsequent transitions and the extracted values examined. For each bit-pair the presence of a set transition bit is used to increment or decrement the corresponding encoder count, depending on the value of the sense bit.

Accepting a Write torque setting request involves, once the encoders have been re-polled, merely reading the subsequent byte from the link and depositing it where the interrupt routine can find it, according to the motor number extracted from the first byte. Replying to an Encoder read request, however, is complicated by the need to poll the encoders between transmission of the two bytes that form the reply. To avoid inconsistencies caused by unfortunate coincidence between transitions and transmissions the entire 16-bit value is read at the start of the operation and stored so that, when the high byte comes to be sent, it is consistent with the low byte, although the value represented may now be obsolete due to encoder transitions between the two transmissions. Sending a byte out via the link-adaptor involves polling WriteStatus until it indicates that the previous transmission has been taken; this may be thought of as an unbounded looping operation which could potentially lead to lost encoder counts. In practice, the transputer's link engines are so much faster than the Z80 and they are always ready to take the proffered data and this situation never arises.

In summary the necessity to develop the IP INTERFACE presented no real problems. This was primarily because of the remarkable simplicity of the architecture of the transputer family. The ease with which the transputer family can be interfaced to a wide range of devices has been demonstrated here and by other work (Barlow *et al.*, 1989). It must be noted that the interface approach used here was chosen so that the existing IP boards were not modified destructively. The development of a 'production' version, however, would be directly along the lines of option 2 given above.

7.4.3 Control Strategies

The 8031 microcontrollers, used on the original RTX IP board, were embedded with a proportional plus integral plus derivative (PID) control algorithm for each motor. This system performed satisfactorily, although there was significant overshoot when the motors were driven near their power limits. Initially this performance degradation was considered to be the consequence of poor control-law design, but early attempts to embed high gain proportional plus integral (PI) and PID control laws (Burge and Bradshaw, 1985) showed this was not the case. Indeed the designs experienced difficulty in rejecting disturbances caused by the link dynamics, friction and end-effector loads (Roskilly *et al.*, 1990). Moreover, despite integrator conditioning (Hanus *et al.*, 1987) it was not possible to drive the motors at their power limits without inducing limit cycle effects (in fact the original control strategy did not permit the motors to reach their power saturation limits, thus avoiding the limit cycle problem but at the expense of performance).

These problems were overcome by implementing an alternative control strategy of the form

$$u\{kT\} = fK_p[u_I\{kT\} - y\{kT\}] \tag{7.1}$$

where

$$u_I\{KT\} = K_I\rho\{kT\} \tag{7.2}$$

$$\rho\{(k+1)T\} = \rho\{kT\} + Te\{kT\} \tag{7.3}$$

$$e\{kT\} = v\{kT\} - y\{kT\} \tag{7.4}$$

$$f = 1/T \tag{7.5}$$

In equations (7.1) to (7.5), $v\{kT\}$ is the demand input, $y\{kT\}$ is the output, $u\{kT\}$ is the controller output, K_p is the proportional gain, K_I is the integral gain and T is the sampling period. The use of this control strategy is fully reported in Roskilly *et al.* (1990), where it is shown that this approach gives robust control in the face of external disturbances such as link dynamics, friction (including the breakout component of coulomb friction), and end-effector loading. Furthermore, the appropriate use of integrator conditioning logic allows the controllers to drive the motors to power saturation without developing unsatisfactory limit cycle behaviour.

The implementation of such control laws (one for each motor) resulted in a significant performance improvement over the original control strategies. The actual implementation was relatively straightforward requiring almost a direct translation of (7.1)–(7.4) into Occam.

7.4.4 Parallelism and Transputer Assignments

The inherent ability of the transputer, or transputer network, to execute processes concurrently is one of its major advances over conventional sequential processors.

Deciding, however, the extent of the parallelism is not an easy task, a point identified by other workers. (Fleming, 1988). For the robotic control system design case study presented herein, the DeMarco specification became invaluable in making decisions as to nature of the parallelism and the final mapping on to a transputer network.

The discipline enforced by the DeMarco method of defining all the data flows and data files in the DD, and the generation of the PSs, provided an overall guide as to the size and complexity of the final executable code. Moreover, the specification gave useful information about the coupling and cohesion between the identified processes. This allowed the decision to be made, at a high level, to implement PROCESSes 1 and 2 (Figure 7.3) in parallel on separate transputers. This choice was also influenced by the decision to include a transputer on the IP INTERFACE board to provide the 'play-back' facility.

At a lower level it was also made clear by the DeMarco analysis that the implementation of the control algorithms, PROCESSes 2.3 to 2.10 could be in parallel. Clearly, from a cost point it would be inefficient to implement each of these algorithms on their own transputer, but implementing these algorithms in parallel on a single transputer significantly eased the problem of real-time scheduling. Indeed, an algorithm process would only be exercised if there were demand data available for it to process.

In summary, this case study indicated that the DeMarco specification provided a useful tool to identify logical routes to parallelism. The method also provided a technique for modelling alternative options, in order that an efficient solution could be found.

7.5 HINTS AND CAVEATS

During the development presented in this chapter a number of general observations were made about the use of transputers and Occam in real-time control systems' design and implementation. While these observations are not really in the spirit of this chapter, which is to present a 'systems type' methodology to the design and development of transputer based control systems, it was considered unfair not to mention some of the 'dodges' which resulted in a successful implementation.

It is important, when designing control systems in 'raw Occam' without the benefit of guaranteed deadline scheduling to think carefully about the priority assigned to processes: the PRI0 level, which interrupts running PRI1 processes and executes to completion, must be used sparingly and is best reserved for off-chip communications, particularly when dealing with slow peripherals. This, and more sophisticated buffering techniques, can be used to ensure that the transputer's link engines, which operate essentially 'for free' by autonomous direct memory access, are utilized as fully as possible.

A second consideration is that the Occam ALT construct, as implemented on the transputer, is not 'fair', although it can be made so (Jones, 1989); if such a piece of code is swamped with messages on several channels, one of them will receive preferential treatment. Close consideration of iteration rates is in order here, although typical message sizes in control applications are very small and thus do not normally cause problems. If the iteration rates in the system are made into compile-time variables it is often quicker (in the spirit of rapid prototyping) to demonstrate that a higher-than-required rate executes correctly and then reduce it, than to perform many tedious calculations as proof of correctness. This will suffice for all but the most safety-critical applications.

Note finally that a late-scheduled event need not affect performance too adversely if it is 'time-stamped' using the transputer's internal clocks. The normal finite difference equations used in discrete-time control can be recast in divided difference form, employing the real measured time intervals between iterations. Such methods exhibit graceful degradation in the face of processor overload and can be used to good effect in designing systems for average loadings which can be shown to recover rapidly from the type of rare, transient occurrence which normally would limit processor utilization.

7.6 CONCLUSIONS

This chapter by means of a case study has shown one potential route to the design and implementation of a real-time robotic control system. Great emphasis has been placed on the use of structured systems analysis and design techniques to ease the development task. It has been proposed that the DeMarco methodology is ideally suited to such an activity when the implementation is on a transputer based system because of the intimate relationship between hardware and software processes on such devices. It has also been shown how relatively easy it is to interface the transputer family to other IC devices.

In summary it has been shown that the transputer family is ideal for implementing real-time control systems, and that by using an appropriate analysis design technique development can be rapid and relatively straightforward.

REFERENCES

BARLOW, M. I., BURGE, S. E. and KONNANOV, P. (1989) A survey of analogue I/O strategies for transputers, *Microprocessors and Microsystems*, **13**(b), 387–95.

BURGE, S. E. and BRADSHAW, A. (1985) 'Design of multi-functional controllers for structural load alleviation', in *Proceedings of IEE International Conference, Control 95, Cambridge*, 440–5.

DEMARCO, T, (1980) *Structured Analysis and System Specification*, Englewood Cliffs, NJ: Yourdon.

FLEMING, P. J. (ed.) (1988) *Parallel Processing in Control*. London: Peter Peregrinus.

HANUS, R., KINNAERT, M. and HENROTTE, J. L. (1987) 'Conditioning technique, a general anti-windup and bumpless transfer method, *Automatica*, **23**(6), 729–39.

INMOS (1988) *Transputer Handbook*, Hemel Hempstead: Prentice Hall.

JONES, G. (1989) 'Carefully scheduled selection with Alt. Occam Users Group', *Newsletter*, January.

KERRIDGE, J. (1987) *OCCAM Programming: A Practical Approach*, London: Blackwell Scientific.

ROSKILLY, A. P. (1991) 'A control system design and implementation strategy applied to an industrial robotic manipulator', PhD Thesis, University of Lancaster.

ROSKILLY, A. P., COUNSELL, J. M. and BRADSHAW, A. (1990) 'Nonlinear modelling of robust controllers for robotic manipulators, in *Proceedings of the I. Mech. E. International Conference on Mechatronics: Designing Intelligent Machines, Cambridge*, pp. 223–30.

WARD, P. T. and MELLOR, S. J. (1985/86) *Structured Development for Real-time Systems*, Vols 1–3, Englewood Cliffs, NJ: Yourdon.

8

Development of a Multiple Target Tracking Algorithm on Transputers

D. P. Atherton and D. M. A. Hussain

8.1 INTRODUCTION

During the last two decades the improved technology available for surveillance systems has generated a great deal of interest in algorithms capable of tracking a large number of objects using information from one or more sensors. Typical sensor systems, such as radar, obtain noisy measurement data from returns received from true targets and possible other bodies. The tracking problem requires the processing of these data to produce accurate estimates of the position and velocity of the targets. There are two types of uncertainties involved with the measurement data: first the positional inaccuracy, as the measurements are corrupted by noise; and second the measurement origin since there may be uncertainty as to which measurement originates from which target. These uncertainties lead to a data association problem and the tracking performance depends not only on the measurement noise but also upon the uncertainty in the measurement origin. Therefore in a multiple target tracking system extensive computation may be required to establish the correspondence between measurements and tracks at every radar scan. After the data association process tracks are normally updated using standard Kalman filter algorithms, and finally tracks whose statistics deviate from the assumed model of the target and all but one of any tracks which can be shown to be following the same target are eliminated. A block diagram for the computation involved in a standard recursive multiple target tracking (MTT) system is shown in Figure 8.1.

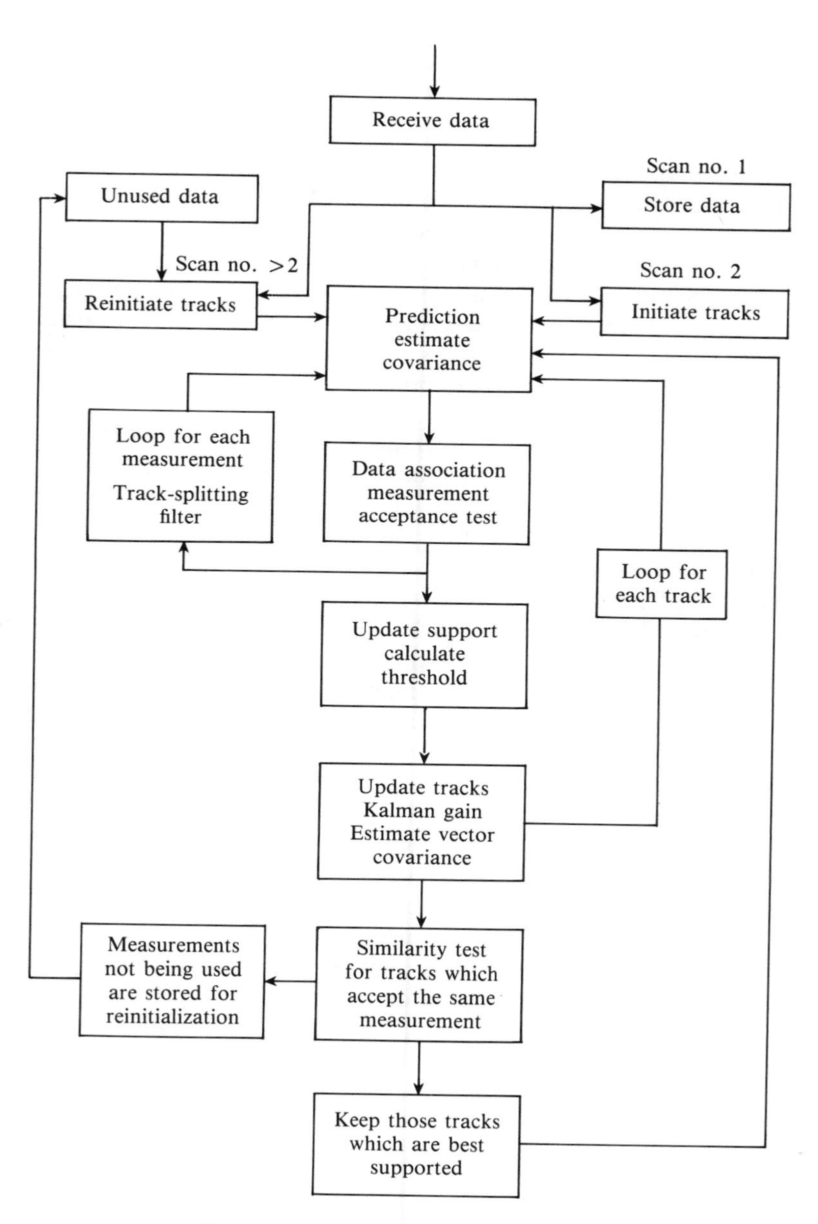

Figure 8.1 Recursive MTT system

8.2 PROBLEM DEFINITION

8.2.1 The Data Association Problem

In a general multiple target tracking situation, as explained in the introduction, a number of measurements from an unknown number of targets are received at each radar scan. Typically the radar sensor measures target position in polar coordinates, that is in range r and bearing θ. Measurement inaccuracy can normally be modelled as additive zero-mean uncorrelated Gaussian noise on r and θ, with given variances σ_r^2 and σ_θ^2 respectively. The multiple target tracking problem requires each measurement received at every radar scan to be associated with an existing or new target track. The associated observations are then incorporated in a state estimation algorithm or initialization procedure to produce updated track position and velocity estimates.

Basically there are two fundamental approaches to deal with the data association problem: first is the deterministic approach in which the most likely of several 'candidate' associations are formed and then treated as if they were certain, ignoring the fact that this may not necessarily be true. The results of the deterministic association are then used in a standard state estimation algorithm. The nearest neighbour filter (NNF) and the track splitting filter (TSF) are two common examples of the deterministic approach. The second method is a probabilistic model, based on a Bayesian approach, which computes the probabilities of individual associations and state estimations are obtained with associated probabilities. The multiple hypothesis tracking (MHT) approach, where a number of hypotheses are generated and evaluated as more data are received, and the joint probabilistic data association filter (JPDAF) are examples of algorithms which use the Bayesian approach.

The fundamental difference between the deterministic approach and the Bayesian approach is that the latter explicitly takes into account the uncertainty in the measurement origin. The Bayesian approach is the optimal solution to the data association problem, but it is computationally very expensive. The JPDAF method is a suboptimal modification of the Bayesian approach and is simpler since it does not require storage of information regarding hypotheses at previous time instants. Here we will be concerned with the implementation of a deterministic approach, the TSF algorithm, which is reasonably efficient computationally and is suitable for implementation on parallel processors.

8.2.2 State Estimation

For the estimation of target position and velocity from the track data, it is common to use a recursive Kalman filtering algorithm (Blackman, 1986; Bar-Shalom and Fortmann, 1988). This requires a model for the motion of the target being tracked and one often assumed is the constant velocity target with random acceleration description given by

$$\mathbf{x}_{n+1} = \Phi\mathbf{x}_n + \Gamma\mathbf{w}_n \tag{8.1}$$

and the corresponding measurement $\mathbf{z}_{n+1}$ is given by

$$\mathbf{z}_{n+1} = H\mathbf{x}_{n+1} + \boldsymbol{\nu}_{n+1} \tag{8.2}$$

The state vector, the state transition matrix, the excitation matrix and the measurement matrix are respectively:

$$\mathbf{x}_{n+1}^T = (x\ \dot{x}\ y\ \dot{y})_{n+1} \tag{8.3}$$

$$\Phi = \begin{bmatrix} 1 & \Delta t & 0 & 0 \\ 0 & 1 & 0 & 0 \\ 0 & 0 & 1 & \Delta t \\ 0 & 0 & 0 & 1 \end{bmatrix} \tag{8.4}$$

$$\Gamma = \begin{bmatrix} \Delta t^2/2 & 0 \\ \Delta t & 0 \\ 0 & \Delta t^2/2 \\ 0 & \Delta t \end{bmatrix} \tag{8.5}$$

$$H = \begin{bmatrix} 1 & 0 & 0 & 0 \\ 0 & 0 & 1 & 0 \end{bmatrix} \tag{8.6}$$

Here Δt is the sampling interval and corresponds to the time interval, assumed uniform, at which radar measurement data are received. The acceleration noise $\mathbf{w}_n$ is zero-mean Gaussian and independent in each Cartesian coordinate with covariance Q_n. The measurement noise $\boldsymbol{\nu}_n$ is assumed Gaussian with covariance R_n. The Cartesian coordinates of the measurements are given from the polar coordinates (r, θ) obtained from the radar by the nonlinear transformation

$$x = r\cos\theta \tag{8.7}$$

$$y = r\sin\theta \tag{8.8}$$

This transformation results in the measurement errors on the Cartesian coordinates being non-Gaussian distributed, but under the reasonable assumption that the measurement errors on the polar coordinates are small compared with the true target coordinates (r, θ), it can be shown that the Cartesian errors are bivariate Gaussian random variables (Farina and Studer, 1986) with zero-mean and covariance R_n given by

$$R_n = \begin{bmatrix} \sigma_x^2 & \sigma_{xy} \\ \sigma_{xy} & \sigma_y^2 \end{bmatrix} \tag{8.9}$$

where

$$\sigma_x^2 = \sigma_r^2 \cos^2\theta + \sigma_\theta^2 r^2 \sin^2\theta \tag{8.10}$$

$$\sigma_y^2 = \sigma_r^2 \sin^2\theta + \sigma_\theta^2 r^2 \cos^2\theta \tag{8.11}$$

$$\sigma_{xy} = (\sigma_r^2 - r^2\sigma_\theta^2) \sin\theta \cos\theta \tag{8.12}$$

Although the analysis in Farina and Studer (1986) is approximate, it is the only way of avoiding a nonlinear tracking filter, because if the measurement errors on the Cartesian coordinates are non-Gaussian distributed the optimum tracking filter in these coordinates is nonlinear. On the other hand, tracking can be performed in polar coordinates to avoid the correlation introduced by the nonlinear transformation but it leads to large dynamic errors when a linear model for target motion is used: even simple constant velocity tracks appear nonlinear in polar coordinates and artificial acceleration components are generated. This problem does not arise if tracking is performed in Cartesian coordinates.

When the target position measurement errors are correlated, a fourth-order full Kalman filter is the optimal tracker in that it minimizes the mean squared error between the estimated and the actual states, and it accounts for the cross-correlation term in the measurement covariance matrix. The standard Kalman filter equations for estimating the position and velocity are

$$\hat{\mathbf{x}}_{n+1} = \Phi\hat{\mathbf{x}}_n + K_{n+1}\boldsymbol{\nu}_{n+1} \tag{8.13}$$

$$K_{n+1} = P_{n+1/n}H^{\mathrm{T}}B_{n+1}^{-1} \tag{8.14}$$

$$P_{n+1/n} = \Phi P_n \Phi^{\mathrm{T}} + \Gamma Q_n \Gamma^{\mathrm{T}} \tag{8.15}$$

$$B_{n+1} = R_{n+1} + HP_{n+1/n}H^{\mathrm{T}} \tag{8.16}$$

$$P_{n+1} = (I - K_{n+1}H)P_{n+1/n} \tag{8.17}$$

$$\boldsymbol{\nu}_{n+1} = \mathbf{z}_{n+1} - H\hat{\mathbf{x}}_n \tag{8.18}$$

where Φ is the assumed target motion model of (8.1), K_{n+1} is the filter gain, $\boldsymbol{\nu}_{n+1}$ the innovations, H is the measurement matrix, P_{n+1} is the state covariance matrix, B_{n+1} is the covariance of the innovations, Γ is the excitation matrix, R_{n+1} the assumed measurement noise covariance matrix, and Q_n the filter acceleration noise matrix. The measurement noise matrix R_n and filter acceleration noise matrix Q_n are parameters which must be estimated in a practical situation. Clearly, when R_n is not diagonal, the recursive updating of the innovation covariance matrix B_{n+1} and the state covariance matrices P_{n+1} and $P_{n+1/n}$ using the above equations, introduces off-diagonal terms. When the measurement errors in each coordinate are independent, that is R_n is diagonal, the Kalman filter may be decoupled into two optimal tracking filters, known as $\alpha - \beta$ filters (Bridgewater, 1978). This filter configuration simplifies the computational requirements considerably, because the states relating to each of the two coordinates can be estimated independently. The equations for an $\alpha - \beta$

filter to estimate x-position and velocity are

$$\hat{x}_{n+1} = x_{n+1/n} + \alpha_{n+1}(z_{x_{n+1}} - x_{n+1/n}) \tag{8.19}$$

$$\hat{\dot{x}}_{n+1} = \hat{\dot{x}}_{n+1/n} + (\beta_{n+1}/\Delta t)(z_{x_{n+1}} - x_{n+1/n}) \tag{8.20}$$

where $x_{n+1/n}$ and $\dot{x}_{n+1/n}$ are the predicted position and velocity respectively, that is $\mathbf{x}_{n+1/n} = \Phi\hat{\mathbf{x}}_n$ for the full Kalman filter, and $z_{x_{n+1}}$ is the x-component of the measurement vector $\mathbf{z}_{n+1}$. The Kalman gain for the $\alpha - \beta$ filter is

$$K_{n+1} = (\alpha_{n+1} \quad \beta_{n+1}/\Delta t)^{\mathrm{T}} \tag{8.21}$$

where α_{n+1} and β_{n+1} are the gain coefficients. The estimation error covariance can be shown to be

$$P_{n+1} = (\sigma_x^2)_{n+1}\begin{bmatrix} \alpha_{n+1} & \beta_{n+1}/\Delta t \\ \beta_{n+1}/\Delta t & \delta_{n+1}/\Delta t^2 \end{bmatrix} \tag{8.22}$$

where $(\sigma_x^2)_{n+1}$ is the variance in the error of the $(n+1)$th x-position measurement. Recurrence relations for α, β and δ can easily be obtained using (8.13)–(8.18).

8.2.3 Initialization of the Filters

One of the basic requirements for any filtering algorithm is satisfactory 'initialization' of the filter, that is providing the initial estimate vector and the initial covariance matrix of the estimate vector. The filter initialization becomes more important in the multitarget environment where additional algorithms such as those for data association are involved The measurement acceptance test (see Section 8.3.1), which is an essential part of the algorithm, is also affected by a poor guess for the initialization of the covariance matrix of the estimate. The initial state is usually assumed to be a normally distributed random variable (Bar-Shalom and Fortmann, 1988), that is

$$\mathbf{x}_0 \sim N[\hat{\mathbf{x}}_{0/0}, P_{0/0}] \tag{8.23}$$

The scheme for track initialization is quite simple. A measurement m is considered to be unused if it has not been used to update a track. This measurement is then stored for possible correlation with another measurement n arriving at a later scan, usually the next scan, and if the actual distance between these two measurements m and n is less than a distance threshold $\delta_d(m,n)$ a new track is initialized. A maximum speed $V_{\max}$ is assumed for a target; thus the maximum distance that can be travelled in j time steps is $V_{\max}\, j\,\Delta t$, where Δt is the time interval between two scans. The distance measurement noise variance $\sigma_T^2(m,k)$ of measurement m at time k is obtained from the measurement noise variances on the (x, y) coordinates as

$$\sigma_T^2(m,k) = \sigma_x^2(m,k) + \sigma_y^2(m,k) \tag{8.24}$$

The distance threshold $\delta_d(m, n)$ between measurements m and n, j time steps apart, is therefore taken as

$$\delta_d(m, k) = V_{max}\, j\, \Delta t + \sigma_T(m, k) + \sigma_T(n, k + j) \tag{8.25}$$

When the above test is satisfied the most recent measurement is taken as the initial track position estimate. The initial velocity of the track is taken by dividing the difference between the two measurements m and n by the time elapsed between the two scans. Thus the x coordinate velocity estimate is given by

$$\hat{\dot{x}}_0 = \frac{z_x(k + j, n) - z_x(k, m)}{j\, \Delta t} \tag{8.26}$$

The initial velocity estimate for the y coordinate can be similarly derived.

The choice of the covariance matrix of the estimate should be such that the expected value of the estimation errors achieved by the filter match the filter calculated covariance, that is

$$E[(x_0 - \hat{x}_{0/0})(x_0 - \hat{x}_{0/0})^{\mathrm{T}}] = P_{0/0} \tag{8.27}$$

The Kalman gain or the weighting given to the predicted estimate is directly proportional to the covariance of the estimate. This means that an optimistic (very accurate) covariance at the initial stage will produce a low gain with the result that a small weighting is given to incoming measurements, which normally results in large errors in the initial estimates. At the other extreme a very high value of gain may have a bad effect on tracking accuracy since a high weighting is placed on the noisy measurements. The covariance of the measurement noise R_n can be taken as an initial uncertainty of the target position; thus, supposing the initial covariance matrix for a two-dimensional Kalman filter is

$$P_{0/0} = \begin{bmatrix} \sigma_x^2 \begin{pmatrix} \alpha & \beta/\Delta t \\ \beta/\Delta t & \delta/\Delta t^2 \end{pmatrix} & \sigma_{xy} \begin{pmatrix} \alpha & \beta/\Delta t \\ \beta/\Delta t & \delta/\Delta t^2 \end{pmatrix} \\ \sigma_{xy} \begin{pmatrix} \alpha & \beta/\Delta t \\ \beta/\Delta t & \delta/\Delta t^2 \end{pmatrix} & \sigma_y^2 \begin{pmatrix} \alpha & \beta/\Delta t \\ \beta/\Delta t & \delta/\Delta t^2 \end{pmatrix} \end{bmatrix} \tag{8.28}$$

where α, β and δ are the gain coefficients. In the target model (8.1), it is assumed that the system noise $\mathbf{w}_k$ is zero-mean Gaussian and independent in each Cartesian coordinate so that

$$E[\mathbf{w}_k] = 0 \tag{8.29}$$

where E is the expectation operator and also that the covariance Q_k of the system noise is a positive semi-definite $(n \times n)$ matrix with

$$E[\mathbf{w}_j\, \mathbf{w}_k^{\mathrm{T}}] = Q_k\, \delta_{jk} \tag{8.30}$$

where δ_{jk} is the Kronecker delta defined by

$$\delta_{jk} = \begin{Bmatrix} 0 & j \neq k \\ 1 & j = k \end{Bmatrix} \tag{8.31}$$

Similar assumptions are made regarding the measurement noise $\boldsymbol{\nu}_k$ sequence so that

$$E[\boldsymbol{\nu}_k] = 0 \tag{8.32}$$

and the covariance R_k of the measurement noise is a positive semi-definite $(m \times m)$ matrix with

$$E[\boldsymbol{\nu}_j \; \boldsymbol{\nu}_k^{\mathrm{T}}] = R_k \; \delta_{jk} \tag{8.33}$$

In simulation studies the target motion is normally generated using these assumptions. The values for the measurement noise covariance and the acceleration noise covariance are required for the tracking Kalman filter equations and these are usually taken in simulations to be equal to the values of R and Q used in the generation model. In a practical situation, these parameters have to be 'guessed': R, being a function of the measurement system, can usually be estimated reasonably well, whereas Q is a parameter associated with the unknown target motion and may be much less accurately known. For this reason simulation investigations are sometimes undertaken using a different value of Q in the Kalman filter from that used in the target motion; when the same values are used we refer to the filter as being '*matched*'. Now rewriting (8.27), which is a symmetrical matrix, in terms of its individual elements, we have

$$P_{0/0} = E \begin{bmatrix} \tilde{x}_0\tilde{x}_0 & \tilde{x}_0\dot{\tilde{x}}_0 & \tilde{x}_0\tilde{y}_0 & \tilde{x}_0\dot{\tilde{y}}_0 \\ \tilde{x}_0\dot{\tilde{x}}_0 & \dot{\tilde{x}}_0\dot{\tilde{x}}_0 & \tilde{x}_0\tilde{y}_0 & \dot{\tilde{x}}_0\dot{\tilde{y}}_0 \\ \tilde{y}_0\tilde{x}_0 & \tilde{y}_0\dot{\tilde{x}}_0 & \tilde{y}_0\tilde{y}_0 & \tilde{y}_0\dot{\tilde{y}}_0 \\ \tilde{y}_0\dot{\tilde{x}}_0 & \dot{\tilde{y}}_0\dot{\tilde{x}}_0 & \tilde{y}_0\dot{\tilde{y}}_0 & \dot{\tilde{y}}_0\dot{\tilde{y}}_0 \end{bmatrix} \tag{8.34}$$

where $\tilde{x}_0 = x_0 - \hat{x}_0$, with x_0 the true state and $\hat{x}_0$ the estimated state at the initial time 0. It is easily shown that this can be written as

$$P_{0/0} = \begin{bmatrix} \sigma_x^2 & \frac{1}{\Delta t}\,\sigma_x^2 & \sigma_{xy} & \frac{1}{\Delta t}\,\sigma_{xy} \\ \frac{1}{\Delta t}\,\sigma_x^2 & \frac{2}{\Delta t^2}\,\sigma_x^2 & \frac{1}{\Delta t}\,\sigma_{xy} & \frac{2}{\Delta t^2}\,\sigma_{xy} \\ \sigma_{xy} & \frac{1}{\Delta t}\,\sigma_{xy} & {\sigma^2}_y & \frac{1}{\Delta t}\,\sigma_y^2 \\ \frac{1}{\Delta t}\,\sigma_{xy} & \frac{2}{\Delta t^2}\,\sigma_{xy} & \frac{1}{\Delta t}\,\sigma_y^2 & \frac{2}{\Delta t^2}\,\sigma_y^2 \end{bmatrix} \tag{8.35}$$

Comparing (8.28) and (8.35) it is seen that the appropriate choices for the gain coefficients are $\alpha = 1.0, \beta = 1.0$ and $\delta = 2.0$.

For a decoupled $\alpha - \beta$ filter it follows that the above covariance matrix becomes the block diagonal matrix:

$$P_{0/0} = \begin{bmatrix} \sigma_x^2 & \frac{1}{\Delta t}\sigma_x^2 & 0 & 0 \\ \frac{1}{\Delta t}\sigma_x^2 & \frac{2}{\Delta t^2}\sigma_x^2 & 0 & 0 \\ 0 & 0 & \sigma^2{}_y & \frac{1}{\Delta t}\sigma_y^2 \\ 0 & 0 & \frac{1}{\Delta t}\sigma_y^2 & \frac{2}{\Delta t^2}\sigma_y^2 \end{bmatrix} \tag{8.36}$$

8.3 THE TRACK SPLITTING ALGORITHM

In this section we discuss the algorithms involved in the implementation of the track splitting filter.

8.3.1 Track Continuation

Measurement acceptance criterion

Following the initial track formation, incoming observations are considered for the continuation of existing tracks. The continuation procedure consists of prediction, measurement association and state estimation (i.e. updating). The predicted target positions in the x–y plane, from the tracking filters updated at the previous scan, are found and the uncertainty associated with these is used to place a measurement acceptance ellipse around the predicted position. A measurement occurring in the acceptance region is assumed to come from the appropriate target and used to update the tracking filter. When targets are very close together, however, more than one measurement may fall within the prediction ellipse and in the TSF the filter is allowed to split to form a number of tracking filters equal to the number of measurements inside the ellipse (Smith and Buechler, 1975). This approach assumes all the measurements falling inside the ellipse are equally probable for that particular track and several filter outputs provide equally probable indications of its future position.

The measurement acceptance criterion uses a simple result. If the dimension of the measurement vector $\mathbf{z}_n$ is M, then the norm d_n^2 of the innovation vector $\boldsymbol{\nu}_n$ at scan n for a filter is given by

$$d_n^2 = \boldsymbol{\nu}_n^{\mathrm{T}} B_n^{-1} \boldsymbol{\nu}_n \tag{8.37}$$

where the M-dimensional Gaussian probability density for the innovation is

$$f(\boldsymbol{\nu}) = \frac{\exp\left(-d^2/2\right)}{(2\pi)^{M/2}\sqrt{|B_n|}} \tag{8.38}$$

with B_n the innovations covariance matrix for the specific filter and $|B_n|$ its determinant. Provided that the filter's model for the track dynamics is accurate and that all the previous measurements used to update the track did indeed originate from one particular target, the quantity d_n^2 is a sum of squares of M-independent zero-mean and unit standard deviation Gaussian random variables. Thus d_n^2 will have a χ^2 distribution with M degrees of freedom. The measurement acceptance criterion for a track is thus defined so that if d_n^2 is less than a threshold J^2, then that particular measurement at scan n is used to update the track.

8.3.2 Track Pruning

Support function criterion

A mechanism for restricting the excess tracks that originate from track splitting under measurement ambiguity is needed and one possibility is the use of the track support function. The likelihood function of a track is the measure of the probability of the track accepting a sequence of measurements in n scans. It is given (Smith and Buechler, 1975) by

$$\Lambda = \prod_{i=1}^{n} f(\boldsymbol{\nu}_i) \tag{8.39}$$

$$= \left(\prod_{i=1}^{n} \frac{1}{(2\pi)^{M/2}\sqrt{|B_i|}}\right)\left(\exp\left(-\frac{1}{2}\sum_{i=1}^{n} d_i^2\right)\right) \tag{8.40}$$

The natural logarithm of the second part of the above equation is called the modified log-likelihood function (or support function S_n) (Bar-Shalom and Fortmann, 1988). The support function S_n is given by

$$S_n = -\frac{1}{2}\sum_{i=1}^{n} d_i^2 \tag{8.41}$$

and it can be calculated recursively from

$$S_{n+1} = S_n - \frac{1}{2}\,\boldsymbol{\nu}_{n+1}^{\mathrm{T}} B_{n+1}^{-1} \boldsymbol{\nu}_{n+1} \tag{8.42}$$

If the support function of a track is smaller then a threshold value, it may not represent a true target in the sense that the measurements it has been using are inconsistent with the assumed target motion. This is used as a criterion in the track splitting algorithm to terminate a track.

The summation of the norm d_n^2 for n scans is χ^2 distributed with N degrees of freedom where $N = n \times M$. Therefore the support function S_n is also χ^2 distributed with N degrees of freedom. If we wish to define a threshold T for a χ^2 distribution so that the probability of the variable exceeding the threshold is P_{r_T} then we can either use χ^2 tables or, if the degree of freedom is large, use the approximate relationship (Abramowitz and Stegun, 1972)

$$T = N + T_g\sqrt{2N} \tag{8.43}$$

which relates the equivalent threshold T_g for a Gaussian distribution of unit variance to that of T for the χ^2 distribution. For example if $N = 30$, $T_g = 2.327$ for a threshold probability of P_{r_T} equal to 0.01, and the above formula yields $T = 48.025$ compared with the value 50.892 from the χ^2 table. For the implementation of the support function criterion, assuming that $M = 2$ and thus $N = 2 \times n$, then

$$T'_n = -(n + T_g\sqrt{n}) \tag{8.44}$$

where T'_n is the threshold for the support function S_n.

Similarity criterion

If the number of filters obtained when using the track splitting algorithm needs to be further reduced, the similarity criterion is used to ensure that not more than one filter is tracking the same target. The similarity criterion provides a measure for the nearness of two tracking filter position estimates and if this is below a certain threshold one filter can be eliminated. Any two tracks i and j are deemed to be similar if

$$(\hat{\mathbf{x}}_i - \hat{\mathbf{x}}_j)^{\mathrm{T}} P^{-1} (\hat{\mathbf{x}}_i - \hat{\mathbf{x}}_j) \leqslant D_{\mathrm{th}} \tag{8.45}$$

where $\hat{\mathbf{x}}_i$ and $\hat{\mathbf{x}}_j$ are the two filter state estimate vectors, P is a weighting matrix chosen to be the sum of the covariance matrices of the two filters with off-diagonal elements set to zero and D_{th} is the chosen threshold. As the estimates $\hat{\mathbf{x}}_i$ and $\hat{\mathbf{x}}_j$ are correlated, P will not be the true covariance of $(\hat{\mathbf{x}}_i - \hat{\mathbf{x}}_j)$ but in many instances a reasonable estimate.

When the two tracks are similar, one obviously wishes to keep the best supported track. The support function of the track is proportional to the life of the track. Therefore if two tracks with different track lives are compared then the track with a longer life may be incorrectly eliminated. To prevent such a situation the support function of a track is normalized where the normalized support function is (Kharbouch, 1990)

$$NS_n = -\frac{S_n + n}{\sqrt{n}} \tag{8.46}$$

Similarity pruning as described above may lead to a different elimination of tracks in a multitarget scenario, depending upon the order in which similarity calculations between tracks are made.

8.4 PARALLEL IMPLEMENTATION

The flowchart of the code for a MTT system using the Kalman filter is given in

```
...LIBRARIES
...LINKS
...CHANNEL DECLARATION
PROC MTT (CHAN OF DATA RECEIVE, SEND,
  CHAN OF INPUT FROM1, FROM2, FROM3
  CHAN OF OUTPUT TRANSP1, TRANSP2, TRANSP3)
  ...VARIABLES
  ...PROC TRACK INITIALIZATION
  ...PROC DISTRIBUTION OF DATA — TO OTHER TRANSPUTERS
  ...PROC TRACK PREDICTION
  ...PROC MEASUREMENT ACCEPTANCE
  ...PROC TRACK UPDATE
  ...PROC TRACK PRUNING
  ...PROC GLOBAL SIMILARITY
SEQ
  WHILE TRUE
    SEQ
      RECEIVE ? FROM B004 BOARD
      IF
        SCAN NO =1
          SEQ
            ...STORE THE DATA
        SCAN NO =2
          SEQ
            ...INITIALIZATION OF TRACKS
        SCAN NO >= 3
          SEQ
            ...IF NECESSARY INITIALIZE THE TRACKS
            ...DISTRIBUTION OF DATA
            PAR
              TRANSP1 ! OUTPUT TO TRANSPUTER 1
              TRANSP2 ! OUTPUT TO TRANSPUTER 2
              TRANSP3 ! OUTPUT TO TRANSPUTER 3
            ...TRACK PREDICTION
            ...MEASUREMENT ACCEPTANCE
            ...TRACK UPDATE
            ...TRACK PRUNING
            PAR
              FROM 1 ? INPUT FROM TRANSPUTER 1
              FROM 2 ? INPUT FROM TRANSPUTER 2
              FROM 3 ? INPUT FROM TRANSPUTER 3
            ...GLOBAL SIMILARITY
            SEND ! TO B004 BOARD FOR PLOTTING
```

Figure 8.2 Occam code flowchart of a MTT system

Figure 8.2. To simulate the output of a radar, a data generator routine was written in Occam to run on a B004 board, and the parameters describing the simulation are entered interactively by the user. Figure 8.3 gives details of these parameters for the situation where the individual targets for the simulation are described by the user. The trajectories for the targets are generated using the kinematics described in (8.1), namely a constant velocity motion with acceleration noise. A single dimensional array is used to store the measurement data and for each target track the data given in Figure 8.4 are stored. This array consists of the noisy position measurements (x_n, y_n) used by the filter, the measurement variances σ_x^2, σ_y^2 and σ_{xy}, and a colour identity used for the graphical display, and it is known as the measurement vector.

The MTT system using the track splitting algorithm was implemented on a B003 board containing four transputers. Since the MTT systems employs multiple filters, a number of strategies can be considered for track and data distribution; some of the possible strategies are explained in the following paragraphs.

	Initial input
1	Initial position (x, y) of the platform/radar
2	Initial velocity of the platform/radar
3	Number of target(s)
4	Initial position(s) (x, y) of target(s)
5	Initial angle(s) of target(s)
6	Initial velocity of target(s)
7	Start/stop time of target(s)
8	Scan interval
9	Number of scans
10	Scan sector
11	Maximum and minimum range
12	Probability of detection
13	Range resolution threshold
14	Angle resolution threshold
15	Clutter density
16	Integer seed to generate Gaussian numbers

Figure 8.3 Radar parameters

x_n	y_n	σ_x^2	σ_y^2	σ_{xy}	colour identity

Figure 8.4 Data for a single target at each radar scan

8.4.1 Track Distribution Configuration

The simplest way to configure the MTT system on a network of transputers is to distribute the individual tracks among the processors as equally as possible. Each individual track consists of an estimate vector (8.13), covariance of the estimate (8.17), track life, that is how long the track has been running since initialization, support function value (8.42) and a colour identity for graphical display all of which are stored in a single dimension array, known as a track vector. Figures 8.5 and 8.6 give two possible task configurations for such an arrangement: one is a tree formation configuration and the other is a pipeline configuration. The difference between these two configurations occurs at the time of data distribution. In the tree formation configuration the data distribution module deals the tracks to the different transputers like cards are dealt in a card game, therefore if there are four tracks each transputer receives one track, but if there are five tracks then one of the transputers is given two tracks. Also all the measurements received at each scan are passed on to all the processors so that each track can be tested for a possible correlation with all the measurements. After each processor finishes its tasks, that is data association, filtering and possibly prediction, where no measurement has been associated with an existing track, and a local similarity routine possibly to prune all but one of those tracks which have accepted a common last measurement and are similar, all the updated tracks are sent back to the first processor where global similarity calculations are carried out. After that the remaining position estimates for the tracks are sent to the B007 board through the B004 board for a real-time display.

In the pipeline approach the data distribution consideration is similar but the tracks are not dealt; instead, each processor keeps its share of tracks and the remaining tracks are passed on to the next processor along with all the measurements received at each scan. After that each processor, similar to the tree configuration, performs data association and filtering. However, to prune the similar tracks on different transputers, the last processor in the network passes the tracks to its previous processor and a cross-transputer similarity routine looks for any similar tracks between the two processors. This process of passing the tracks and cross-transputer similarity calculations continues until all the tracks reach the first processor (Gul and Atherton, 1989; Atherton *et al.*, 1990). In both configurations the first processor is responsible for the track initialization and data distribution calculations. Thus from the time of receiving the data until data distribution, this processor is a serial deadlock as for the whole of this time the other three processors in the system are idle. There is no other alternative to this problem as the data have to be distributed among the different processors. The time taken by the first processor for initialization and data distribution is known as the scheduling time. The individual processors possess the same modules, therefore ideally if the number of tracks divides equally among the processors they will finish their tasks approximately at the same time, but under data ambiguity it is never certain which processor will finish its task first.

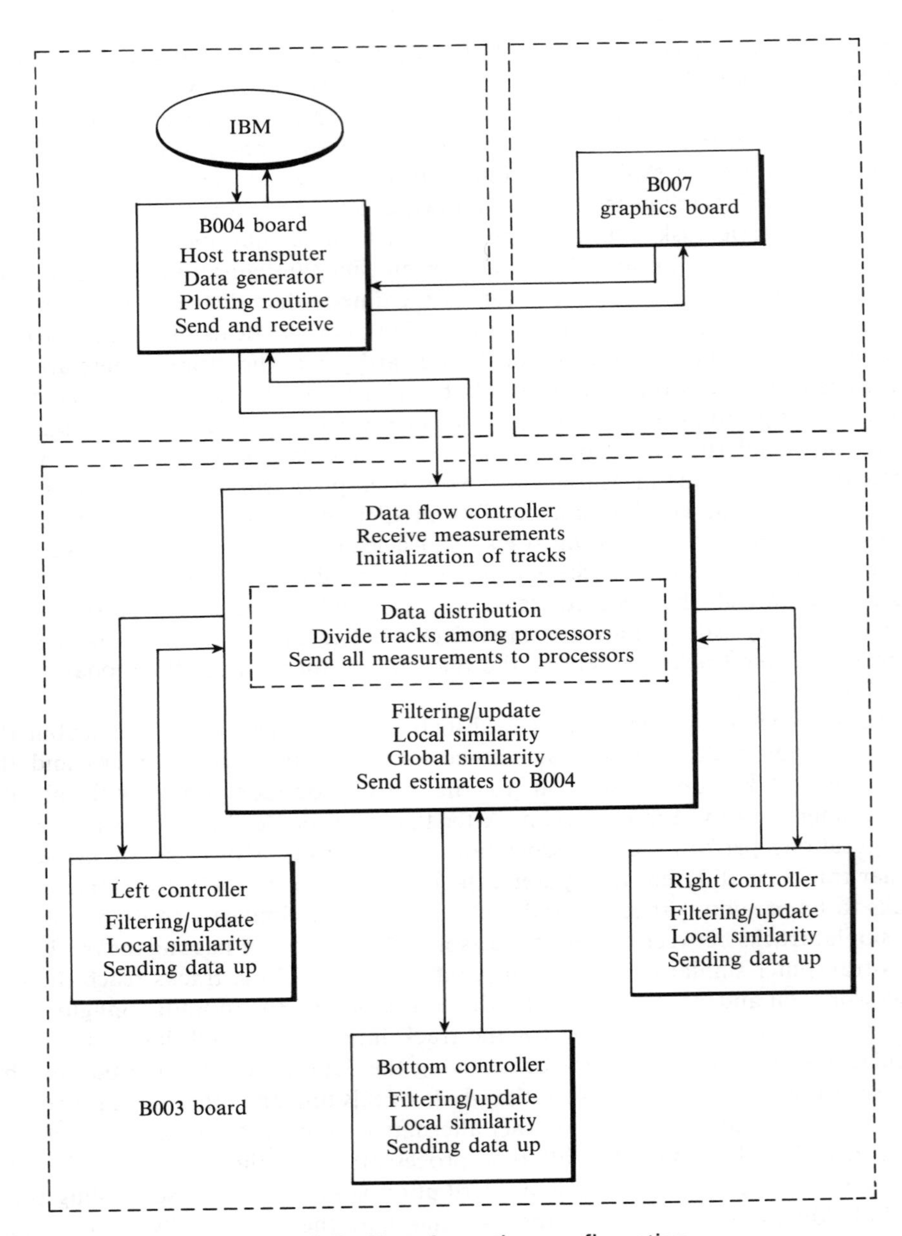

Figure 8.5 Tree formation configuration

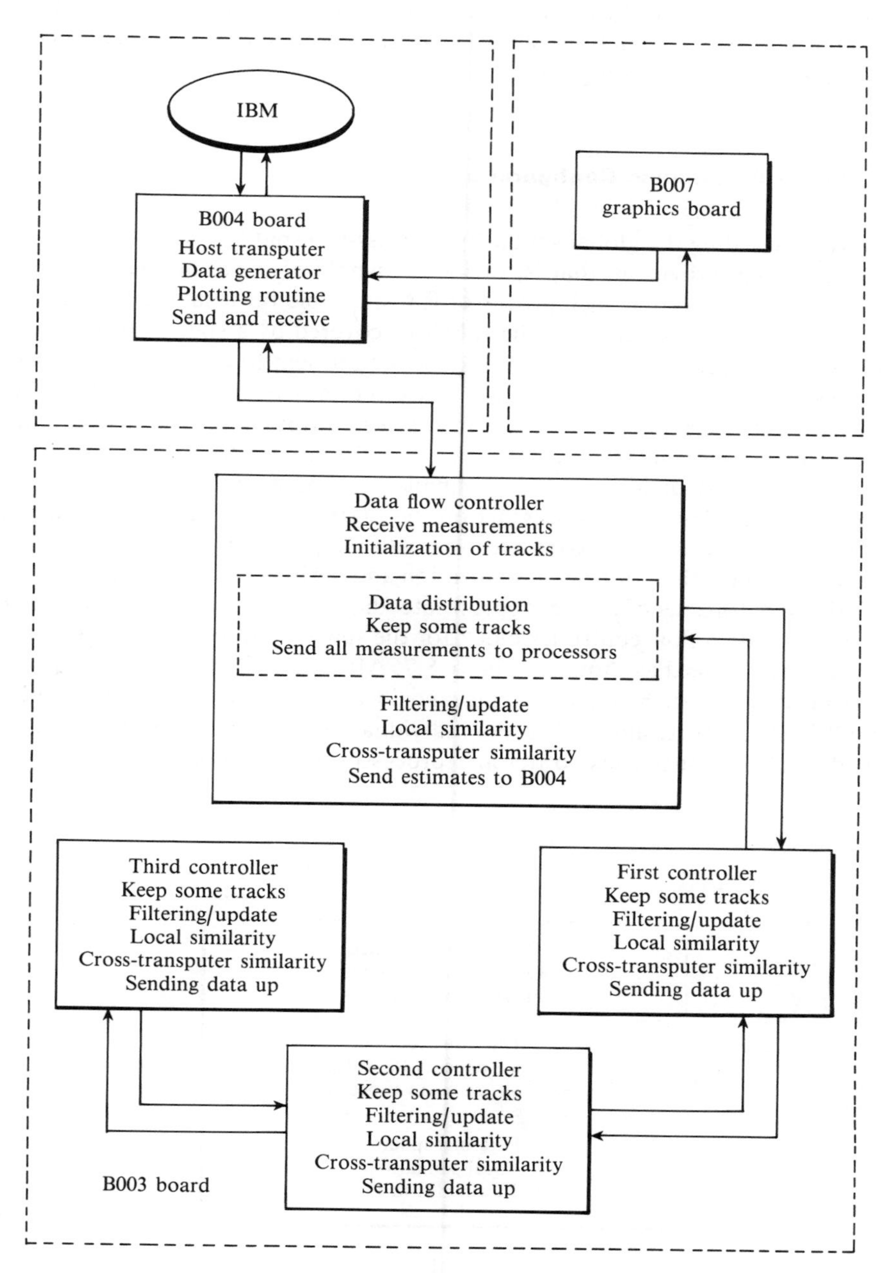

Figure 8.6 Pipeline configuration

In both the above configurations the need to perform global similarity or cross-transputer similarity calculations slows down the computation as some processors remain idle while this is being done.

8.4.2 Space Partitioning Configuration

One way of avoiding the global similarity or cross-transputer similarity comparison is to make a space division; that is, divide the whole of the scan area into smaller regions (subspaces) so that each processor is responsible for a subspace. Only those measurement vectors which lie in the subspace covered by a specific transputer are then sent to that processor, but the track vectors are sent to all transputers. All the track vectors are sent to each processor since the target estimated positions for a track near a subspace boundary may lie in another subspace, and this also eliminates the need for global similarity calculations. For the transputer implementation the global window was divided into four smaller subspaces and each subspace was allocated to one of the four processors on the B003 board. At each radar scan the first processor on the B003 board receives data via one of the hard links (Inmos, 1988) and is responsible for track initialization and performing the subspace division calculations before sending the appropriate data to the other processors. The changes to the data flow controller block for the space partitioning configuration in Figure 8.5 are changed as shown in Figure 8.7. After that all four processors, which have data association, Kalman filtering and local similarity routines as shown in Figure 8.5, start processing. When the individual processors finish their tasks, the updated tracks are sent back to the first processor and the position estimates and

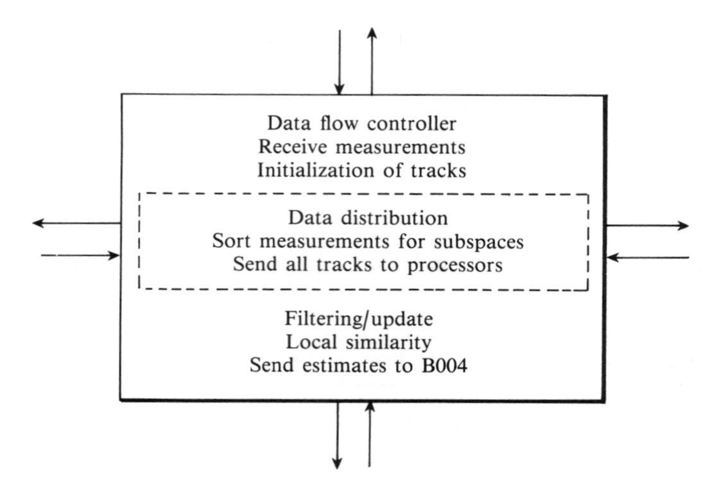

Figure 8.7 Data flow controller for space partitioning configuration

other necessary information are passed on to the B004 board for the real-time graphics display. The procedure is then repeated at the next scan. It is evident for this configuration that if the number of targets in each subspace are not equal, that is the destiny of the targets is not uniform in the global space, the individual processors will finish their tasks at different times and a bottleneck will occur. In an attempt to improve the concurrency further two different strategies have been adopted and are explained in the next sections.

8.4.3 Track Clustering Configuration

An alternative approach for distribution of the data between the processors is to use a track clustering algorithm. Here all the tracks which accepted the same last measurement are taken to define a cluster. Sorting into clusters is done by the first processor which then distributes the clusters, beginning with the largest cluster to the first processor, then the next largest to the second processor and so on. Since each processor is allowed to handle a maximum number of tracks, a cluster will be divided between transputers if it is too large for one transputer. If the number of clusters is greater than four then the fifth cluster is placed in the first transputer and so on. If the total number of tracks is greater than the number allowed for the number of transputers available then tracks are removed by successively removing the worst supported tracks from the largest cluster. The changes to the data flow controller block for the track clustering configuration in Figure 8.5 are changed as shown in Figure 8.8.

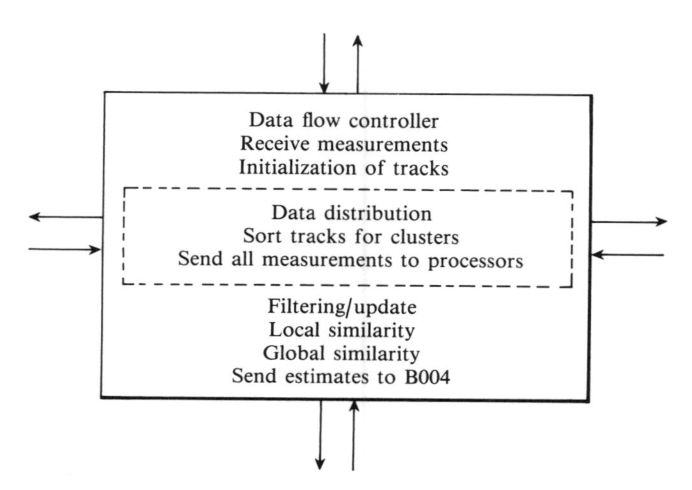

Figure 8.8 Data flow controller for track clustering configuration

8.4.4 Intelligent Tracking Algorithm

The main feature of the intelligent tracking algorithm is that it makes expert decisions in a very short time. These eliminate the need for global similarity or cross-transputer similarity calculations and provide approximately balanced data loading among the processors. The changes to the data flow controller block for the intelligent tracking algorithm in Figure 8.5 are changed as shown in Figure 8.9. The decision maker which is the new feature of the algorithm uses the following two conditions:

1. Condition A: this condition is basically an AND (logical) operation of two criteria. The first criterion is that all the distances $D_{i,j}$ between two measurements must be greater than a given threshold D_p, which can be a function of the radar resolution. The formula for calculating the distance is given by the expression

$$D_{i,j} = \sqrt{[(x_i - x_j)^2 + (y_i - y_j)^2]} \quad (8.47)$$

$$i = 0, \ldots, (n-1)$$

$$j = (i+1), \ldots, n$$

where n is the total number of measurements. The second criterion is that the number of tracks must be equal to or less than the number of measurements. These two criteria, for example, apply when the tracks are distinct. When condition A is satisfied, the tracks are divided (like cards are dealt in a card game) to the processors and all the measurements are sent to all the processors.

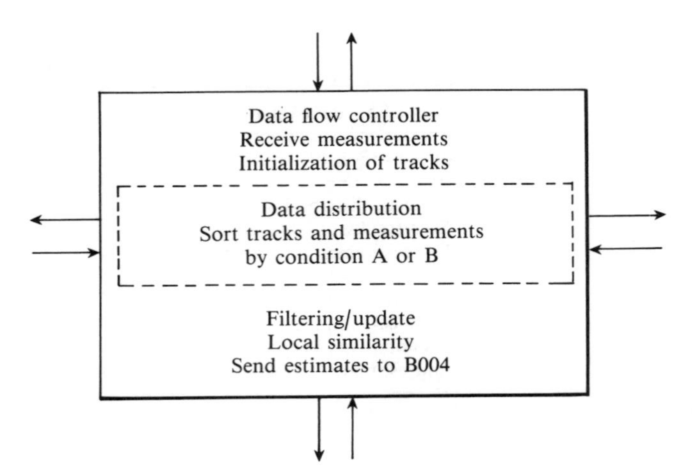

Figure 8.9 Data flow controller for intelligent tracking algorithm

2. Condition B: this condition is the OR (logical) operation of two criteria. The first criterion is that one or more distances $D_{i,j}$ must be less than or equal to the given threshold D_p. The second criterion is that the number of tracks must be greater than the number of measurements. Therefore either of these two criteria will invoke this condition in which case the measurements are divided (like cards are dealt in a card game) to the processors and all the tracks are sent to every processor. This case typically represents either that branching is about to start or that branching is already in progress. Thus to keep similar tracks on the same processor, measurements should be divided among the processors.

To demonstrate how the intelligent algorithm works, consider as an example the tracking of four crossing targets which are initially well separated. The decision-maker routine has to execute six (loops) possible combinations (as there are four measurements) to calculate the distance $D_{i,j}$ and to make comparisons with the given D_p. Assume that the time taken for a single loop calculation of $D_{i,j}$ and comparison with D_p is T_w. This results in a total time of

$$T_{\max} = 6T_w \tag{8.48}$$

which will be small compared with the time required if a clustering algorithm is used. After execution of the decision process, since condition A is valid, each processor will be sent a single track with all the four measurements. Now suppose that later when the targets are near to each other that each track has been associated with every measurement and that none of the tracks is pruned by either the support or local similarity criteria. When the new cycle begins there will be 16 tracks and four measurements, but this time the distances $D_{i,j}$ between any two measurements will normally be less than the chosen threshold D_p. Therefore since condition B is valid, measurements will be distributed and all the tracks will be sent to each processor. Even though the number of tracks has increased, the intelligent algorithm takes the same amount of time to make its decisions since the number of measurements remains the same. This is in contrast with clustering techniques, which take more time as the number of tracks grows, since every track is tested for association with each measurement. Thus every processor will be given all 16 tracks and one measurement each. There is a slight increase in the communication time as all the tracks are sent to all the processors but this is relatively small. Since all the tracks are present in the individual processors and they are individually associated with the given measurement, the cross-transputer similarity calculation is not required.

8.5 PERFORMANCE COMPARISON

In order to test the performance of the different algorithms several target scenarios were simulated. Two simulation programs were written: one which may be described as a deterministic scenario, where the initial positions of the targets and their velocities can be input by the user, as illustrated in Figure 8.3, and the other a random scenario where the initial positions are selected randomly but uniformly distributed in the global tracking window. The user specifies the target density, the mean and standard deviation for the Gaussian distribution for the initial speed, and the initial direction is selected from a uniform distribution between 0° and 360°. If a target reaches the boundary of the global window the target density is kept constant by replacing it with another target. Four scenarios are described below, where in all cases matched values were used in the tracking filter for the variances of the acceleration and measurement noises. In all cases full Kalman filters were used. The simulation results given are also averages for five simulation runs using different random number seeds to generate the measurement noise.

The first scenario chosen consisted of ten targets all moving left to right but with initial positions such that the ideal paths of two sets of five, that is with no measurement noise added, each crossed at a single point. The crossing points chosen which were reached after 30 seconds were (0.0, 0.0) and (0.0, 2.0), respectively. The processing times for the four algorithms are shown in Figure 8.10. When the targets are well separated and no track splitting takes place the simple track distribution algorithm with its small scheduling time is best, but around the target crossing time where there is significant track splitting the intelligent algorithm is superior. For this scenario the space partitioning algorithm is not particularly good owing to the location of the target crossing points.

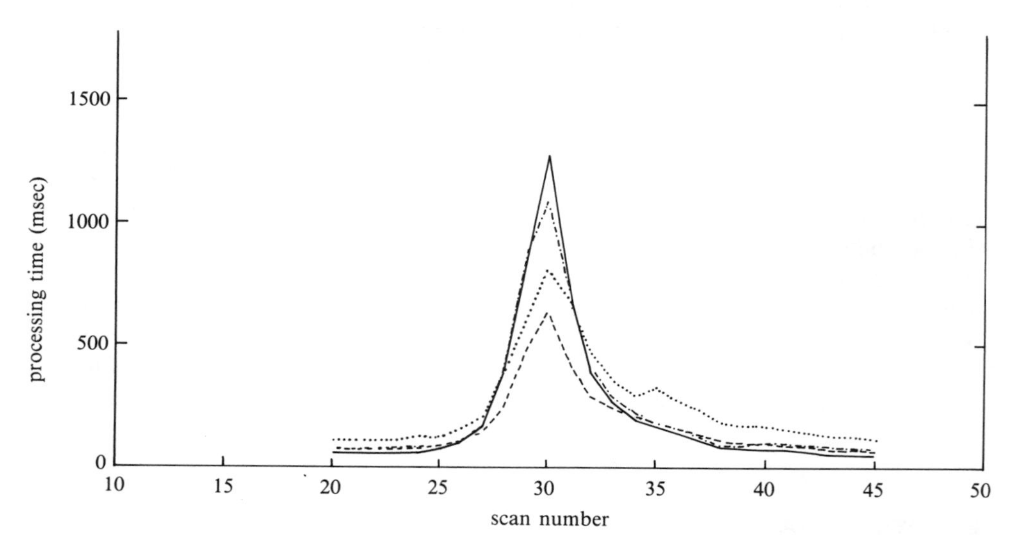

Figure 8.10 Processing times for two groups of five crossing targets: — track distribution; ... track clustering distribution; --- intelligent tracking distribution; --- space partitioning distribution

Figure 8.11 shows results from another deterministic simulation with more targets. For this case 25 target paths were simulated. There are three sets of four targets all moving from left to right and the ideal paths for these three sets of targets cross at (0.0, −2.0), (0.0, 0.0) and (0.0, 2.0) respectively. The other 13 targets, each with a path which did not interfere with any of the other targets, were distributed approximately equally in each of the four windows of the global space. The results are seen to be very similar to the previous case with the intelligent algorithm giving a 50% improvement in the computation time relative to the basic algorithm in the region of the target crossing time. In both examples outside the small time region where the targets cross, all four algorithms provided satisfactory tracking of all the targets with essentially the same tracking error.

Figures 8.12 and 8.13 show results for tracking random scenarios, first in Figure 8.12 for a total of 10 targets and second in Figure 8.13 for a total of 40 targets. For the lower target density results of Figure 8.12, where the number of tracks is such that the probability of track splitting taking place is quite low, the basic algorithm with its low scheduling time performs slightly better than the intelligent algorithm. However, as expected, as the target density increases, the intelligent algorithm achieves superior performance as shown in Figure 8.13. The space partitioning algorithm has no real advantages even though the targets are basically uniformly distributed in the global window, since it was found in practice that the number of targets in each of the five windows was never equal in any of the five simulation runs. Thus in the high-density scenario it is only marginally better than the basic

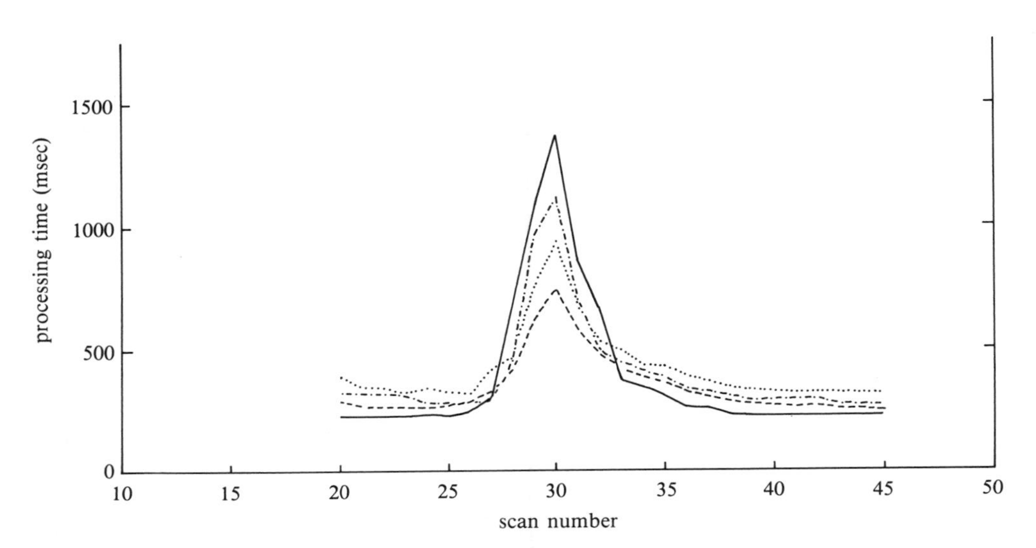

Figure 8.11 Processing times for three groups of 4 crossing targets and 13 separate targets going in different directions: — track distribution; ··· track clustering distribution; --- intelligent tracking distribution; -·- space partitioning distribution

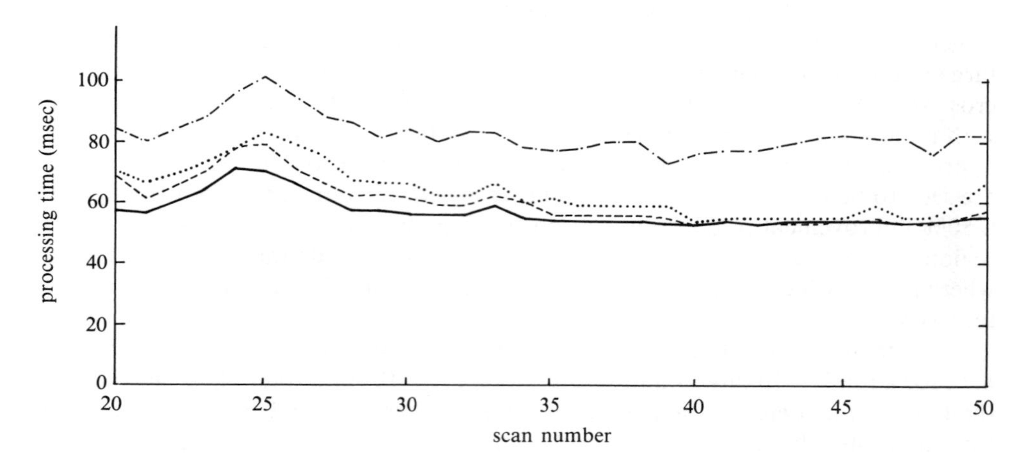

Figure 8.12 Processing times for 10 targets (random scenario): — track distribution; ··· track clustering distribution; --- intelligent tracking distribution; -·- space partitioning distribution

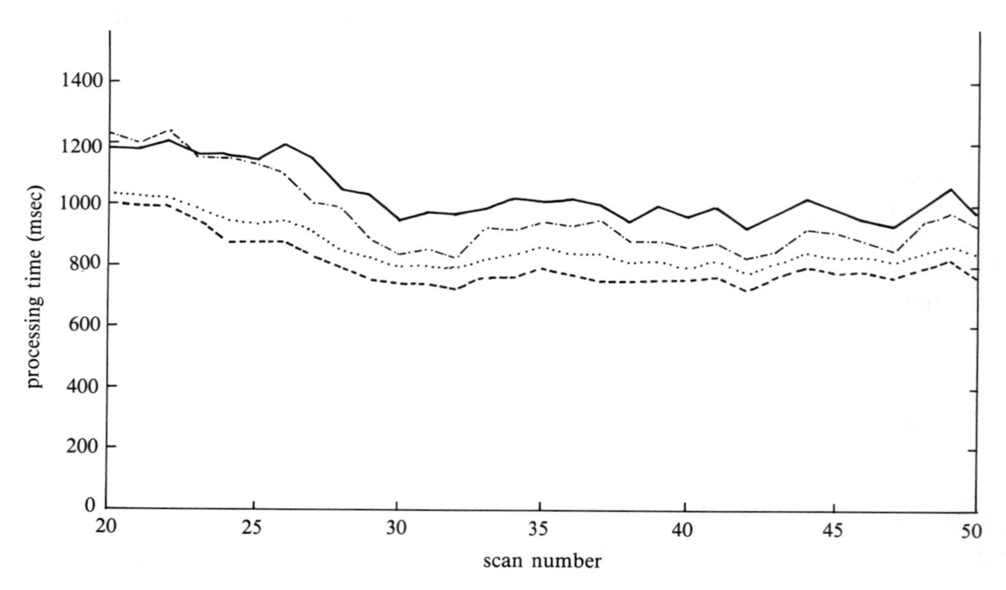

Figure 8.13 Processing times for 40 targets (random scenario): — track distribution; ... track clustering distribution; --- intelligent tracking distribution; -·- space partitioning distribution

algorithm and because of its relatively high track scheduling time it shows the worst performance for the low density scenario. Again, the target tracking accuracy for all the algorithms was essentially the same.

8.6 CONCLUSIONS

This chapter has discussed the problem of tracking multiple targets in two dimensions and outlined in detail one specific approach, namely the track splitting algorithm. When there are a large number of targets, and in particular when track splitting occurs, a large number of tracking filters are required. To speed up the computations the filters can be distributed on several transputers and we have discussed four approaches to doing this. Results of simulations for the four algorithms running on a four transputer system have been presented for several target motion scenarios from which it has been seen that in general the intelligent tracking algorithm gives the best results when a reasonable amount of track splitting takes place. The basic algorithm, which takes the smallest time for the scheduling, is satisfactory when the probability of track splitting is low. Although it has not been discussed here it should also be pointed out that this basic algorithm gives a speed up of 60–70% (Gul and Atherton, 1989) of the theoretical maximum of four when run on four transputers rather than one.

REFERENCES

ABRAMOWITZ, M. and STEGUN, I. A. (eds) (1972) *A Handbook of Mathematical Functions*, New York: Dover.

ATHERTON, D. P., GUL, E., KOUNTZERIS, A. and KHARBOUCH, M. M. (1990), 'Tracking multiple targets using parallel processing', *IEE, Pt D*, **4**, 225–31.

BAR-SHALOM, Y. and FORTMANN, T. E. (1988) *Tracking and Data Association*, London: Academic Press.

BLACKMAN, S. S. (1986), *Multiple Target Tracking with Radar Applications*, London: Artech House.

BRIDGEWATER, A. W. (1978) 'Analysis of second and third order steady state tracking filters', in *Proceedings of AGARD Conference No. 252, Monterey, CA*, pp. 9.1–9.11.

FARINA, A. and STUDER, F. A. (1986) *Radar Data Processing*, Vol. 1, Chichester: Research Studies Press.

GUL, E. and ATHERTON, D. P. (1989) 'A transputer implementation for multiple target tracking', *Micro-processors and Microsystems*, **13**(3), 188–94.

INMOS, (1988) *Transputer Reference Manual*, Hemel Hempstead: Prentice Hall.

KHARBOUCH, M. M. (1990) 'Some investigations of multiple target tracking', PhD Thesis, University of Sussex, UK.

SMITH, P. L. and BUECHLER, G. (1975) 'A branching algorithm for discriminating and tracking multiple objects', *IEEE Trans Auto-Control*, **AC-20**, 101–4.

9

IMS A100/Transputer Based Heterogeneous Architectures for Embedded Control Problems

Y. Li and E. Rogers

9.1 INTRODUCTION

The performance demands placed on modern control systems often mean the use of advanced schemes and techniques such as adaptive control, rule-based and neural network systems, and the provision of a fault-tolerance capability. These complex processes are almost always implemented using digital computing techniques and are often required to operate in real-time, i.e. the complete control algorithm must be computed within the loop sample time. This in turn means that, owing to the requirement of synchronization, there is a direct dependence between processing time and the maximum input frequency which can be adequately processed.

This dependence can still cause a problem despite the vastly increased computing power now available. The basic reason for this is the fact that increased performance demands require shorter sampling times. For example, cases have been reported of self-tuning controllers, implemented on a DSP (digital signal processing) chip with an instruction cycle of 200×10^{-9} seconds and fixed-point arithmetic, where the input signal is restricted to signals below 250 Hz even when a (standard) second-order model is used. Of course, this problem becomes more severe or even impossible as the model order is increased (often a necessary consequence of high-performance control) and/or the process is multivariable. Li (1990) gives a full treatment of this general point.

The structural links between algorithms used in control and digital signal processing problems is well known. One obvious approach to the general problem outlined above is to attempt to exploit the new generations of programmable DSP chips. This has been investigated by, for example, Rees and Whitting (1988) and the general conclusion is that these chips perform 'poorly' on the more general computational tasks arising in applications. Reasons for this include the following:

1. Coefficient resolution which is limited to, for example, 16 bits and fixed-point arithmetic.
2. Sensitivity to number representation.

Work has also been reported by, for example, Ganesan (1991) on using DSP chips to build parallel processing architectures. If N processors are used then the theoretical improvement in data throughput relative to a single (sequential implementation) processor is (of course) N. In practice, however, this is not possible owing to increased communication/process management overheads as N increases. Such architectures can, however, be successfully used for problems composed of 'small' regular tasks.

The transputer can efficiently compute irregular tasks and coordinate parallel applications. Its application to real-time control problems has been reported by, for example, Li *et al.* (1991) and in Irwin and Fleming (1992). One problem (which is somewhat surprising given the claims made in respect of real-time embedded systems) highlighted by this work has been the shortcoming revealed by the special demands of real-time control software. In particular, some cases considered strongly suggest that the architecture granularity – basically a measure of computation power to interprocessor communications overheads – is not appropriately matched. This is because the communications overhead is too dominant or, alternatively, the task size is not 'sufficiently large'.

Work by Fleming and Garcia Nocetti (1992), for example, has used performance analysis software to undertake a detailed study of this problem. One of the major conclusions from this work is that a granularity reduction of one order of magnitude should be more suitable. Hence it is to be expected that fusing (or merging) the finer granularity of parallel DSPs with the transputer's ability to handle irregular computations and manage parallel operations should prove highly effective. This is the so-called heterogeneous system approach to parallel control and simulation problems which is the subject of this chapter based on transputers and the IMS A100 DSP chip.

To provide continuity, the starting point here is the same general problem as Chapter 1. The relevant details are reviewed in Section 9.2, concluding with the two semi-systolic architectures which are the effective starting point of this chapter. Section 9.3 then develops transputer based architectures from these two semi-systolic 'building blocks' and critically analyzes their expected performance. This then leads to the heterogeneous (or mixed granularity) architectures of Section 9.4 and their application to self-turning/adaptive control schemes. Finally, Section 9.5 summarizes the results obtained here and briefly outlines some future research objectives.

9.2 A TYPICAL CONTROL COMPUTATION AND BASIC ARCHITECTURES

In common with conventional hardware, a control algorithm is naturally written in sequential code, since there are dependence relationships between the computation steps. Further, the order in which data are to be processed must be specified in the algorithm. If there is no data dependence in the operations for a particular special case, full parallelism is possible. In general, however, this is impossible and a minimum amount of serialism is required. Potential concurrency in these algorithms does, however, often exist and appropriate reorganization of them could lead to an implementation with maximum concurrency realized by pipelining.

To provide continuity, this chapter considers the same general problem as Chapter 1. The required results are now summarized, starting with the single-input single-output (SISO) plant state-space model

$$\dot{\mathbf{x}}_p(t) = A_p\mathbf{x}_p(t) + B_p u(t) + w_1(t) \tag{9.1}$$

$$y(t) = C_p\mathbf{x}_p(t) + w_2(t) \tag{9.2}$$

Here $\mathbf{x}_p(t) \in R^n$ is the plant state vector, $u(t)$ the plant input, $y(t)$ the plant output and $w_1(t)$ and $w_2(t)$ are Gaussian white noise disturbances. The matrices A_p, B_p and C_p are of appropriate dimensions with elements which (for the remainder of this chapter) are assumed to be either known or computed on-line. Suppose also that this plant is to be controlled as per the scheme of Figure 9.1. Then, without loss of generality, the following transfer function description of the controller for the plant of (9.1) is assumed

$$\frac{U(z)}{E(z)} = H(z^{-1}) = \frac{\sum_{i=1}^{n} a_i z^{-i}}{1 - \sum_{i=1}^{n} b_i z^{-i}} \tag{9.3}$$

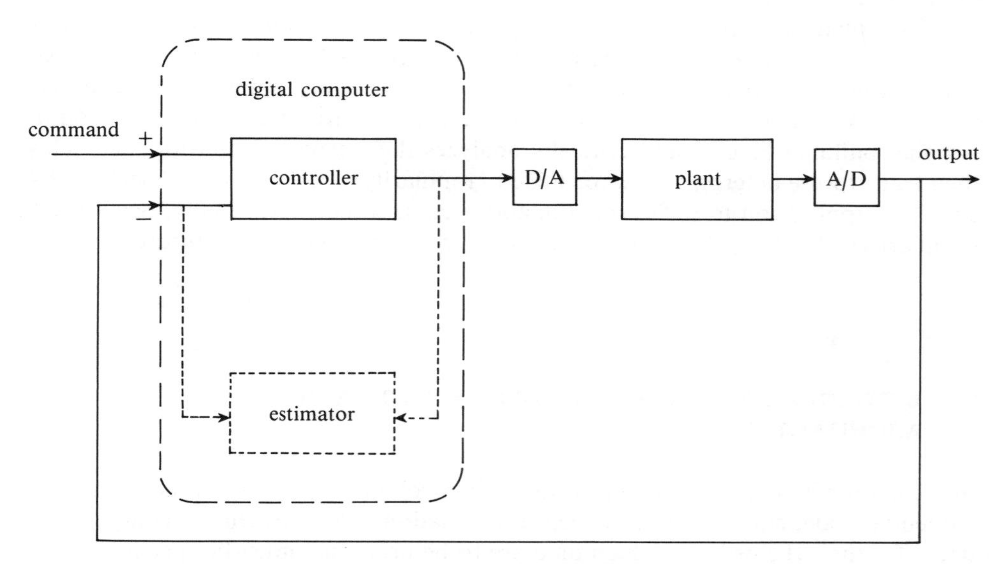

Figure 9.1 Block diagram of a typical digital feedback control scheme

or, in difference equation terms,

$$u(k+1) = a_1 e(k) + a_2 e(k-1) + \dots + a_n e(k-n+1) \\ + b_1 u(k) + b_2 u(k-1) + \dots + b_n u(k-n+1) \quad (9.4)$$

where

$$e(k) = r(k) - y(k) \quad (9.5)$$

and $r(k)$ is the command or reference signal.

In this context z^{-1} denotes either the argument of the z transform or the shift operator of a one cycle delay as appropriate. The controller structure used here includes many well-known choices as special cases. Further, in the case of adaptive/self-tuning control schemes, treated in detail in Section 9.4.3, it is assumed that the identification/estimation phase is completed using any one of the numerous well-known techniques.

Consider now real-time computation of the controller output. Then it follows immediately (see Li, 1990) that this is, in effect, equivalent to applying an infinite impulse response (IIR) or auto-regressive moving average (ARMA) filter. Here, however, the filter is embedded within the feedback loop of the overall control scheme.

Serial implementation of this system would require $2n$ multiplications and

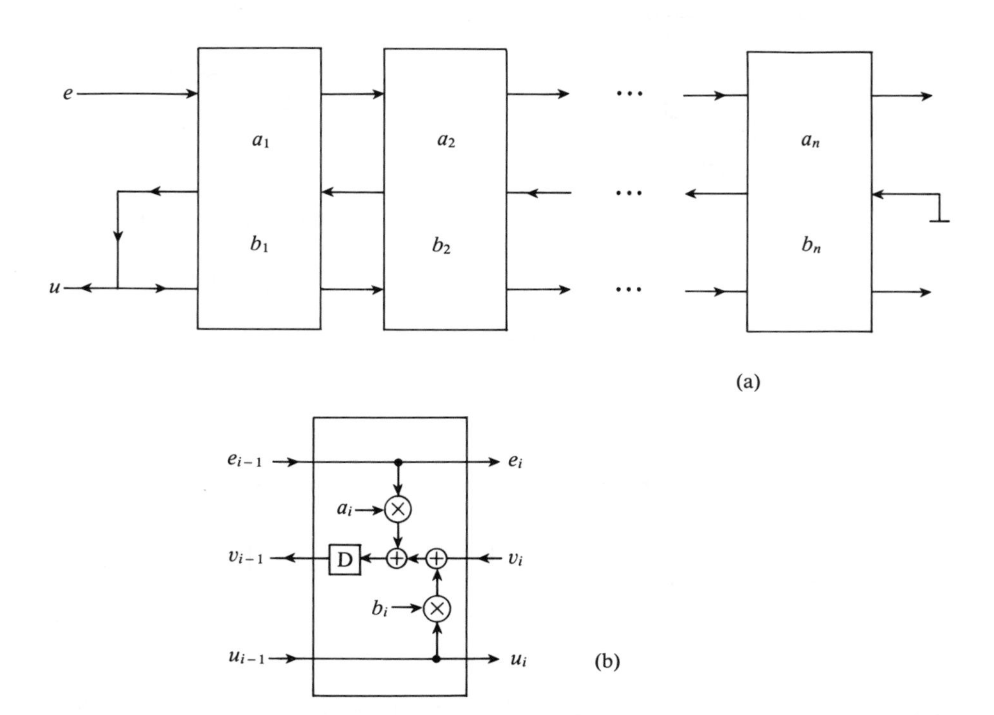

Figure 9.2 Type 1 semi-systolic IIR filter based on its reversed computation graph (RCG): (a) array configuration; (b) cell configuration

$(2n - 1)$ additions. Further, it is clear that a totally serial implementation of (9.4) could require the processor to fetch and decode the data plus instructions for every multiplication or addition and store the intermediate results. Hence it could be subject to program overheads and inefficient software management. Further, use of a global shared memory system would heavily restrict the processing speed due to limited processor/memory communication bandwidth.

High arithmetic operation speed can, however, be achieved by partitioning the computation into several subtasks suitable for basic multiplication plus accumulation units. Further, a solution to this communications bottleneck problem is to use processors, such as transputers, which have their own (or local) memories and direct interprocessor communication channels. Using this approach, excessive memory accessing can be avoided by using parallel and pipelined implementations which make use of the delay operations inherent in the transfer function.

Candidate architectures for the computation of (9.4) which satisfy these requirements are the semi-systolic architectures developed in Chapter 1. These are redrawn here as Figures 9.2 and 9.3 respectively, and note again that they have been

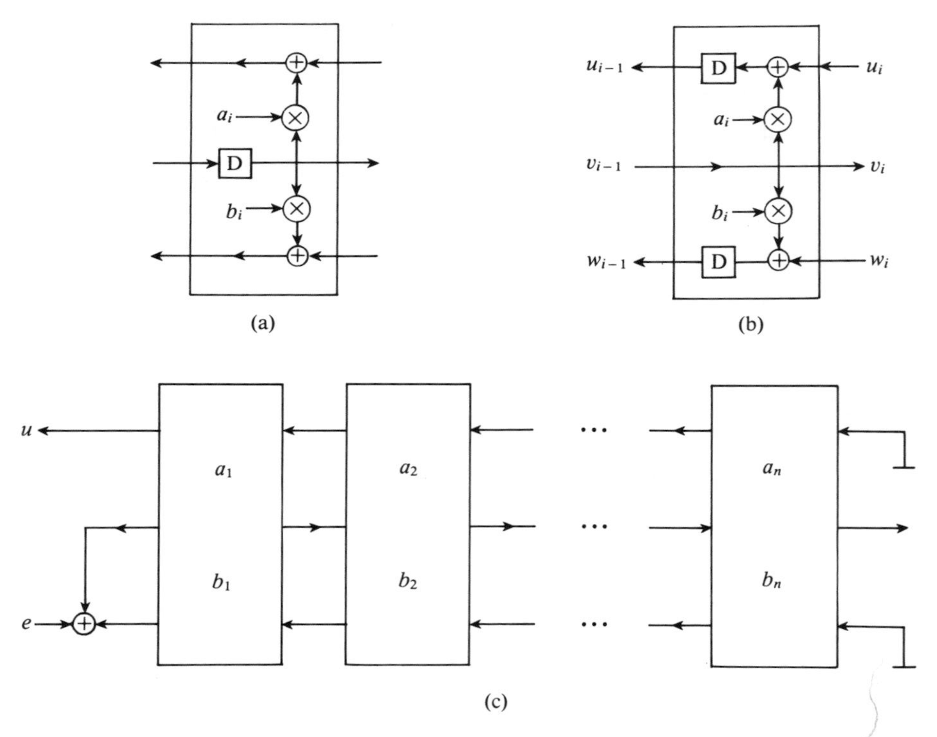

Figure 9.3 Type 2 semi-systolic array: (a) RCG of the cell of Figure 9.2(b); (b) cell configuration and (c) array architecture

derived from the canonical signal flow graph of the underlying IIR filter structure and its reversed computation graphs. Note also that these semi-systolic arrays are pipelined for word-level multiplication plus accumulation operations and match the computational requirements of (9.3) with very fine and balanced partitions.

9.3 TRANSPUTER BASED COARSE-GRAIN ARCHITECTURES

The transputer is now well established and, after a very brief overview of the essential basics, this section considers candidate networking architectures constructed solely from transputers to provide increased speed. Hence mappings of architectures based on multiplication plus accumulation elements will be considered. An alternative using transputer supervised IMS A100 parallel processor based architectures is considered in the next section with the objective of improved efficiency and performance.

9.3.1 Transputer Basics

In effect, all members of the transputer family share the same architecture, combining on one silicon chip the processor, local memory, external memory interface, intertransputer links, internal interrupt and timers for real-time processing, and system services (including a hardware scheduler for concurrent programs). Complete details can, for example, be found in manuals produced by Inmos (1985). The transputer operates within minimum local data transmission time due to single-chip VLSI integration and can therefore be used as a building block uniprocessor of a concurrent system. These components define a family of programmable systems with a powerful processing capability when networked with transputers.

The ad hoc transputer language Occam is high level, makes programming tasks relatively easy, and is also the lowest code which transputers understand. Further, the instruction set permits efficient implementation of high-level coding and provides direct support for the Occam model of concurrency. Overall, the system provides a very rich matched mapping capability between hardware architectures and software modules.

The first release, the IMST414, is a 32-bit architecture with excellent uniprocessor performance including 10 MIPS (million instructions per second) performance, 20 MHz data rate for on-chip RAM, up to 20 Mbits s^{-1} serial links, and the capability to directly address 4 Gbytes at 26.6 Mbytes s^{-1}. The IMST800 and the recently announced T9000 are architecturally similar to the T414 but offer a much higher peak performance of up to 200 MIPS and an on-chip 64-bit floating point unit (FPU). Further, the FPU and arithmetic logic unit (ALU) of the T9000

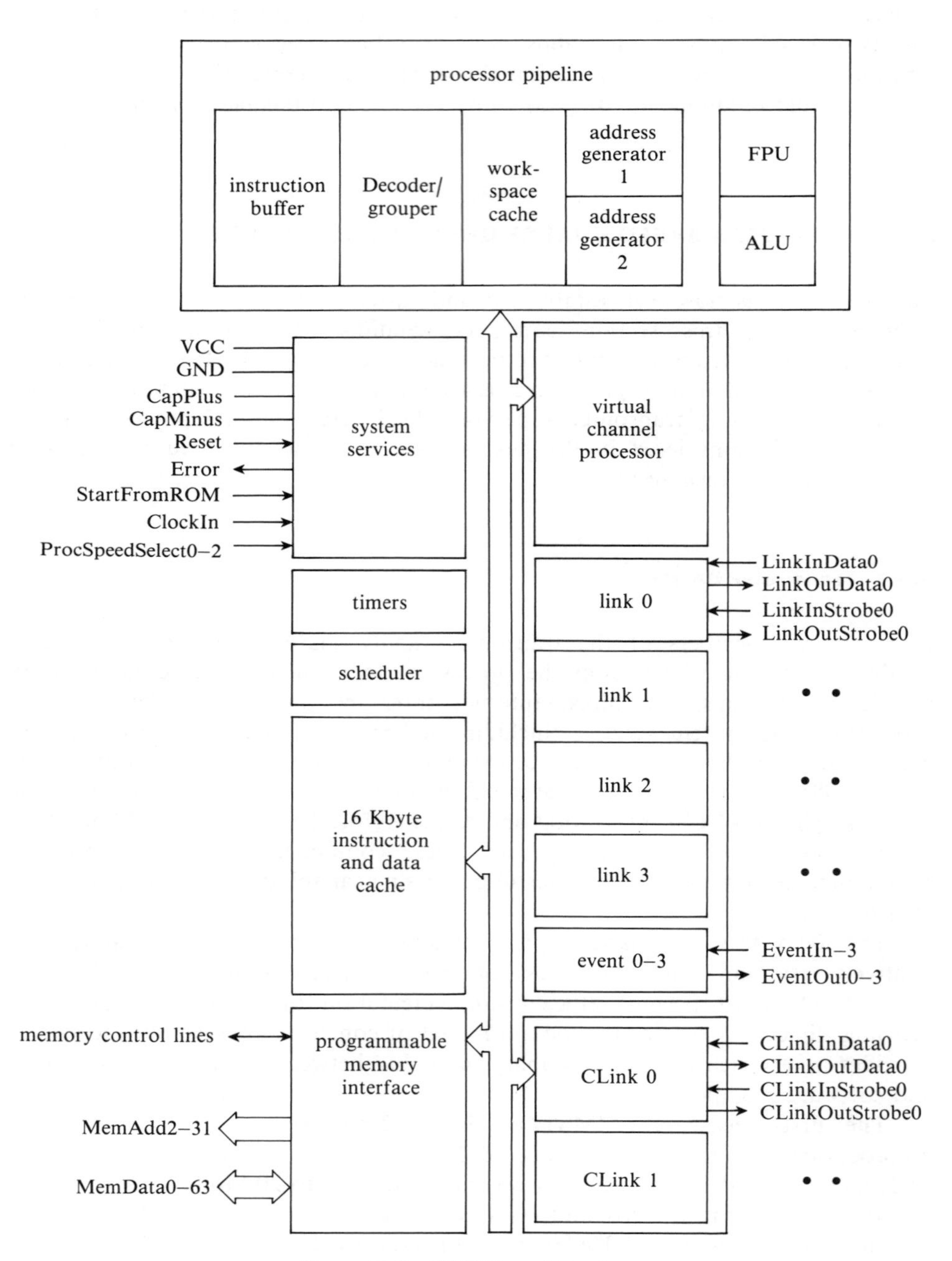

Figure 9.4 T9000 architecture

are pre-pipelined with support from a workspace cache and hence higher speed and processor efficiency.

In order to support an effective and high-speed local memory access, the T9000 is integrated with an upgraded 16 Kbyte on-chip data cache. Further, this architecture also includes a virtual channel processor (VCP) to speed up off-chip communications and to multiplex software channels on to each of its hardware links. Overall this gives a better balanced concurrent networking. Note also that, for embedded real-time applications, this architecture has been designed with an on-board kernel which provides hardware support for multitasking, multiple interrupts and fast interrupt response, and context switching. Figure 9.4 shows a schematic block diagram of this device and a more in depth treatment of the aspects discussed briefly here can be found, for example, in Jones (1991).

9.3.2 A Synchronous Architecture

It is a straightforward exercise to reconfigure the semi-systolic architectures of Figures 9.2 and 9.3 into transputer based concurrent networks. For example, each cell in these arrays could be executed by a T800 with an architecture similar to that shown in Figure 9.4 for the T9000. In the case of the Type 1 array of Figure 9.2, this results in the simply connected linear network for the computation of (9.3)–(9.4) shown in Figure 9.5.

The PC in this arrangement is only used for system development and setting up the required reference and coefficient values and is not involved in the computations. A host T800 transputer is used to supervise analog to digital (A/D) and digital to analog (D/A) conversions and also performs the data processing, the results of which are broadcast into every cell of the array which is also implemented by a T800 transputer. Once the array receives input data, each transputer runs in parallel on its own program and data and this complete operation is begun by the initialization

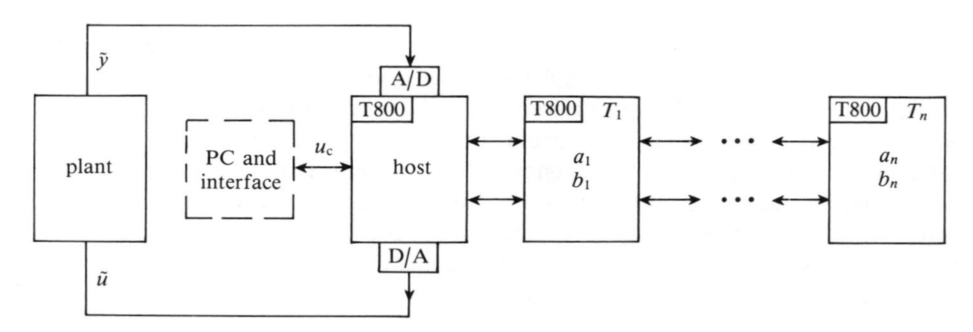

Figure 9.5 Transputer based multiple instruction multiple data (MIMD) architecture

procedure which sets the state output to zero and also signals that these values have been output.

At this stage it is also necessary to set the controller coefficients. Once these are in place, each cell requests the data, which are broadcast from the host serially, and the state output from its right-hand neighbour. A simplified Occam structure which demonstrates the operation of this system is listed in Appendix A.

9.3.3 A Data-driven Architecture

The simple algorithm given above shows how parallelism in the transputer array is realized but only considers the case of synchronized implementation. Here data must be transmitted serially to every cell of the linear array and computations can only begin when this is completed. Hence such an array will suffer from the serial communication bottleneck within the network when the transputer processing elements (PEs) compete for data transmission from and to the host. Note, however, that it is also implied in the coding that synchronization is not necessary, since the operation of individual transputers can be triggered by arrival of data, or equivalently, the array can be programmed in a data-driven mode and hence this communication bottleneck can be avoided. For example, the execution in the ith processor (transputer), denoted T_i, can be arranged to start when the data from its neighbours T_{i-1} and T_{i+1} arrive. This is an advantage of the asynchronous mode over the synchronous one. The output to the D/A converter, however, must still be synchronized for normal sampled-data systems. In the data-driven mode, programming of each transputer could be made easier by using the 'Pri PAR' process of Occam. This is, however, difficult in implementations using T414 or T800 transputers and may be easier if a T9000 is used instead. (A more costly option at current prices!)

In a similar manner to the Type 1 semi-systolic array of Chapter 1 (Figure 1.4), this realization has a theoretical cycle time (for work-level operations) of

$$T_w = T_m + 2T_a + T_b \tag{9.6}$$

where T_m and T_a are the word-level multiplication and addition times respectively. Note that here the data transmission time T_b will, in general, be longer than that for the semi-systolic single chip implementations developed in Chapter 1.

This implementation has the desired system latency of one cell cycle. The sample Occam code also shows that many procedures which are represented by the 'PAR' statements have potential concurrency. Only those which are implemented by corresponding individual transputers are, however, actually run in parallel in this architecture. If additional transputers are used in the implementation, each of these 'PAR' processes could run in parallel. Note, however, that this case still requires a minimum amount of serialism unless these processors are fully pipelined and that the total cost would be increased.

The Type 2 semi-systolic array (Figure 1.5) can also be implemented by an array of transputers configured in a similar manner to that described above. Here, however, the host transputer must also perform the additions of the outside adder, i.e. calculate and broadcast (by suitable software modifications) $r - y + w_0$ instead of $r - y$. This realization therefore has the same structure as that shown in Figure 9.5 and has similar performance in terms of speed and response times. Hence further details are omitted here.

In this realization of purely transputer based architectures, $n + 1$ transputers are required. Further, the timing control of the network is complex and, despite the large number of processors used, it is still difficult to exploit the potential concurrency in the controller computation fully. This leads to the conclusion that networking this type of general-purpose processor to implement a simple PE based regular architecture is very much less than ideal.

9.4 TRANSPUTOR BASED HETEROGENEOUS ARCHITECTURES

A more system-specific device in the Inmos parallel processor range is the IMS parallel DSP element termed the IMS A100 processor array (Inmos, 1989). This element, or device, is, in effect, a transversal filter with a 32-stage multiply–accumulation array as shown in Figure 9.6. It is capable of delivering 80 MIPS for 16-bit multiplication and 32-bit addition and of performing subtasks in parallel on its 32 processing units.

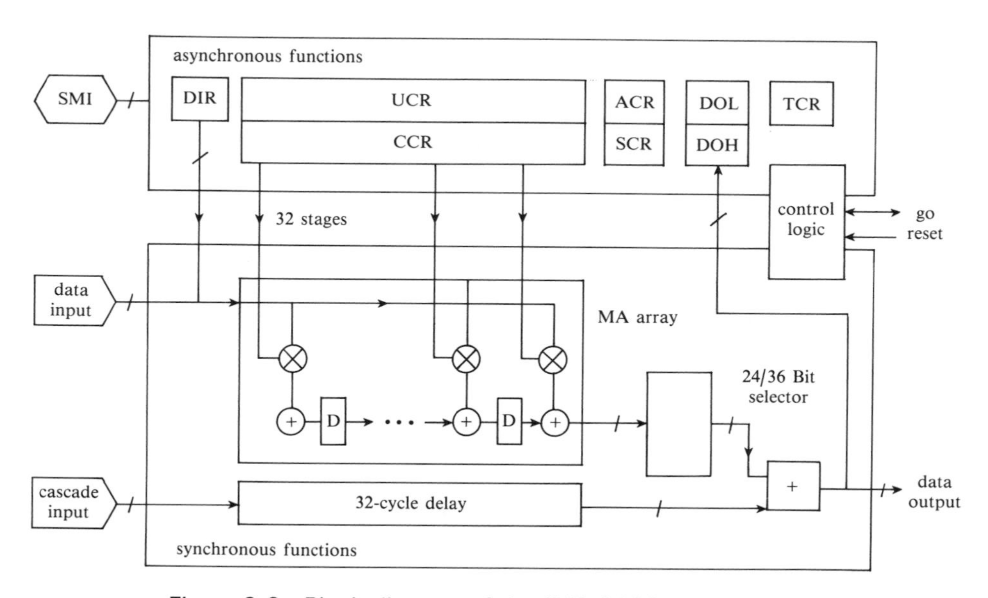

Figure 9.6 Block diagram of the IMS A100 processing core

Each processing cycle is triggered, or 'clocked', by a 'GO' signal from external circuitry, which, for example, can be generated by a host transputer. Further, if a particular application requires more stages (for example, 'high'-order controllers) the A100 can be used as a building block, i.e. several can be cascaded together without degradation in throughput. Since it is a special-purpose device, it is always supervised by a general-purpose processor, such as transputers communicating data through their links, in order to provide greater flexibility.

Another important feature of the A100 is that there are two coefficient registers, namely, the current coefficient register (CCR) and the update coefficient register (UCR). These enable the filter coefficients to be program-controlled and updated on each 'GO' cycle by the host transputer during operation. Hence the A100 can be applied to applications such as adaptive/self tuning control where the implemented controller coefficients must be updated on-line. Section 9.4.3 gives a full treatment of this area. Note also that, in architectural terms, this device performs as a fixed structure semi-systolic FIR (finite impulse response) filter and requires only a very simple supervisory program, with a consequent reduction in instruction time and program overhead.

As shown in Figure 9.6, the internal architecture of the IMS A100 is that of a FIR filter. The controller computation of (9.4), however, has an IIR filter structure. Hence structural modifications are required in order to configure the A100 to implement this controller. The key to this is the fact that a first-order section of an IIR filter can be regarded as constructed from two sections of a FIR filter. Hence two stages of a single A100 processor or, alternatively, two such processors, are employed.

The progammability of the A100 data and coefficients offers a degree of flexibility sufficient to meet the requirements of (9.4). Instead of the n transputers used in the purely transputer based design of Section 9.3.2, an IMS A100 and transputer based heterogeneous architecture offers improved performance at reduced cost with lower space requirements for real-time on-line operations. This approach is considered in detail in the next two subsections.

9.4.1 Type 1 Architecture

The starting point for this section is the Type 1 semi-systolic architecture whose cells are shown in Figure 9.2. In each cell of this array, two multiplications and two additions are performed. Hence two stages of an A100 processing core will be used to realize such a first-order section and clearly the cell arithmetic operations must be correctly timed.

To detail these timings, first note that the feedback coefficients, b_i, are initially stored in the CCR of the A100 and the feedforward coefficients, a_i, in the UCR. Then, on the next 'GO' cycle the a_i coefficients in the UCR are swapped with the b_i coefficients in the CCR and passed into the processing array as illustrated in

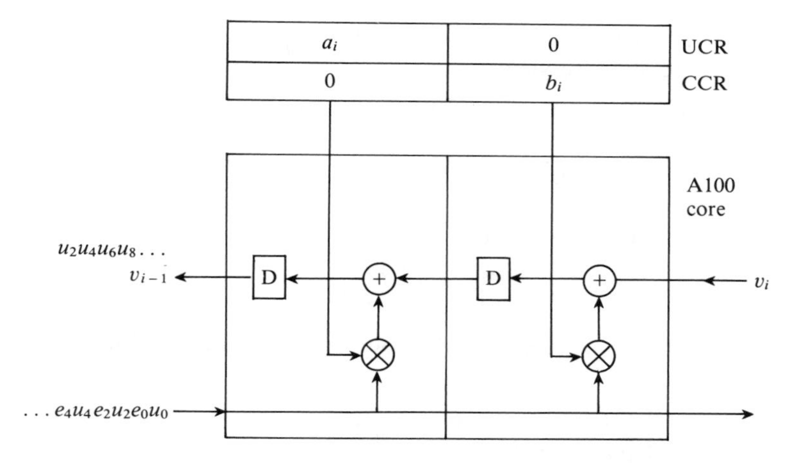

Figure 9.7 Type 1 cell using two stages of A100 processing core

Figure 9.7. Further, the contents of these two coefficient banks are continuously swapped on each cycle. In order to obtain the correct timing, and hence a correct transfer function, the coefficients are interleaved with zeros.

The feedback and feedforward coefficients here are alternately positioned. Hence the current computed control signal, $u(k)$, must be delayed one cycle and then interleaved with the input error data $e(k)$ (including the initialized state values). The time period of this one cycle delay is also used to compute $e(k)$ which is required by (9.4). Similarly the input data to the cell or filter section must also be interleaved, i.e. the computed control signal data, $u(k)$, and the error signal $e(k)$ are interleaved with each other as shown in Figure 9.7.

This latter requirement means that the plant output, $y(k)$, must be sampled and $u(k)$ collected once every two 'GO' cycles. Hence, in comparison to the original Type 1 cell shown in Figure 9.2, this realization adds an additional unit into each cell. Further, the results of the multiplications in each stage, which are associated with data $u(k)$ and $e(k)$, are added together at the end of each even 'GO' cycle.

In order to guarantee correct timing and an invariant transfer function it is clearly necessary to verify the implemented transfer function. This problem also arose in Chapter 1 where methods based on linear time invariant systems were developed. These methods can also be applied (Li, 1990) to the architecture developed here but, since it is inherently time varying, this verification exercise must be applied to two complete 'GO' cycles. Here only a summary of the main results is given, with complete details in Li (1990).

First note that the data corresponding to $e(k)$ are delayed by one cycle before input to the array. Hence, in frequency domain terms a single cell or filter section is governed by

$$E^1 = z^{-1}E \tag{9.7}$$

where E^1 denotes the state of the error signal delayed by one 'GO' cycle. Applying this to the cell architecture yields

$$\begin{bmatrix} E^1 \\ V_{i-1} \\ U \end{bmatrix} = \begin{bmatrix} 1 & 0 & 0 \\ a_i z^{-1} & z^{-1} & 0 \\ 0 & 0 & 1 \end{bmatrix} \begin{bmatrix} 1 & 0 & 0 \\ 0 & z^{-1} & b_i z^{-1} \\ 0 & 0 & 1 \end{bmatrix} \begin{bmatrix} E' \\ V_i \\ U \end{bmatrix}$$

$$= \begin{bmatrix} 1 & 0 & 0 \\ a_i z^{-1} & z^{-2} & b_i z^{-2} \\ 0 & 0 & 1 \end{bmatrix} \begin{bmatrix} E' \\ V_i \\ U \end{bmatrix} \tag{9.8}$$

Cascading n of these cells to obtain the complete array description and using (9.7) yields

$$\begin{bmatrix} E \\ U \\ U \end{bmatrix} = \prod_{i=1}^{n} \begin{bmatrix} 1 & 0 & 0 \\ a_i z^{-2} & z^{-2} & \mathrm{b}_i z^{-2} \\ 0 & 0 & 1 \end{bmatrix} \begin{bmatrix} E \\ 0 \\ U \end{bmatrix} \tag{9.9a}$$

$$= \begin{bmatrix} 1 & 0 & 0 \\ \sum_{i=1}^{n} a_i \zeta^{-i} & \zeta^{-n} & \sum_{i=1}^{n} b_i \zeta^{-i} \\ 0 & 0 & 1 \end{bmatrix} \begin{bmatrix} E \\ 0 \\ U \end{bmatrix} \tag{9.9b}$$

or

$$U = \sum_{i=1}^{n} a_i \zeta^{-i} E + \sum_{i=1}^{n} b_i \zeta^{-1} U \tag{9.10}$$

where $\zeta^{-1} = z^{-2}$. This last equation is, in effect, equivalent to (9.3), provided the plant output is sampled and the controller output collected at every even cycle. Further (structurally similar) time domain analysis of this array yields

$$v_{i-1}(2k') = v_i(2k' - 2) + a_i e(2k' - 2) + b_i u(2k' - 2) \tag{9.11}$$

which converges (by iteration) to (9.4) provided $2k' = k$, where k' denotes the new discrete time-scale for the interleaved data streams.

Two transputers are employed in this implementation where the host T800 is used to calculate $e(k')$ and supervise the A/D and D/A operations and communications, with the development system residing in the IBM PC. A 'slave' transputer – the T212 (16-bit) – is especially employed to broadcast $e(k')$ and the alternately interleaved $u(k')$ data into the A100 array on each 'GO' cycle, i.e. half the new sampling period. This 'slave' also supervises the continuous swapping of the two coefficient banks.

Overall, this system provides a high data broadcasting rate and hence a high throughput rate for the control system. Note also that the continuous bank swapping mode of the A100 is set in advance. Further, one A100 realizes at maximum a 16-order controller and higher orders (if required) can be provided by

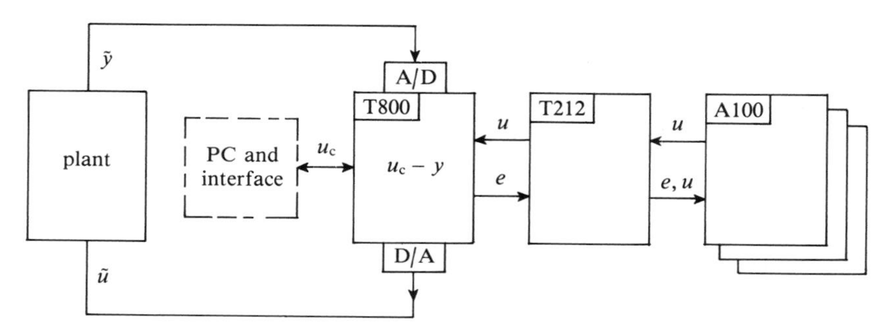

Figure 9.8 Types 1 and 2 mixed architectures using the transputer/A100

cascading several A100s. The final architecture is shown in Figure 9.8 and termed a Type 1A mixed (transputer/A100) architecture.

An Occam structure to demonstrate the concurrent operation of this implementation is given in Appendix B. Note that here the minimum period time of a 'GO' signal is bounded by the longest 'WHILE' process. This, in turn, determines the half sampling period time which (in theory) is

$$T_w = 2(T_m + T_a + T_b) \tag{9.12}$$

where T_m and T_a are defined as before and the data transmission time T_b includes the program overhead time used in the implementation.

9.4.2 Type 2 Architecture

Mixed transputer/A100 based coarse-grain architectures can also be developed from the Type 2 semi-systolic array of Figure 9.3. Here, however, using two stages of the same A100 processing core to realize on first-order section is difficult since two computed results, $u_{i-1}(k)$ and $w_{i-1}(k)$, need to be collected and their initial values, $u_i(k)$ and $w_i(k)$, need to be added. Hence realizations based on this type of architecture require two complete IMS A100 components.

Returning to Figure 9.3, it is easily seen that if the $v(k)$ data bus is separated into two connected buses then separate chips can be used to realize the upper and lower halves of the array as shown in Figure 9.9. The accumulation of the outside adder can, however, be completed by a T800. Consequently this realization has the same structure as the Type 1 architecture of Figure 9.8 and is termed a Type 2 mixed architecture.

In this case both $u(k)$ and $w_0(k)$ must be broadcast to the A100 arrays. Further, the coefficients swapping facility of this array is not employed for non-adaptive applications. Note also that the throughput rate of this architecture is twice that of the Type 1 architecture. Further, this implementation can realize a controller with

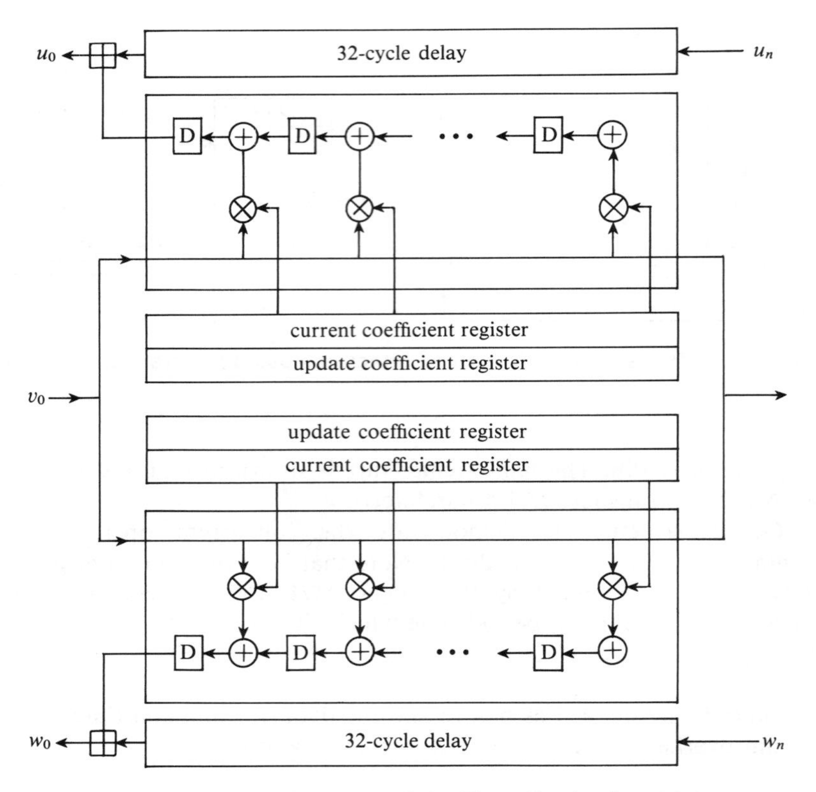

Figure 9.9 Processing core of the Type 2 mixed architecture

order up to 32 and (if required) higher-order controllers can be realized by cascading several A100 chips without degrading performance.

An Occam structure for this implementation is shown in Appendix C. Note also that the minimum period time of the 'GO' signal is again bounded by the longest 'WHILE' process which (in theory) is given by (9.5).

9.4.3 Application to Self-tuning/Adaptive Control

In this section it is shown that, with appropriate modifications to the software, the architectures developed in Sections 9.4.1 and 9.4.2 can be used to implement self-tuning/adaptive control schemes. In these cases, the control task, i.e. computation of the controller output, consists of two major subtasks. These are parameter estimation (or reference model based computation) and control signal provision (Goodwin and Sin, 1984; Åström and Wittenmark, 1989).

In this work, it is simply assumed that the estimation phase is completed using any one of the numerous well-established methods. Similarly, the control law used can be designed using any one of the equally well-known methods. To provide a basis for this section, the case of a real-time pole placement self-tuning control scheme is considered with the plant described by

$$Y(k) = \frac{C(z^{-1})}{D(z^{-1})} u(k) + w(k) \tag{9.13}$$

where

$$C(z^{-1}) = \sum_{i=1}^{n} c_i z^{-i}, D(z^{-1}) = 1 - \sum_{i=1}^{n} d_i z^{-i} \tag{9.14}$$

and $w(k)$ is an uncorrelated random disturbance sequence with zero mean.

Note that the analysis that follows generalizes in a natural manner (see Li *et al.*, 1991) to the case of coloured noise. In this arrangement, z^{-1} denotes the shift operator of a one cycle delay. Further, the coefficients of the plant model, c_i and d_i, are estimated on-line and then used to calculate the coefficients of the regulator which is assumed to have the IIR filter structure:

$$\frac{A(z^{-1})}{B(z^{-1})} = \frac{\sum_{i=1}^{n} a_i z^{-i}}{1 - \sum_{i=1}^{2n} b_i z^{-i}} \tag{9.15}$$

Suppose also that desired closed-loop characteristic polynomial is defined by

$$T(z^{-1}) = 1 + \sum_{i=1}^{2n} t_i z^{-i} \tag{9.16}$$

and that the polynomials $A(z^{-1})$ and $B(z^{-1})$ satisfy

$$D(z^{-1})B(z^{-1}) + C(z^{-1})A(z^{-1}) = T(z^{-1}) \tag{9.17}$$

Then, with $r(k)$ defined as in (9.5), the closed-loop system is

$$y(k) = \frac{C(z^{-1})A(z^{-1})}{T(z^{-1})} r(k) + \frac{D(z^{-1})B(z^{-1})}{T(z^{-1})} w(k) \tag{9.18}$$

At this stage, return (briefly) to the estimation of a_i and b_i. Here, the algorithm used is based on minimizing the output variance. Further, the so-called direct method (Åström and Wittenmark, 1989) is preferred, because the alternative indirect method would take longer to complete (Li, 1990).

Consider now the implementation of this scheme in real-time. The computations are performed by the supervisory transputer – here a T800 (or an array of these). The results are then immediately sent to the A100 chips through the associated T212 via the links and Occam channels. This mode of operation is based on the assumption that the speed of updating the plant, and hence the regulator, parameters is 'much lower' than the plant dynamics (Åström and Wittenmark, 1989) and hence these computations are not necessary every sampling period.

The effect of this assumption is that the load is eased on the T800. Further, it offers a match of the two computation phases, i.e. parameters and control signal, which now operate at an asynchronous speed. Suppose, however, that it is required to operate the parameter updating at a faster speed, i.e. the above assumption is violated. Then a number of T800s can be networked together subject to appropriate task partitioning and use of the T800 host to compute the error signal $r(k) - y(k)$. Both the Type 1 and 2 architectures of Sections 9.4.1 and 9.4.2 respectively can be used here, but note that the Type 2 architecture doubles the speed. A block diagram of the resulting scheme is shown in Figure 9.10.

Simulation studies have been undertaken for the implementation based on the Type 1 architecture. In these studies (Niessen, 1990; Li *et al.*, 1991) the coefficients a_i and b_i were written to the CCR and UCR banks at even (including the starting time point) and odd 'GO' cycles respectively once they are computed. Further, the CCR and UCR banks were continuously swapped in the cycles following until the next update of the coefficients was available.

These simulations yielded a throughput rate (and hence sampling rate) of 47 kHz for 16-bit data and coefficients which is judged to be sufficient for most applications. This implementation is based on the IMS B009 board, which, in effect, is an IBM PC board containing four cascaded IMS A100 components controlled by an on-board IMS T272 transputer as shown in Figure 9.11 (Inmos, 1989). Hence one such board can implement a controller of up to 32nd order, which should be sufficient for most applications. Further, an IMS B404-8 board containing one 25 MHz T800 and IBM RAM is plugged onto the B909 board for the general-purpose processing detailed above.

An Occam structure for this realization is given in Appendix D. Note also that similar architectures can be used to realize other self-tuning and adaptive control schemes, with the software (routinely) modified to meet any particular requirements of these schemes. This is because the IMS A100 is always ready to accept parameter updates and can be controlled by software (Inmos, 1989). Niessen (1990) and Li (1990) detail these and numerous other related aspects.

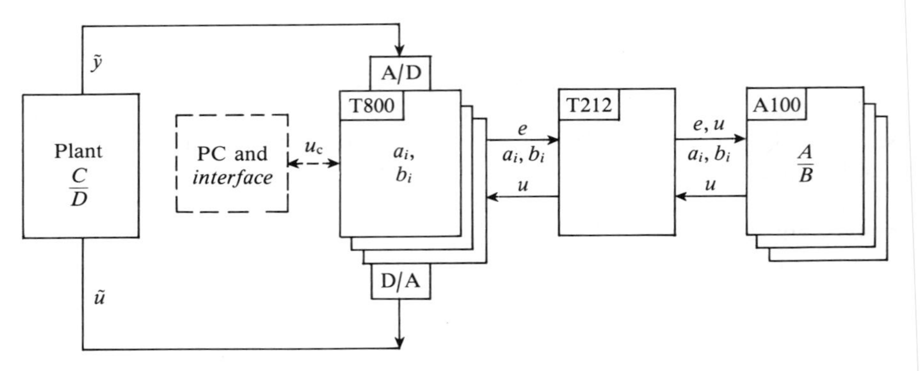

Figure 9.10 Types 1 and 2 mixed architectures for self-tuning/adaptive control

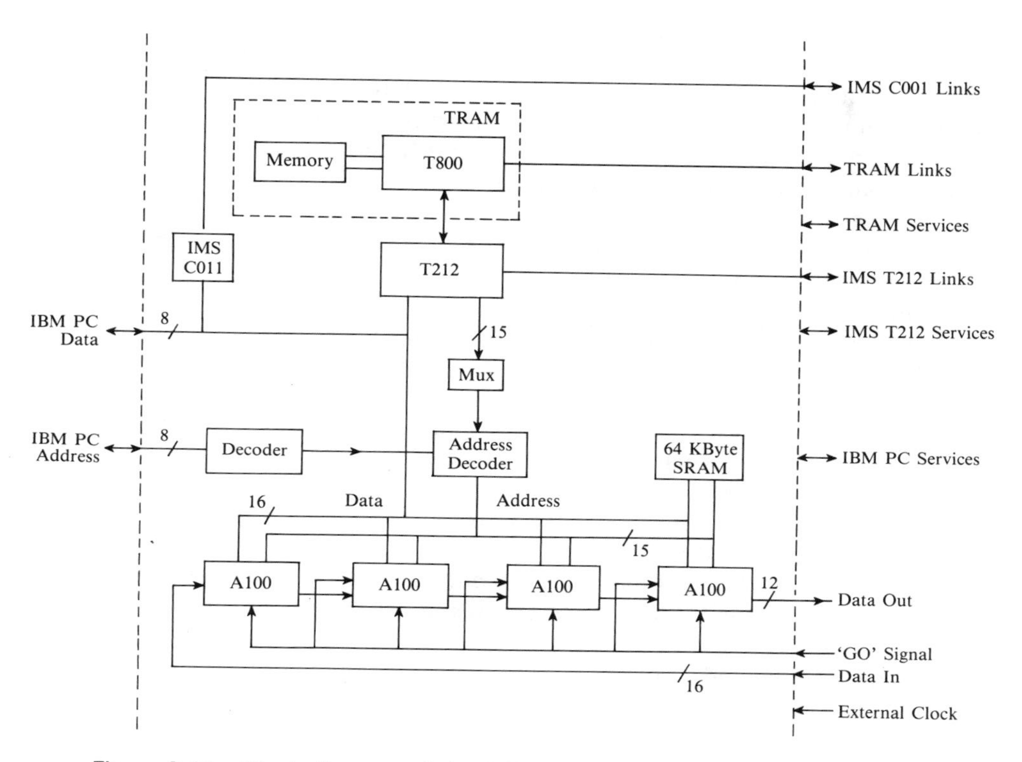

Figure 9.11 Block diagram of the IMS B009 and B404 boards employed

9.5 CONCLUSIONS

The effective starting point for this chapter has been the shortcomings revealed in other work in terms of the transputer's ability to deal with the special demands of real-time control applications. This is because the communications overhead is too dominant or, alternatively, the task size is not sufficiently large. As a result, the architecture granularity (the ratio of computation power to interprocessor communications overheads) is not appropriately matched.

Other work has used performance-related software to study this problem and one of its main conclusions is that a granularity reduction of one order of magnitude should be more suitable. This chapter has investigated a solution to this general problem based on merging, or fusing, the finer granularity of parallel DSPs with the transputer's ability to handle irregular computations and manage parallel operations. The basis of the work reported here has been transputers and the IMS A100 DSP chip and this is one version of the so-called heterogeneous system approach to parallel control and simulation problems.

Basically, two transputer-supervised semi-systolic architectures have been developed on the A100 array, starting from some of the work reported in Chapter 1. Further, it has been shown that these new architectures can be easily modified for application to self-tuning/adaptive control schemes. In this mode of operation, the control signal computation is executed by the A100 devices and data acquisition, updating and broadcasting is undertaken by two transputers. Some predictions of expected performance have also been given.

In practical applications of these architectures, the speed of the data streams will be very high and consequently the effects of this high rate clocking must be taken into consideration. For example, most A/D converter circuits are based on a comparison with a digital counter and would be considered 'slow' in comparison to the clock speed and hence a 'long' response time. Consequently, parallel A/D convertors should be used in order to reduce this time. If, however, this is still deemed 'too long', then the conversion time must be compensated for, particularly in cases where the performance specifications are high in terms of, for example, the accuracy of the feedback signal or the return time for the plant input signal. Methods for achieving this based on state augmentation or predictive control techniques can be found in Li (1990).

The work reported here has clearly demonstrated the basic feasibility of the heterogeneous approach. Much work remains to be done, however, before its full potential can be assessed. Areas include (see Li, 1990, for a complete list) benchmark testing and prototype development, including the advantages and disadvantages of replacing the A100 with other parallel chips such as the TMS and Motorola.

APPENDIX A

9.A1 Main Process

```
PROC Data.Driven.Array ()
    #USE userio
    ...{{{declarations}}}
    CHAN Input, Output, T0.to.T1,
    T1.to.T0, ..., Ti.to.Ti+1, Ti+1.to.Ti, ...
    --(i=1,2,...,n-1)
    SEQ
        ...{{{Initialization}}}
                        ...set state values zero and
        output them and set controller coefficients
        Input?r
        WHILE NOT End
            SEQ
                Host (Input, Output, T0.to.T1, T1.to.T0)
```

```
Broadcast (T0.to.T1,
T1.to.T0, ..., Ti.to.Ti+1, Ti+1.to.Ti, ...)
PAR
   T1(T0.to.T1, T1.to.T0, T1.to.T2,
   T2.to.T1)
   ......
   Tn(...)
Output!v0                          --then D/A
...{{{Await until elapsed.time=Tw}}}
```

9.A2 Subprocesses

```
PROC Host
     SEQ
          T1.to.T0?vo
          PAR
               T0.to.T1!v0
               SEQ
                    Input?y                      --A/D first
                    e := r-y
                    T0.to.T1!e

PROC Broadcast
     SEQ                                         --(i=1,2,n-1)
          ...
          PAR
               Ti+1.to.Ti?vi
               SEQ
                    Ti-1.to.Ti?v0
                    PAR
                         Ti.to.Ti+1!v0
                         Ti-1.to.Ti?e
                    Ti.to.Ti+1!e
          ...
          PAR
               vN := 0
               SEQ
                    Tn-1.to.Tn?v0
                    Tn-1.to.Tn?e

PROC Ti                                          --(i=1, 2, ..., n)
     SEQ
          PAR
                    vi := vi+bi*v0
                    vi-1 := ai*e
          vi-1 := vi-1+vi
          Ti.to.Ti-1!vi-1
```

APPENDIX B

9.B1 Main Process

```
PROC Type.1.Array ()
     #USE userio
     ...{{{declarations}}}
     CHAN Input, Output, To.T212, From.T212, To.A100,
     From.A100
     SEQ
          ...{{{Initialization}}}   --set state values
          and CCR=0, bᵢ; and UCR=aᵢ, 0
          Input?r
          WHILE NOT End
               PAR
                   T800 (Input, Output, To.T212,
                   From.T212)
                   T212 (To.T212, From.T212, To.A100,
                   From.A100)
```

9.B2 Subprocesses

```
PROC T800 (Input, Output, To.T212, From.T212)
     ...{{{declarations}}}
     SEQ
          PAR
               From.T212?v₀
               Input?y   --after A/D
          Output!v₀   --and D/A
          ...{{{Await until elapsed.time=Tw}}}
          e := r-y
          To.T212!e
          ...{{{Await until elapsed.time=Tw}}}

PROC T212 (To.T212, From.T212, To.A100, From.A100)
     ...{{{declarations}}}
     SEQ
          ...{{{set 'GO' high}}}
          From.A100?v₀
          PAR
               T.A100!v₀
               From.T212!v₀
          ...{{{Set 'GO' low, signal to swap CCR and UCR
          and await until elapsed.time=Tw
then set 'GO' high}}}
          To.A100!e
          ...{{{Set 'GO' low, signal to swap CCR and UCR
          and await until elapsed.time=Tw}}}
```

APPENDIX C

9.C1 Main Process

```
PROC Type.2.Array ()
     #USE userio
     ...{{{declarations}}}
             CHAN Input, Output, To.T212, From.T212,
             To.A100A, From.A100A, To.A100B,
             From.A100B
        SEQ
             ...{{{Initialization}}}  --set state
             values
             Input?r
             WHILE NOT End
                PAR
                   T800 (Input, Output, To.T212,
                   From.T212)
                   T212 (To.T212, From.T212 To.A100A,
                   From.A100A, To.A100B, From.A100B)
```

9.C2 Subprocesses

```
PROC T800 (Input, Output, To.T212, From.T212)
     ...{{{declarations}}}
     SEQ
        PAR
             SEQ
                From.T212?u0
                PAR
                   From.T212?w0
                   Output!u0   --and D/A
             SEQ
                Input?y   --after A/D
                v0 := r-y+w0
                To.T212!v0
                   ...{{{Await until
                   elapsed.time=Tw}}}

PROC T212 (To.T212, From.T212, To.A100A, From.A100A,
     To.A100B, From.A100B)
     ...{{{declarations}}}
     SEQ
        ...{{{Set 'GO' high}}}
        PAR
             SEQ
                From.A100A?u0
                From.T212!u0
             SEQ
                From.A100B?w0
                From.T212!w0
```

```
            SEQ
               To.T212?v0
               PAR
                  To.A100A!v0
                  To.A100A!v0
          ...{{{Set 'GO' low and await until
          elapsed.time=Tw}}}
```

APPENDIX D

9.D1 Main Process

```
PROC Type.1.Self.Tuning.Array ()
     #USE userio
     ...{{{declarations}}}
     CHAN Input, Output, To.T212, From.T212, To.A100,
     From.A100
     SEQ
          ...{{{Initialization}}}  --set state values,
          initial coefficients of ai and bi and CCR=0, bi;
          and UCR=ai, 0
          Input?r
          WHILE NOT End
               PAR
                  T800 (Input, Output, To.T212,
                  From.T212)
                  T212 (To.T212, From.T212, To.A100,
                  From.A100)
```

9.D2 Subprocesses

```
PROC T800 (Input, Output, To.T212, From.T212)
     ...{{{declarations}}}
     PAR
          SEQ
               PAR
                  From.T212?v0
                  Input?y   --after A/D
               Output!v0   --and D/A
               ...{{{Await until elasped.time=Tw}}}
               e := r-y
               To.T212!e
               ...{{{Await until elapsed.time=Tw}}}
          ...{{{Keep ai and bi unchanged and estimate new
          ai and bi
               SEQ
                  To.T212!bi
                  To.T212!ai}}}
```

```
PROC T212 (To.T212, From.T212, To.A100, From.A100)
    ...{{{declarations}}}
    SEQ
        SEQ k=0 FOR Ne/2-1
            ...{{{Set 'GO' high}}}
            From.A100?v0
            PAR
                To.A100!v0
                From.T212!v0
            ...{{{Set 'GO' low, signal to swap CCR and
            UCR and await until elapsed.time=Tw}}}
            ...{{{Set 'GO' high}}}
            To.A100!e
            If k=Ne/2-1
                SEQ
                    To.T212?bi
                    PAR
                        To.A100!bi   --set CCR=0, bi
                        To.T212?ai
                    To.A100!ai   --set UCR=ai, 0
            ...{{{set 'GO' low, signal to swap CCR and
            UCR and await until elapsed.time=Tw}}}
```

REFERENCES

ÅSTRÖM, K. J. and WITTENMARK, B. (1989) *Adaptive Control*, Reading, MA: Addison-Wesley.

FLEMING, P. J. and GARCIA NOCETTI, D. F. (1992) 'Real-time control and simulation performance analysis tools', in *Transputers for Real-Time Control* (eds G. W. Irwin and P. J. Fleming), Taunton: Wiley Research Studies Press. To appear.

GANESAN, S. (1991) 'A dual-DSP microprocessor system for real-time digital correlation', *Microprocessors and Microsystems*, **15**(7), 379–84.

GOODWIN, C. G. and SIN, K. S. (1984) *Adaptive Filtering, Prediction and Control*, Englewood Cliffs, NJ: Prentice Hall.

INMOS (1985) *IMS T414 Reference Manual*, Bristol: Inmos.

INMOS, (1989) *Digital Signal Processing Handbook*, Bristol: Inmos.

IRWIN, G. W. and FLEMING, P. J. (eds) (1992) *Transputers for Real-Time Control*, Taunton: Wiley Research Studies Press. To appear.

JONES, G. (1991) *IMS T9000 Transputer: Benchmark Performance*, Inmos Technical Note 72.

LI, Y. (1990) 'Concurrent architectures for real-time control', PhD Thesis, University of Strathclyde, Glasgow, UK.

LI, Y., NIESSEN, K. and ROGERS, E. (1991) 'Mapping systolic structures onto transputer/A100 based parallel processors for adaptive/self-tuning control', *Int. J. Control*, **54**(6), 1399–411.

NIESSEN, K. (1990) 'Transputer operated array processors for adaptive filtering and control', Diploma paper, University of Stuttgart, Germany.

REES, D. and WHITTING, P. (1988) 'Controller implementations using novel processors', in *Industrial Digital Control Systems* (eds K. Warwick and D. Rees), London: Peter Peregrinus, pp. 423–52.

10

Fault-Tolerant Parallel Systems for Control Applications

A. M. Tyrrell

10.1 INTRODUCTION

Distributed systems offer a number of advantages over single processor systems, including distributed functionality, higher speed of computation through concurrency, piecewise growth capability and fault tolerance through redundancy and reconfiguration. However, these advantages are gained at the expense of increased system complexity (Randell *et al.*, 1978) which can often lead to a decrease in system reliability. The demand for reliable computer systems which provide the intended service despite possible faults in hardware or in software has led to interest in making computer systems fault-tolerant (Ayache *et al.*, 1982; Wensley *et al.*, 1978).

The different methods used in obtaining high reliability systems can be divided into a number of stages (Anderson and Lee, 1981):

1. *Fault-avoidance techniques.* One method of increasing the reliability of a computer system is to try to reduce the possibility of faults. With hardware this could include careful routeing of signals, shielding, input-line filters. With software systems this includes structured design techniques and the use of structured design and programming languages.
2. *Fault-tolerance techniques.* If the techniques used to avoid faults cannot be 100% successful, which is the case, then we must accept that faults are there and try to tolerate them. Such techniques include the duplication of hardware modules in the system, and the inclusion of multiple algorithms to perform the same function.

Despite the use of fault prevention methods, it is not unusual for latent software design faults to remain in systems delivered for use. Software faults, being design faults, are inherently unanticipated; it follows that software has failures modes which are unknown. Furthermore, in harsh environments, the devices and interconnecting buses that make up the computing system will be susceptible to induced faults, for example caused by electrical noise, which again are inherently unpredictable. Hence for high reliability the designer must also provide a degree of tolerance to unanticipated faults (Leveson and Shimeall, 1983). Hardware faults, as components age, may occur; such faults have predictable failure rates and discrete failure modes can be tolerated using redundancy in hardware and/or test-and-recovery routines in software. However, it can be argued that modern VLSI devices are so complex that all possible failure modes and failure rates are not known; therefore these devices may be subject to unpredictable failures as well.

Software fault tolerance entails the use of special software structures which attempt to increase the reliability of a computing system by either masking an error, *N*-version programming (Avizienis, 1985), or by allowing the system to recognize the occurrence of an error and to take action to recover from that error (Randell, 1975; Leveson, 1983). *N*-version programming offers arguably the most effective fault tolerance, but is costly in software design, development and testing and in software execution overheads. On the other hand, error detection and recovery methods are characterized by the economical use of redundant software and can be effective in protecting against many types of fault (Hecht, 1979).

Error recognition and recovery techniques are well established for sequential and deterministic software executing on single processor systems. However, the techniques require substantial modification if they are to be adapted for use on a distributed system. Interprocess interaction is central to the control and coordination of the component processes in a distributed system, yet this interprocess interaction provides a medium for error propagation. The techniques must limit this promulgation of errors and make all the interacting processes cooperate in the error recognition and recovery process. If appropriate design methods are not used, then the final design will not improve the reliability of the system.

Software for a distributed system may be designed as discrete processes, executing in parallel on spatially distinct computers, which interact by message passing. It can be argued that the methods of CSP (Hoare, 1985) and the notation of Occam (Inmos, 1984) are appropriate for the design of such loosely coupled distributed systems. Occam provides the minimum set (Carpenter and Tyrrell, 1990) of nine constructs (sequential and parallel execution of processes, variable assignment, selection, iteration, synchronous communications for input/output, guarded processes and time), which are necessary for the design of software for loosely coupled distributed systems. These features encourage a rigorous approach to software design by making the designer consider the properties of each component process very carefully so that interactions with other processes are fully understood. Its use leads to concise specifications and designs, based on the parallelism inherent in a system (Newport, 1986); these can be implemented naturally as a set of distributed processes, capable of asynchronous execution, which are coordinated by synchronous interprocess communications.

The process of coordinating and synchronizing a distributed system is non-trivial and interprocess dependencies may lead to deadlock. To ensure that the system is free from such problems, the design process must be augmented to include a study of the process interactions by simulation and mathematical analysis. For example, the states of the software and the sequences of allowable state transitions, including interprocess communications, can be modelled using Petri net techniques (Tyrrell, 1987) to produce a graph which can be used to simulate and analyze process execution. The designer can then solve many state reachability problems concerning the dynamic behaviour of the system. Recent work (Tyrrell and Holding, 1986) on the systematic design and placement of fault-tolerant structures in distributed systems has shown the advantages of this approach.

A loosely coupled distributed system will operate without error only if the software is designed and implemented properly. This will be achieved only if each component process is designed and implemented correctly and the complete set of such processes are coordinated and synchronized correctly. Despite the proper use of appropriate design methods and the careful use of fault avoidance techniques throughout the software life cycle (Anderson, 1985), design faults often remain undetected until a system is in operational use.

For high-reliability applications, a system must also perform satisfactorily in the presence of certain classes of fault, whether caused by hardware or software. The performance of a computer system under fault conditions requires detailed analysis. Many types of fault (such as sensor failure) can be anticipated during the design phases and the designer can incorporate structures which will minimize the effect of a fault (Hecht, 1976). However, it is not feasible to anticipate all software and hardware faults and consideration may be given to introducing software fault tolerance.

10.2 ERRORS, FAULTS AND FAILURES

Designing a fault-tolerant system requires finding a way to ensure that the logical fault that arises from a physical failure does not cause an error. A failure can be defined as occurring when the behaviour of a system first deviates from that required by the specification of the system. A fault is the erroneous state of hardware or software resulting from failures of components, physical interference from the environment, such as electromagnetic interference or radiofrequency interference, failure of power supplies, operator error, or incorrect design of software and/or hardware. An error is the manifestation of a fault within a program or data structure. The error may occur some distance from the fault point. There are many causes also for systematic errors. These include: faulty analysis of customer requirements; failure to spot a key hazard; mistakes in specification; mistakes in high-level designs; errors in computer programs; and errors in hardware design (Campbell *et al.*, 1983).

Note that the only difference between a fault and an error is with respect to the structure of the system; a fault in a system is an error in a component or in the design of the system. A fault is the cause of an error and an error is the cause of a failure.

A system contains an error when its state is erroneous, whereas a system failure is the event of not producing behaviour as specified.

10.3 ACCEPTANCE TEST DESIGN

The recovery mechanisms rely on errors being detected successfully (Kane and Yau, 1975). This is the object of the acceptance tests. These tests must use prior knowledge of the data, the data transformations and the interrelationships between the data to construct a number of tests that will identify error conditions. As the complexity of these tests increases, and hopefully their effectiveness to identify errors, so the execution time required to perform these tests will increase. There is therefore a compromise to be struck between the complexity, and thus effectiveness, of the tests and the time taken to perform them. This compromise is very much dependent on the application area: real-time systems obviously require a faster response time than batch type systems.

Three criteria have been identified to enable checks to be useful (Anderson and Lee, 1981): (a) the check(s) should be derived solely from the system specification; (b) the check should be a complete check on the behaviour of the system; and (c) the system and its check should be independent. It is usually impossible, in practice, to achieve such stringent aims. Thus acceptability checks are used which will identify many errors, but cannot guarantee the absence of undetected errors.

The majority of error checking measures for computer systems can be broadly classified in one of the following: (a) replication checks, as the name suggests, an alternative implementation of the system (or subsystem) is created, and the system results are checked for consistency against these; (b) timing checks, if timing constraints are applied to the system, timing checks can be incorporated into the system to 'flag' errors if results are not produced within the specified time; (c) reversal checks, which take the output from a system and calculate what the inputs should have been to give these outputs (these are, however, heavy on time); (d) coding checks, a well-known example being parity checking; (e) reasonableness checks, where the objects within a system can be checked to see if their value is reasonable, with reference to their intended purpose, such as range checks; (f) structural checks, which can be applied to data structures within a system to check for structural integrity; and (g) diagnostic checks, where a component in a system is tested by applying inputs and monitoring the output response, and comparing this response to known answers.

Deciding upon the type of acceptance test for a given application is nontrivial.

10.4 HARDWARE METHODS FOR FAULT TOLERANCE

As components age, hardware faults may occur; such faults have predictable failure rates and discrete failure modes and can be tolerated using redundancy in hardware and/or test-and-recovery routines in software. However, it can be argued that modern VLSI devices are so complex that all possible failure modes and failure rates are not known; therefore these devices may be subject to unpredictable failures. More importantly, despite the use of fault prevention methods, it is not unusual for latent software design faults to remain in systems delivered for use. Software faults, being design faults are inherently unanticipated; it follows that software has failure modes which are unknown. Furthermore, in harsh environments, the devices and interconnecting buses making up the computing system will be susceptible to induced faults, for example caused by electrical noise, which again are inherently unpredictable. Hence for high reliability the designer must also provide a degree of tolerance to unanticipated faults. Fault tolerance is achieved by including in the system design special mechanisms which recognize and automatically override the effects of the fault, giving a satisfactory performance even in the presence of the fault.

Fault tolerance can be provided by including structures which attempt to *mask* the effect of a fault so that erroneous actions are not taken. For example, in an N-modular redundant system, all computations are performed on 'N' computational modules of diverse design using the same input data; each module relays its results to a further 'voting' module which determines the 'correct' decision by a majority vote. The concept was first outlined by von Neumann and since that time many systems of this sort have been proposed and built, principally using triple modular redundancy (TMR; $N = 3$) and with diversity in both hardware and software (Avizienis, 1985). Their effectiveness is well documented and extremely high reliability can be achieved but at high cost.

The transputer provides the ideal facilities required to build parallel systems, with parallelism and communication designed into its model of concurrent systems. A module can be designed that uses a network of transputers which perform the functions required to produce high integrity systems. It is envisaged that these modules will provide a standard building block that can be used in a system requiring high integrity. The use of a network of transputers enables the fault-tolerant mechanisms to be executed at high speed, thus making the module suitable

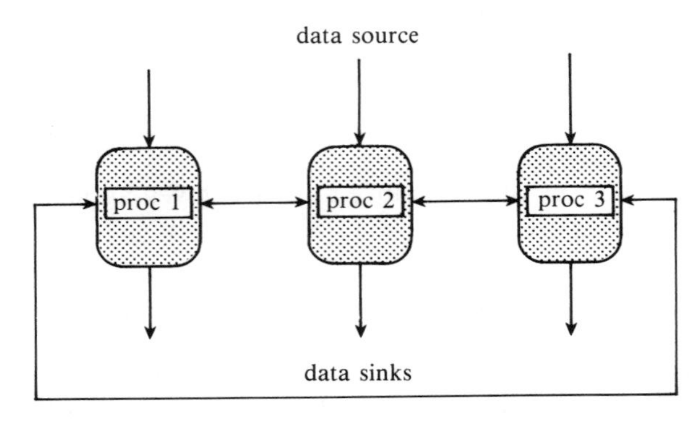

Figure 10.1 Hardware fault-tolerant module

for real-time applications. An additional advantage of the standard module approach is that once the hardware has been tested the same design is used for all applications.

The module consists of three processors, which are connected as shown in Figure 10.1. The processors all receive the same data from the source, each processor communicates its copy of the input data to the other two processors, each processor performs a voting procedure on the input data to check for integrity of data. Each processor then executes its function and communicates its resulting data to the other two processors. Again voting takes place and a consensus on the correct result is arrived at. Procedures are included in the module design to cope with interprocessor communication failures, such that if one, or more, processors or links fail the module will continue to operate, possibly in a degraded manner.

10.5 SOFTWARE METHODS FOR FAULT TOLERANCE

High reliability can be achieved by the use of massive redundancy to mask faults. For example, the whole system may be triplicated such that an error in a single channel is masked by the majority fault-free channels. Common mode failures can be avoided through diversity.

10.5.1 Fault Masking by *N*-version Redundancy

The main aim of *N*-version redundancy is to hide any faults that may be present. This is achieved by using multiple copies of the software (and/or hardware) and voting on their outcome. If one is found to be 'out-voted' by the others, then its results will be ignored and the consensus result taken. The fault in the single process has thus been masked out. This is illustrated in Figure 10.2.

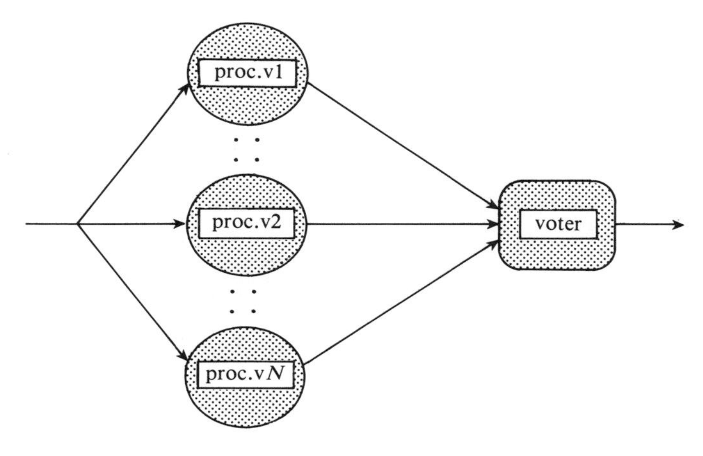

Figure 10.2 *N*-version redundancy

10.5.2 Fault Tolerance by Backward Error Recovery

A degree of fault tolerance can be achieved at less cost by partitioning the software into recovery blocks, see Figure 10.3. These are special software structures which recognize when a system state is in error (by assessing partial results for acceptability and consistency) and which initiate an error recovery mechanism (by backtracking and restoring a previously recorded high integrity state). Further progress then requires the system to resume operation through inbuilt software redundancy or hardware redundancy (Lee and Shin, 1984; Campbell and Randell, 1986). Many faults, ranging from sensor failure to software design faults, can be tolerated effectively without necessarily resorting to total system replication (Koo and Toueg, 1987).

10.5.3 Fault Tolerance by Forward Error Recovery

A second recovery method is 'forward' error recovery. The system substitutes predefined states for those which it deems to be erroneous, thereby allowing the system to continue satisfactorily (Taylor, 1986; Cristian, 1982). Thus, in Figure 10.4, if an error is recognized by means of an acceptance test (at state s1), then the process goes into a recovery procedure which attempts to restore normal execution (at state s2). If it cannot recover, then an abnormal exit occurs. Since forward error

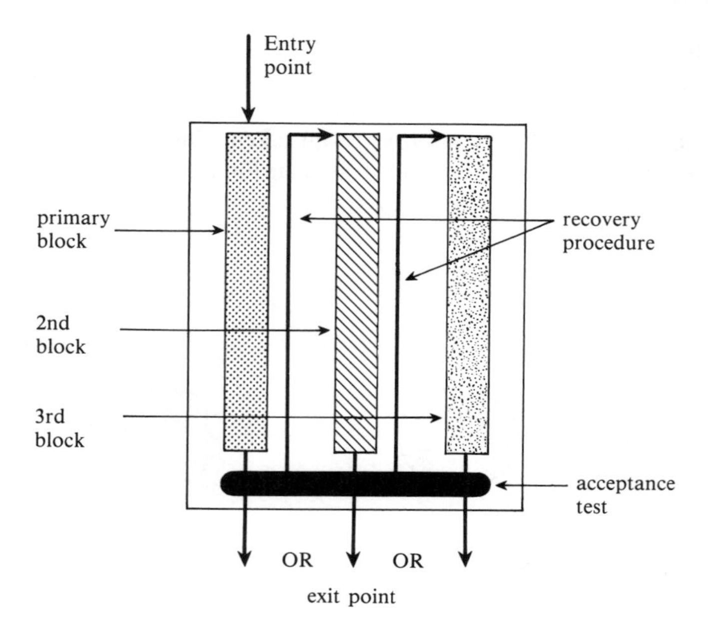

Figure 10.3 Structure of backward error recovery block

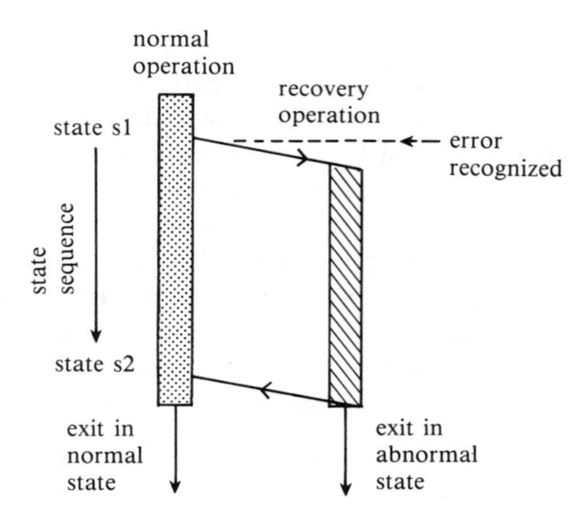

Figure 10.4 Illustration of forward error recovery

recovery relies on changing system states which are in error, accurate assessment of the likely extent of the errors is essential (Tyrrell and Carpenter, 1989; Leveson, 1983).

10.6 FAULT TOLERANCE IN PARALLEL SYSTEMS

When considering distributed computer systems, interprocess communications are central to the operation of the system (Campbell *et al.*, 1983; Campbell and Randell, 1986). If the system is truly decentralized, it comprises discrete processes executing in parallel on spatially distinct computers where no single process has a complete record of the overall system state. In such a system, individual processes can cooperate in the provision of the correct overall system function only if there is a reliable method of communication between the processes. While the communication channels are essential to the proper functioning of the system, they also provide a mechanism for the propagation of errors through interprocess interactions which cannot be revoked. Fault tolerance can be achieved in such a system provided error recognition and error recovery is coordinated between all interacting processes, such as in the conversation scheme (Randell, 1975). Recent work in this area using Petri nets has resulted in a systematic method for the design of such conversations.

10.6.1 Parallel Systems

In many respects the computational performance limits of the conventional microprocessor have already been reached. However, there remain many applications where computational support would be highly desirable but cannot be provided by a single processor, executing a sequential program. Often, the necessary improvements in computer performance can be achieved by exploiting whatever degree of parallelism is inherent in the application problem and implementing the software using a set of parallel processors (Hockney and Jesshope, 1989).

The design of software intended for a distributed implementation is demanding, requiring the understanding and application of proper design methods including techniques which exploit and control the parallel nature of the system. The software will comprise a set of processes in asynchronous concurrent execution where coordination is provided by synchronizing interprocess communications. The designer is therefore presented with three interrelated problems:

1. Partitioning the software into appropriate discrete processes.
2. The design of the interprocess communication structure.
3. The internal design of each individual process.

Established 'stepwise refinement' design methods for conventional, sequential software can be applied effectively in (3). However, (1) and (2) are outside the scope of software engineering methods for sequential software. Partitioning software into a set of discrete processes requires care since an inappropriate partitioning strategy can lead to a complicated interprocess communication structure, in which it is difficult to avoid dynamic faults, such as deadlock, when processes interact.

10.6.2 Fault-tolerant Structures for Parallel Systems

Error recovery mechanisms require the software to detect the presence of a fault by the errors which it generates and to take action to recover from those errors. The design of a recovery mechanism for a set of distributed communicating processes is nontrivial. If one process is faulty, recovery will affect that process and every other process to which the error could propagate through interprocess communication. It is necessary to limit the extent of error propagation if an intolerably long sequence of recovery actions is not to occur (Campbell *et al.*, 1983). As a consequence the recovery procedure must be properly coordinated within the interacting processes.

An effective method requires the set of communicating processes to be partitioned into atomic actions. Within the atomic action, interprocess interactions are permitted, but *no* interaction is allowed through the boundary of the atomic action. This guarantees that an error within the atomic action will not be propagated further (Jalote and Campbell, 1986). Recovery can then be applied at the level of the atomic action.

10.6.3 Backward Error Recovery

In order to implement backward error recovery in a distributed system, the set of intercommunicating concurrent processes must be partitioned into atomic actions; such a partition is conventionally referred to as a conversation (Randell, 1975); Figure 10.5.

An implementation of backward error recovery at the level of a conversation is presented using Occam later. The ability to test asynchronous inputs without imposing timing constraints (through the use of PAR and ALT constructs) provided the main strategy for selecting between alternative processes. Furthermore, since an Occam design is deterministic, state-space simulation methods can be used to demonstrate freedom from synchronization pathologies, such as deadlock, in a communication-intensive design. The one-to-one nature of Occam channels makes it easy to eliminate information smuggling.

The conversation consists of its constituent processes and a test line coordinator for the conversation. The processes included in a conversation can be enclosed within a fold associated with the conversation. This clarifies the program structure and facilitates checks for conversation side wall violations, i.e. interactions with processes outside the conversation.

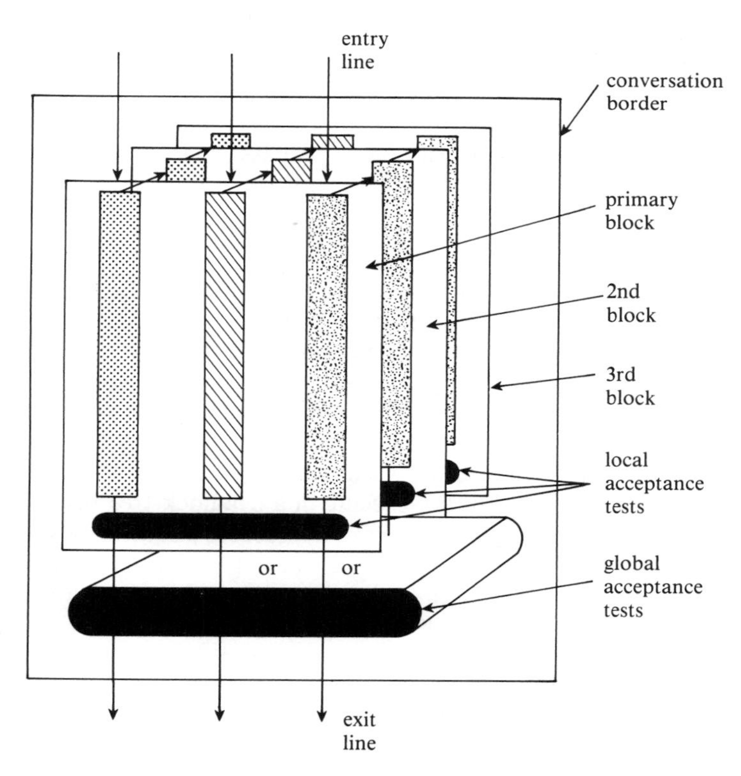

Figure 10.5 Conversation structure

The structure of the control part of the conversation including the recovery procedure is as follows (Tyrrell and Sillitoe, 1991):

```
{{{conversation
PROC example(CHAN OF INT chana, chanb)
    SEQ
        PAR
            SEQ
                ... primary block
              ... acceptance test
            SEQ
                ... 2nd block
              ... acceptance test
            SEQ
                ... 3rd block
              ... acceptance test
        ... check acceptance tests
        ... choose highest priority block which passed
:
}}}
```

The primary and alternative blocks are folded away to reveal the structure of the recovery mechanism. On entry, the recovery variables are passed to each of the blocks for them to start executing. On exit from each block the acceptance test is executed on the results from the blocks. If an acceptance test is failed the process's results are not used. The results of each block's acceptance test is checked, in priority order, and the highest priority block to pass its acceptance test is used.

10.6.4 Forward Error Recovery

In a system of concurrent processes in asynchronous execution, interprocess interactions again introduce complexity into the recovery procedure. There are two problems. Firstly, errors may be propagated to a fault-free process by interprocess interactions; as with backward error recovery the extent of this error migration must be contained. Secondly, synchronization violations during recovery could lead to deadlock.

Consider the poorly designed forward recovery in Figure 10.6. In normal execution three processes, Pa, Pb, Pc, execute in a deadlock-free fashion. Suppose an error is detected at state s1 by process Pa. This process will attempt state recovery which, if successful, will bring it back to a normal permitted state at state s2. However, if process Pa has missed the communication at A, process Pb will deadlock. Similarly, Pc will deadlock at B awaiting a communication with Pb.

It can be seen that forward error recovery in a distributed system has to be coordinated amongst the participating processes. As with backward error recovery, atomic actions must be identified among the interacting processes and recovery performed at that level. If an error is recognized in one process, all other processes within that atomic action have to undergo recovery.

Two implementations of forward error recovery have been investigated (Tyrrell

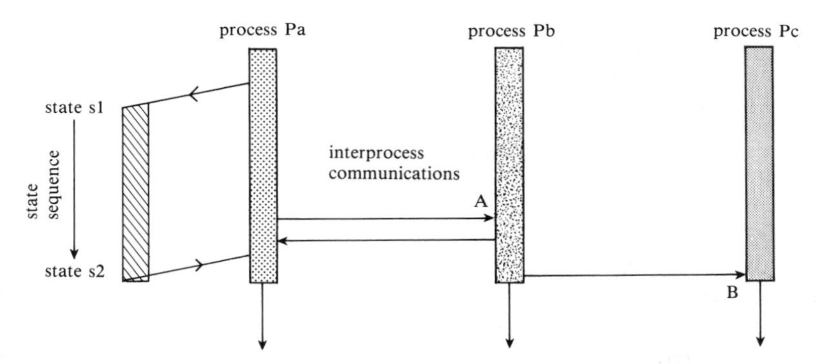

Figure 10.6 The consequences of poorly designed forward error recovery

and Carpenter, 1989). The first implementation requires a process to raise an exception flag if an error is recognized; the exception is then handled when the protected atomic action has terminated. This method is termed a *static* exception test and can be achieved with only a small amount of additional programming. The second method, termed a *dynamic* exception test, permits recovery as soon as the error is recognized, but requires considerably more code. Both methods require a centralized process to coordinate recovery.

Static exception test

Consider a set of processes which perform a function that has to be protected. During the performance of the protected function the processes interact as necessary within a properly constituted atomic action. Within the atomic action, each constituent process may assess its local results for acceptability (either as its results are generated or when all its results have been assembled) and if the data are unacceptable will set an exception flag and record the error type but take no action to terminate the atomic action. To provide protection, the designer must include the following control structure for execution in each process as soon as the atomic action has terminated.

In each process:

```
...prior code
SEQ
  ...protected function (atomic action)
  IF
    (exception = TRUE)
      to.coordinator[i] ! error.type
    TRUE
      to.coordinator[i] ! no.error
  from.coordinator[i] ? action
  IF
    (action = null)
      SKIP
    TRUE
      ...perform recovery (substitute predefined state)
```

```
Coordinator:
   PARi = 0 FOR all
    to coordinator[i] ? error.type
   SEQ
    ...decide on action
    PAR i = 0 FOR all
     from.coordinator[i] ! action
```

Each process sends to the coordinator an assessment of the exceptions that have been raised (or a no-exception signal) and then awaits a command from the coordinator process. The coordinator process instigates any necessary recovery and may select from a hierarchy of recovery actions. The recovery mechanism can be extended to allow the coordinator process to monitor the recovery actions.

The structure of the communications which effect the error recovery mechanism is shown in Figure 10.7 for two processes. Error recognition (by acceptability testing) is integral to the constituent processes in the atomic action, but the coordinator process effectively controls the onward promulgation of acceptable data. A simple extension of this mechanism includes a further phase of acceptability testing whereby the local results are assembled into a global set and tested for global acceptability.

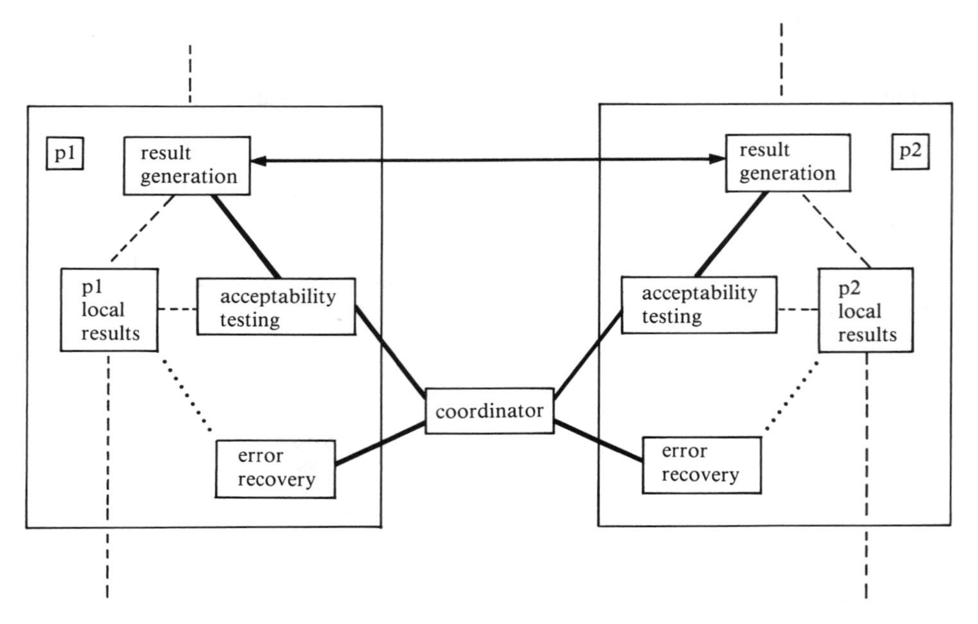

Figure 10.7 The communication structure underlying the static exception method of forward error recovery (heavy lines indicate the flow of control, dashed lines the flow of data and dotted lines the action of error recovery)

Dynamic exception test

Waiting for process termination before instigating error recovery may be undesirable. It is possible to act on an exception shortly after it is raised at the expense of additional coding. The method described here is similar to that above but does not require output guards and can therefore be implemented in Occam.

The implementation takes advantage of existing interprocess communications. An input statement can be considered as guarding the next phase of processing. It can therefore be enclosed within an ALT construct where the alternative process is the recovery process required for that level of processing, guarded by a trigger input from the coordinator:

```
chan.x ? variable
... next phase of
processing
```

```
PRI ALT
  from.coordinator[i] ? action
    ... perform recovery
  chan.x ? variable
    ... next phase of processing
```

An output statement requires a related approach being preceded by an ALT construct with two guards, one from the coordinator, the other a timeout mechanism:

```
chan.y ! expression
... rest of code
```

```
TIMER clock:
SEQ
  clock ? now
  PRI ALT
    from.coordinator[i] ? action
      ... perform recovery
    clock ? AFTER (now PLUS delay)
      SKIP
  chan.y ! expression
  ... rest of code
```

Consider a set of processes which perform a function that has to be protected. During the performance of the protected function the processes interact as necessary within a properly constituted atomic action. For dynamic forward error recovery, the atomic action is required to also enclose a recovery coordination process.

Each process within the atomic action is required to report an error to the coordinator process immediately it is recognized. The coordinator then commands all other processes in the atomic action to recover; the command will be accepted by each process at the next communication point and recovery will take place. The recovery action will normally impose a predefined system state and terminate the atomic action.

The overall structure of the coordinator is as follows:

```
ALT i=0 FOR ALL
 to. coordinator[i] ? exception.type
  SEQ
   ... decide on error. type
   PAR i=0 FOR all
    from.coordinator[i] ! action
```

As in the static forward error recovery implementation, this structure can be extended to allow for a hierarchy of errors and a diverse set of recovery actions.

10.7 REAL-TIME METHODS

In real-time applications, real-world synchronism of the overall system is usually imposed by a 'real-time clock' driven schedule. In such systems it is common to place critical timing requirements on the system's execution. To ensure that a system complies with its timing requirements, the performance of a control process can be monitored by a real-time counter and appropriate actions can be taken should it appear that a time-critical event will not meet its critical timing requirement. For example, in the design notation (based on Campbell *et al.*, 1979; Upadhyaya and Saluja, 1986):

```
ensure acceptance test AT
 within time t by process P
 else by default;
```

The timing performance of the application process 'P' is monitored by setting a 'watch-dog timer' to trip at a predetermined period somewhat less than the time-critical time '*t*'. The watch-dog timer will run concurrently with the control process 'P' and the first of these processes will determine subsequent actions; the other process is aborted. When no fault is present, the control process 'P' will be designed to produce results well within the critical time '*t*'. Usually this process will be required to execute every *T* units of time in Figure 10.8.

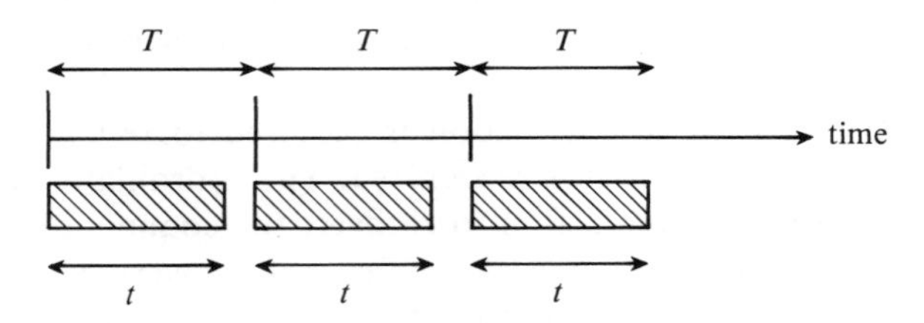

Figure 10.8 Watch-dog mechanism

The structure assumes that the watch-dog timer, the acceptance test mechanism and the default process are immune to faults. The default actions to be taken in the event of a timeout should be detailed properly in the system specification.

This structure does not constrain the implementation of the watch-dog mechanism and several strategies have been proposed for sequential implementation (Upadhyaya and Saluja, 1986). Nor does the structure imply that process 'P' must be executed before the 'default' process: both could execute in parallel if the system design allows this.

Time-critical real-time distributed systems must rely on the integrity of interprocess communications for data interchange and the proper synchronization of the constituent processes. Since the communications medium is likely to be vulnerable and could itself be a source of errors, it is essential that faults occurring in the interprocess communications channel must be recognized promptly so that recovery actions can be initiated within the critical time period.

10.7.1 Parallel Watch-dog Mechanism

Since such systems rely on interprocess synchronization, it is difficult to provide each distributed process with its own, independent, embedded watch-dog mechanism. A more structured design can be developed by using a separate watch-dog process with process termination being handled by a coordinating process.

The design comprises the following:

1. A (centralized) process coordinating all participating processes.
2. The time-critical control process 'P'.
3. A 'timer' process.
4. A 'default' process.
5. A buffer process for collecting appropriate results and outputting them to the real world.

The coordinating process initiates all participating processes in parallel. The process 'P' and the timer processes are then made to race against each other. The outcome of the race is communicated to the coordinating process which then aborts the losing process.

The implementation of such a race is not trivial. For example, a simple PAR construct cannot be used because the PAR construct is terminated synchronously. (This precludes the constituent process which wins the race from aborting the losing processes since they are fellow constituents of the PAR construct.) The use of a simple ALT construct to terminate the race also leads to problems because, in the event of a timeout, process 'P' can not be aborted without the risk of deadlock; nor can process 'P' reliably abort the timer process (since, in both cases, the termination of the timeout and the abort transaction are not atomic). Alternative approaches, such as leaving the losing processes to run gracefully to completion are equally troublesome because of the need to 'mop-up' the communications of the losing processes. This leads to complexity because the point-to-point nature of Occam communications prevents the use of simple garbage collectors. This approach also prevents the PAR construct from terminating until the mopping-up process is complete.

The following approach circumvents this problem. The coordinating process includes a termination process which comprises two parallel subprocesses, each of which forces the other to abort before terminating. One subprocess is activated primarily by the completion of 'P' and will abort the timer; the other is activated primarily by a timeout trip and aborts 'P'. Both subprocesses are guaranteed to run to completion because the inactive subprocess is activated by the process which aborts. The 'result buffer' process must, of course, absorb outputs from both subprocesses. Thus, sub-process (A) is activated by either the timer tripping or the timer being aborted on the completion of 'P'. Similarly, subprocess (B) is activated by either 'P' completing or 'P' being aborted by timeout:

```
PROC P ()
 INT status, any:
 BOOL exit :
 SEQ
  ... initialization.
  exit :=FALSE
  status :=no.abort
  WHILE NOT exit
   SEQ
    PRI ALT
     p.abort ? any
      SEQ
       status := abort
       exit := TRUE
       P. complete ! abort
     SKIP
      ... compute results
  IF
   status=no.abort
    SEQ
     P.complete ! no.abort
     P.abort ? any
     P. request ? any
     P.data ! data
   status=abort
    SKIP
:
```

```
PROC TIMER ()
 INT start.time :
 TIMER clock :
  SEQ
   clock ? start.time
   ALT
    timer.abort ? any
     time.trip ! no.trip
    clock ? AFTER start.time+t.out
     SEQ
      time.trip ! trip
    timer.abort ? any
     SKIP
:
```

```
PROC coordinator ()
 ... declare and initialize
 PAR
  P
  timer
  default
  result.buffer
  SEQ -- (A)
   timer.trip ? status
   to.result.buffer1 ! status 1
   P.abort ! any
  SEQ -- (B)
   P.complete ? status2
   to.result.buffer2 ! status2
   timer.abort ! any
:
```

10.7.2 Busy Polling Watch-dog

An alternative approach is to provide a separate timeout process, which is polled frequently by the process 'P':

```
PROC P ()
  ... declare and initialize
  SEQ
    exit := FALSE
    go := TRUE
    WHILE NOT exit
      SEQ
        P.command ! not.complete
        P.ack ? state
        IF
          state=abort
            SEQ
              go := FALSE
              exit := TRUE
          state=no.abort
            SKIP
        ... compute partial results.
        ... set exit TRUE on completion.
    IF
      go
        SEQ
          P.command ! complete
          P.ack ? any
          go := FALSE
      NOT go
        SKIP
:
```

Each control process performs only a limited amount of work before confirming that further execution is required (a 'busy' polling solution). When 'P' completes, it signals the coordinating process which responds by sending back an acknowledge signal; data exchange can then take place, typically on the next clock signal.

10.7.3 Scheduling

The situation often occurs where there are a number of 'control processes' executing on a single processor. These control processes having different deadlines to meet. How can we make sure that the correct processes are scheduled at the correct times to give all control processes a chance to finish 'on-time'? A simple and efficient method was proposed in Welch (1990). It is a rate-monotonic scheduler. This type of scheduler allocates the highest priority to the process which has the highest rate, and thus the shortest time to execute. To implement such a scheme on transputers some additional code must be written since it only has two priority levels built into it.

Each control process must send to a *kernel* process an *active* message while it is performing its control law. When it has completed its task, it sends an *asleep* message to the kernel process. If a control process is being pre-empted by a process with higher priority then the kernel process just does not accept its active message. Once there are no more higher-priority control processes to complete their execution the kernel will accept the active message. The framework of the control processes and the kernel are given as in Welch (1990):

```
PROC control-process (CHAN OF SIGNAL out, .....)
 WHILE TRUE
  SEQ
   ... wait for start of time period
   BOOL active:
   SEQ
    ... input data
    active := TRUE
    WHILE active
     SEQ
      out ! awake
      ... work
    out ! asleep
    ... output data
:

PROC kernel ([]CHAN OF SIGNAL in)
 VAL INT n IS SIZE in:
 INT j:
 SEQ
  j := n
  WHILE TRUE
   PRI ALT i=0 FOR j
    in[i] ? CASE
     awake ? CASE
      awake
       j := i PLUS 1
      asleep
       j := n
:
```

The kernel will look for control processes in order, from 0 to j. Thus, processes must be connected in the correct order. Initially all processes will be scanned. The process with the lowest number will always be accepted. Other, lower priority, processes will be blocked; if they were running they will be pre-empted. When an asleep message is received, all processes will then be scanned again, and the highest priority process which requires time is chosen. A number of other interesting scheduling methods are given in Welch (1990).

10.8 A CASE STUDY

10.8.1 Introduction to the Problem

Most control strategies employed on industrial manipulators simplify the control by approximating the manipulator to a set of uncoupled SISOs, the control of which can then be achieved using second-order system theory. This approximation is only possible by the use of very large damping factors within the joint controls, to mask the effects of centrifugal, coriolis and coupled forces, and results in a much reduced performance. On the other hand, the more complex techniques given in the literature (which for the most part are simulated) use a model of manipulator dynamics to design a more precise control. These techniques provide higher performance but have two major problems. Firstly, in general the time taken to evaluate the model of the manipulator's inverse dynamics is too large for it to be completed at the required rate to control the manipulator, when implemented on a traditional computational architecture (Khosla and Neuman, 1985). Secondly, the coefficients used within the model are determined from measurements; these have large tolerances associated with them (Armstrong *et al.*, 1986), they vary between manipulators and with time and some are impractical to measure (e.g. frictional).

Hence, most of the more practical schemes adapt to the characteristics of the manipulator, one such technique is the *computed torque* method, Figure 10.9. This uses as much of the model of the inverse dynamics as is practical in order to decouple the control of the joints and then places a proportional-derivative controller about this reduced system in order to control the effects of the

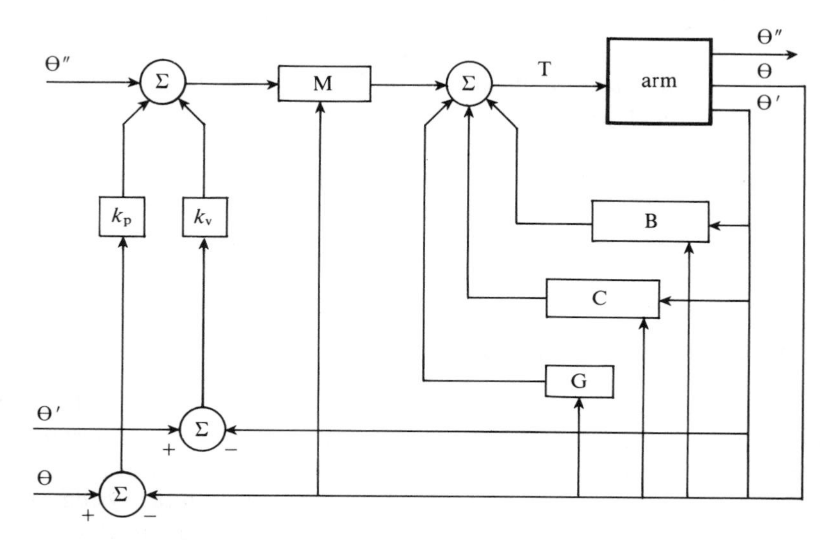

Figure 10.9 Robot arm controller block diagram

unmodelled and time-varying parameters. Although, this technique strikes a balance between the two approaches of control, the difficulty of evaluating the inverse dynamics within the necessary sample period still remains. One possible solution to achieve the desired execution rate would be by the use of suitable parallel architectures (Mirab and Gawthrop, 1990).

For a robot arm with six joints, each angle vector has six elements:

$$\Theta = [\theta_0, \theta_1, \theta_2, \theta_3, \theta_4, \theta_5] \tag{10.1}$$

In the laboratory the control processes are exercised firstly with software which simulates the operation of the robot arm. The simulation involves the computation of:

$$\Theta_n'' = \mathrm{M}^{-1}(\Theta)[T - \mathrm{B}(\Theta)[\Theta'\Theta'] - \mathrm{C}(\Theta)[\Theta'^2] - G(\Theta)] \tag{10.2}$$

$$\Theta_n' = \Theta_{n-1}' + \Theta_{n-1}'' \ \Delta t \tag{10.3}$$

$$\Theta = \Theta_{n-1} + \Theta_{n-1}' \ \Delta t + 0.5\Theta_{n-1}'' \ \Delta t \tag{10.4}$$

where Δt is derived from the sample rate. The arm process itself has been modelled using the data flow simulator. (In fact, copies of the M, B, C and G processes are required to perform the calculations.)

10.8.2 Distribution of Tasks

The system was partitioned on to a three transputer network. There was a requirement to partition the most complex process (i.e. B), while retaining as much of the term reduction as is feasible (Sillitoe and Tyrrell, 1991). The possible term reduction in B and the orthogonal distribution of complexity terms in B point to two possible partitions of B; namely major joint partition and minimum dependency partition. Major joint partition makes use of the facts that the most complex terms will belong to the first joint in the chain and that the first joint will necessarily provide the least-term reduction within B. On the other hand minimum dependency partition looks for a partition which maintains maximum term reduction on either side of the partition. The two forms of partition mapped as in Figure 10.10 result in very similar load ratios (i.e. 1:0.8:0.84) and reductions in complexity (approximately 1.2).

10.8.3 Design of Acceptance Tests

Three different types of acceptance tests were investigated. The first, and easiest test, was simply to check the value of certain critical constants within the system against the 'true' value. This worked well, but only detected a very limited set of faults. Reversal checks were carried out on some of the blocks within the system. Here the output of a particular block was taken and the reverse process to that performed by the block was applied to these data. The result of this was then compared to the

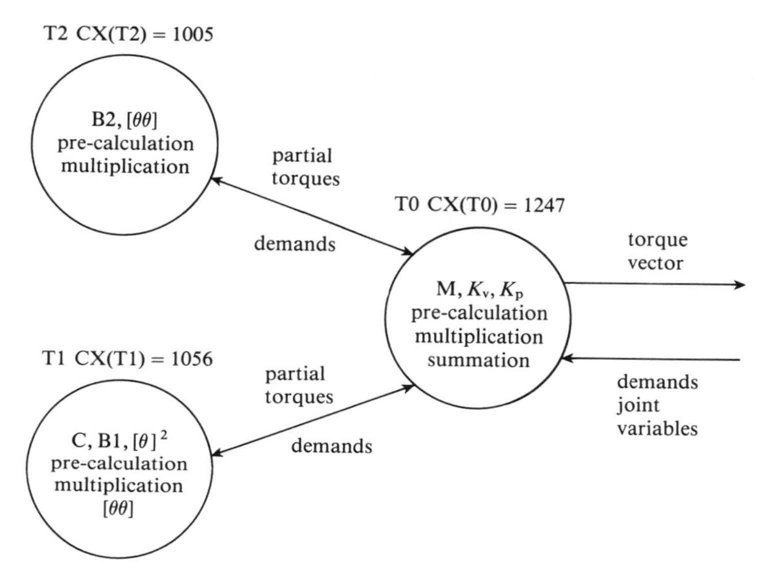

Figure 10.10 Mapping for three transputers

input data. The third type of test used a simple form of predictive tests. The combined function of the blocks M, B, C and G within the control is to decouple the nonlinear characteristics of the manipulator; hence, the outputs from these blocks are complex and a function of the state of the manipulator. Predictive techniques based on the knowledge of the previous two samples of the block's output was developed. The test makes the assumption that the output from each of the blocks is continuous within the sample interval.

10.8.4 Faults Injected

Excluding any unpredictable errors which were present within the system (from the previous discussions it is likely that there would be some), the faults included in the system were 'designed' into it. These tended to be of a static nature, that is, a decision was made to place an incorrect value into one of the variables and this was left at that value for the duration of the test. Some randomness was introduced in a few of the tests by using a random number generator to determine (a) the value of a variable and (b) if the error value should be included in that particular execution or not. Another type of fault placed in the system was to change the function of a part of the system by altering the arithmetic operations involved, i.e. change an addition for a subtraction.

The tests reported here are limited to a single fault in any given execution. This enabled predictions of the errors caused by the fault to be made more easily than would have been the case with multiple faults.

10.8.5 Results

Performance of system in normal operation

Under normal conditions, that is with no injected faults, the control loop based upon the 10% model given in Armstrong *et al.* (1986) cycles at 250 Hz. The system was tested using a series of step inputs to the path generator, which provides the demands [Θ Θ' Θ'']. The input signal and output response are shown in Figure 10.11.

Performance of system with injected faults

The set of results shown in Figures 10.12–10.14 shows the response of the system when faults were injected into the system and no fault-tolerant mechanisms were included to cope with these. It can be seen from Figures 10.12–10.14 that the output response of the system with such faults present is unacceptable for the control of any safety critical system. The output for Figure 10.13 with no fault tolerance was too large to fit on the graph!

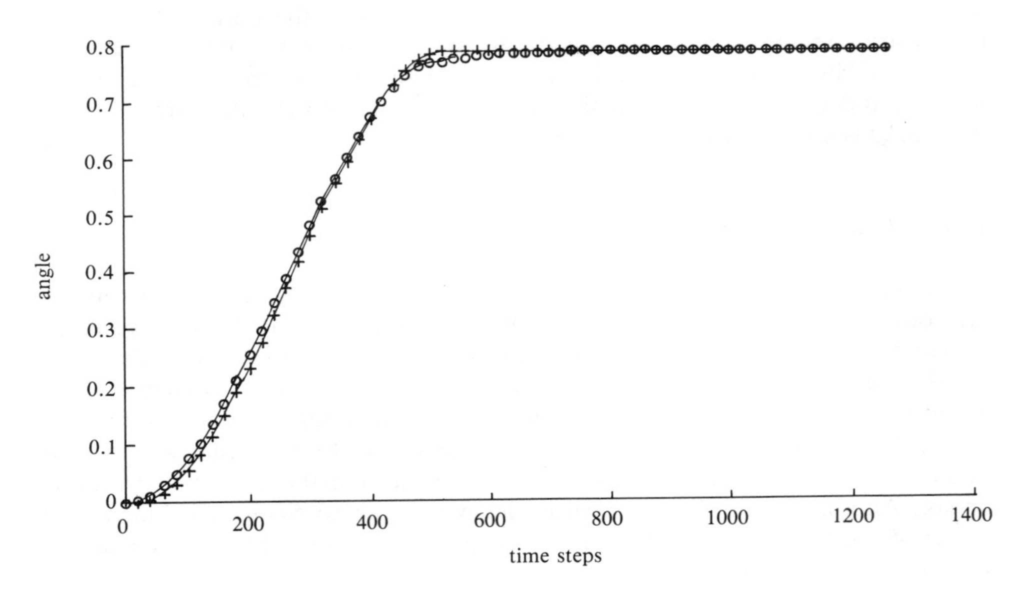

Figure 10.11 Robot response with no errors: ○ input demand; + output response

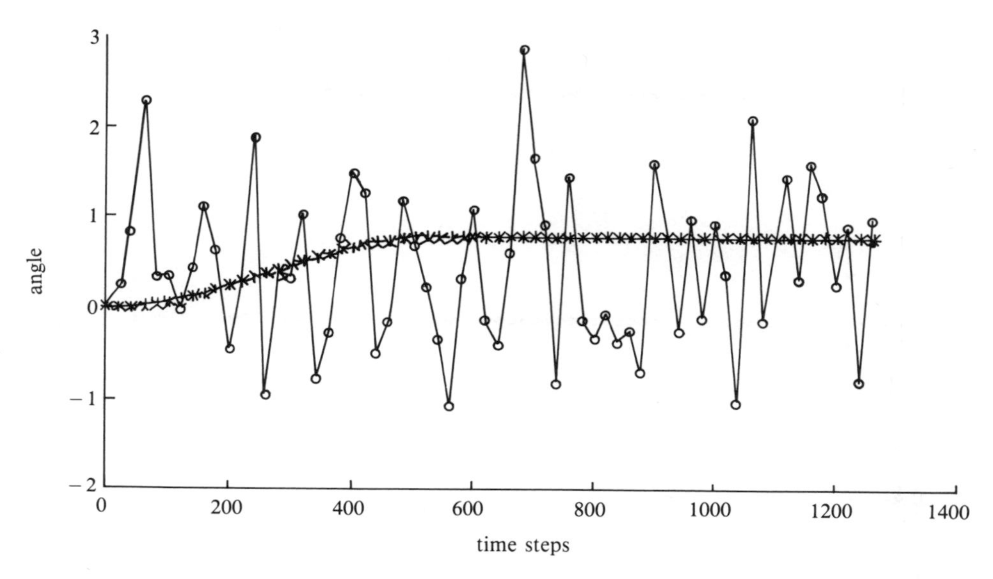

Figure 10.12 Robot response with errors: × input demand; ○ output with errors included and no fault tolerance; + output response with fault tolerance

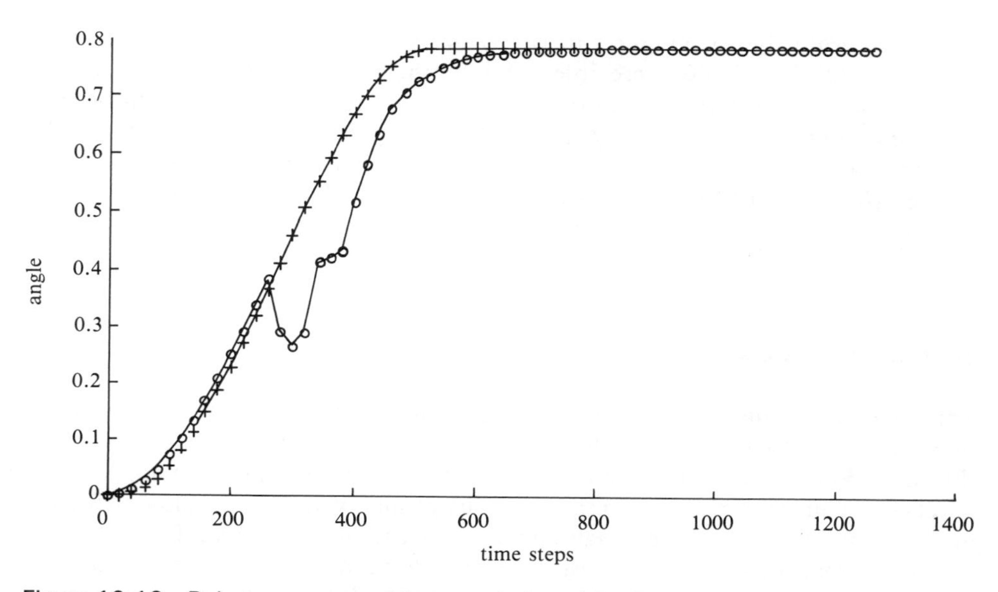

Figure 10.13 Robot response with errors induced for limited time with forward error recovery: + input demand; ○ output response with fault tolerance

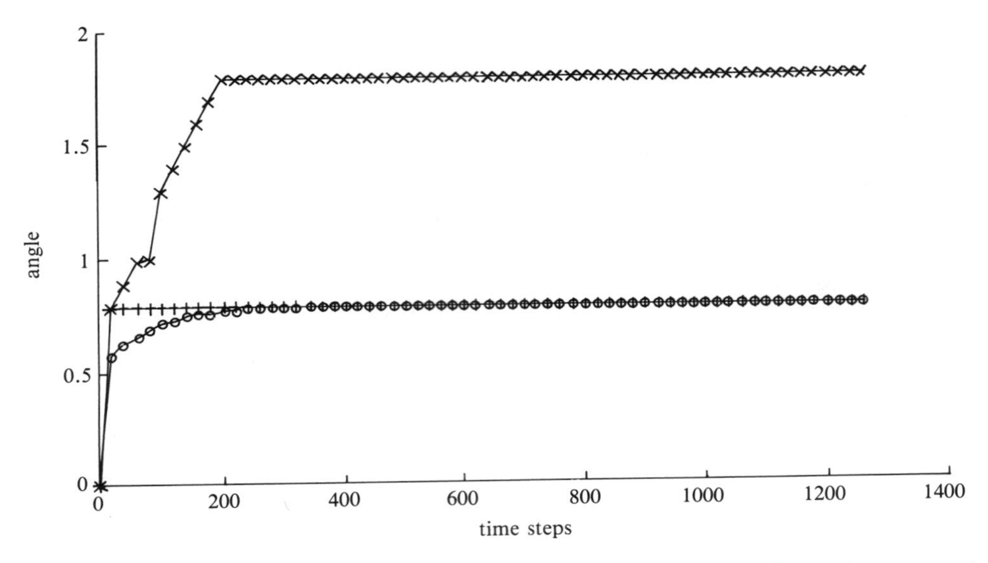

Figure 10.14 Robot response to step input with errors: + input demand; × output response with no fault tolerance; ○ output response with fault tolerance

Performance of system with injected faults and fault-tolerance structures

The set of results shown in Figures 10.12–10.14 also show the response of the system when the different faults were injected into the system and the fault-tolerant mechanisms were included in the system design. From these results it can be seen that the output response of the system, with such faults present but with the fault-tolerant structures included, is considerably better than in the previous cases. Figures 10.12 and 10.14 use backward error recovery. Figure 10.13 uses forward error recovery.

10.9 CONCLUSIONS

This chapter has introduced a number of structures, both hardware and software, that can be used to increase the reliability of parallel computer systems. It has shown the effectiveness of certain software fault-tolerance mechanisms when applied to a parallel system, with real-time responses. The results of this study show that such mechanisms are useful when high-integrity systems are required. However, the major problem with such mechanisms is in the detection of the faults. In many of the tests carried out in this work the acceptance tests did not identify an error state and thus the fault-tolerant mechanisms were never executed.

It has illustrated, using a number of examples, that the parallel programming language Occam and the microprocessor or the transputer are well suited to implementing a number of these fault-tolerant structures. In addition, the nature of Occam makes for clear, easy to understand system designs, which have a mathematical background.

REFERENCES

ANDERSON, T. (1985) 'Can design faults be tolerated? *Software and Microsystems*, **4**(3), 59–62.

ANDERSON, T. and LEE, P. A. (1981) *Fault tolerance: principles and practice*, Englewood Cliffs, NJ: Prentice Hall.

ARMSTRONG, B., KHATIB, O. and BURDICK, J. (1986) 'The explicit dynamics model and inertial parameters of the Puma 560 arm', in *IEEE Conference on Robotics and Automation*, pp. 510–18.

AVIZIENIS, A. (1985) 'The *N*-version approach to fault tolerant software', *IEEE Trans. Software Eng.*, **SE-11**, 1491–501.

AYACHE, J. M., COURTIAT, J. P. and DIAZ, M. (1982) 'REBUS: a fault-tolerant distributed system for industrial real-time control', *IEEE Trans. Computers*, **C-31**(7), 637–47.

CAMPBELL, R. H. and RANDELL, B. (1986) 'Error recovery in asynchronous systems', *IEEE Trans. Software Eng.* **SE-12**(8), 811–26.

CAMPBELL, R. H., HORTON, K. H. and BELFORD, G. G. (1979) 'Simulations of a fault-tolerant deadline mechanism', in Digest of papers. *Fault-tolerant Computer Systems, Madison*, pp. 95–101.

CAMPBELL, R. H., ANDERSON, T. and RANDELL, B. (1983) 'Practical fault tolerant software for asynchronous systems', *Safety of Computer Control Systems*, (Safecomp 1983), IFAC, 59–65.

CARPENTER, G. F. and TYRRELL, A. M. (1990) 'Occam: A second language', *Int. Journal Elec. Eng. Educ.*, October Issue.

CRISTIAN, F. (1982) 'Exception handling and software fault tolerance', *IEEE Trans. Computers*, **C-31**(6), 531–40.

HECHT, H. (1976) 'Fault-tolerant software for real-time applications', *Computing Surveys*, **8**(4), 391–407.

HECHT, H. (1979) 'Fault tolerant software', *IEEE Trans. Reliability*, **R-28**, 227–32.

HOARE, C. A. R. (1985) *Communicating Sequential Processes*, Hemel Hempstead: Prentice Hall.

HOCKNEY, R. W. and JESSHOPE, C. R. (1989) *Parallel Computers*. Bristol: Adam Hilger.

INMOS (1984) *Occam Programming Manual*. Hemel Hempstead: Prentice Hall.

JALOTE, P. and CAMPBELL, R. H. (1986) 'Atomic actions for fault tolerance using CSP', *IEEE Trans. Software Eng.*, **SE-12**, 59–68.

KANE, J. R. and YAU, S. S. (1975) 'Concurrent software fault detection', *IEEE Trans. Software Eng.*, **SE-1**(1), 87–99.

KHOSLA, P. K. and NEUMAN, C. P. (1985) 'Computational requirements of customised Newton–Euler algorithms', J. Robotic Systems, **2**, 309–27.

KOO, R. and TOUEG, S. (1987) 'Checkpointing and rollback-recovery for distributed systems', *IEEE Trans. Software Eng.*, **SE-13**(1), 23–31.

LEE, Y. H. and SHIN, K. G. (1984) 'Design and evaluation of a fault-tolerant multiprocessor using hardware recovery blocks', *IEEE Trans. Computers*, **C-33**(2), 113–24.

LEVESON, N. G. (1983) 'Software fault tolerance: the case for forward error recovery', in *Proceedings of the AIAA Conference on Computers in Aerospace*, pp. 50–4.

LEVESON, N. G. and SHIMEALL, T. J. (1983) 'Safety assertions for process-control systems', *FTCS 13, Fault Tolerant Computing*, pp. 236–40.

MIRAB, M. and GAWTHROP, P. J. (1990) 'Transputers for robot control', in *DTI/SERC Transputer Initiative Mailshot*, pp. 70–7.

NEWPORT, J. R. (1986) 'An introduction to Occam and the development of parallel software', *Software Engineering J.* July, 165–9.

RANDELL, B. (1975) 'System structure for software fault tolerance', *IEEE Trans. Software Eng.*, **SE-1**, 220–32.

RANDELL, B., LEE, P. A. and TRELEAVEN, P. C. (1978) 'Reliability issues in computing', *Computing Surveys*, **10**, 123–65.

SILLITOE, I. P. W. and TYRRELL, A. M. (1991) 'Evaluation of cost effective transputer architectures for the implementation of the computed torque method for robotic manipulators', in *Proceedings of the IEE International Conference of Control 91, Edinburgh*, pp. 855–60.

TAYLOR, D. (1986) 'Concurrency and forward recovery in atomic actions', *IEEE Trans. Software Eng.*, **SE-12**, 69–78.

TYRRELL, A. M. (1987) 'Design of fault tolerant software for loosely coupled distributed systems', PhD Thesis, Aston University, UK.

TYRRELL, A. M. and CARPENTER, G. F. (1989) 'Forward error recovery using Occam', in *Proceedings of the 12th International Conference on Fault Tolerant Systems and Diagnostics, Prague, Czechoslovakia*.

TYRRELL, A. M. and HOLDING, D. J. (1986) 'Design of reliable software in distributed systems using the conversation scheme', *IEEE Trans. Software Eng.*, **12**(9), 921–8.

TYRRELL, A. M. and SILLITOE, I. P. W. (1991) 'Evaluation of fault tolerant software structures for parallel systems in industrial control', in *Proceedings of the IEE International Conference of Control 91*, Edinburgh, pp. 393–8.

UPADHYAYA, J. S. and SALUJA, K. K. (1986) 'A watchdog processor based general rollback technique with multiple retries', *IEEE Trans. Software Eng.*, **SE-12**(1), 87–95.

WELCH, P. H. (1990) 'Multi-priority scheduling for transputer-based real-time control', in *Proceedings of the 13th Occam User Group Technical Conference, York, UK*, IOS Press, pp. 198–214.

WENSLEY, J. H. *et al.* (1978) 'SIFT: design and analysis of a fault-tolerant computer for aircraft control', *Proc. IEEE*, **66**(10), 1240–55.

Part IV

Parallel Machines and Algorithms

An alternative to VLSI-oriented systems or the construction of user-configured networks using 'building block' type processors, is to employ existing parallel machines. An early example of these systems is the vector processor, of which the Cray is an example; but such systems do not fully exploit parallelism. A single-instruction multiple-data (SIMD) system does permit full parallelism and examples here include the Illiac IV and the ICL Distributed Array Processor. Further, in the CDC Cyber 205, pipelining techniques are combined with parallelism to increase efficiency and yield higher performance. They are, however, less suitable for general-purpose computing where the data are not inherently in the form of 'large' uniform arrays.

In contrast to SIMD systems, multiple-instruction multiple-data (MIMD) machines are best suited for general-purpose parallelism. This type of 'shared memory multiprocessor' is more cost effective but less scaleable than 'distributed memory multicomputers', since adding processors to a shared memory system can suffer from bus saturation. Further, the scaleability of a distributed memory system is strongly dependent on the topology used.

One of the most common and successful topologies is the hypercube architecture, which provides the best trade-off between the longest path between processors and the number of physical connections each processor must have. In the case of an n-dimensional machine, for example, the architecture typically consists of 2^n processors, each of which is nearest-neighbour connected by n bidirectional and asynchronous communication channels. Examples include the Intel iPSC, Ametek System 14, NCUBE/TEN and the Connection Machine from Thinking Machines.

Of the various parallel machines, user reconfigurable systems provide better flexibility for different applications. Further, they also allow topology reconfiguration by programs and are best suited for the design and simulation of VLSI-oriented architectures, such as those developed in Parts I and II. An early example is the Accelerated Processors Model 10 system which has between 4 and 12 grooves of 8 arithmetic logic units. Recent examples based on transputers include the Microway Quadputer, the Meiko Computing Surface System and the Parsytec clusters, and Giga Cube machines. These machines are also well suited to the development of the architectures of Part III.

Chapter 11 presents an overview of the hardware of various parallel machines from a control applications standpoint. Strong emphasis is placed on interprocessor connection/communication schemes and, in particular, the hypercube case. This is supported by brief case studies on process and vehicle control systems applications.

In Chapter 12, the subject is the use of currently available parallel machines to develop software for control systems design. This takes the form of a tutorial-style survey of the current 'state of the art', plus some open research problems using, as an illustrative basis, some algorithms and programming techniques for use with hypercubes. A very important point emerging here is that the development of parallel algorithms for (commonly used) control systems design algorithms is not simply a case of modifying existing (sequentially based) software to run in parallel.

11

Parallel Computing Architectures and Machines for Time-critical Control Applications

K. J. Hunt

11.1 INTRODUCTION

This chapter focuses on the use of parallel processing machines for time-critical applications. A particular interest is the problem of real-time control. A broad overview is given of the parallel architectures and machines which are commercially available, concentrating on those machines having true real-time capabilities.Two case studies are presented where the systems described were applied to complex real-time simulation and control problems.

A very wide range of computational tasks is met throughout the field of real-time control. In process control, for example, there is a requirement for time-critical control. This includes real-time feedback and sequence control at the loop level. At a higher level, supervisory control and process monitoring are required to ensure safe operation. Coupled with these requirements is the common need for high-volume data acquisition and analysis, often using remote telemetry stations. In many cases safety and training requirements demand the availability of high-fidelity real-time simulation. Indeed, in aerospace and nuclear power systems this is mandatory. Safety issues aside, it has also been demonstrated in many control situations that operator training through simulation can lead to optimization of production and subsequent economic benefits. Finally, the recent emergence of applied artificial intelligence (AI) techniques, such as expert (or knowledge based) systems, has generated much interest in the process community. Although knowledge based systems typically require substantial computing resource, in some cases important benefits have been realized.

In short, it is clear that the aggregate computing power in a typical real-time control situation is very high (Hunt, 1991). Moreover, as the demand for improved safety margins, better training and optimized production increases, the need for very high and cost-effective computing power will grow. Historically, the development of computing systems has gone from conventional sequential processing machines through to high-cost, high-performance supercomputers. Supercomputing machines are typically equipped with vector processors which have the ability to process instructions simultaneously on a number of data streams in order to achieve very high throughput. These machines are also characterized by a very high cost due in part to the advanced processing units used and the extensive cooling mechanisms needed.

Recently, a number of vendors have introduced multiprocessor *parallel processing* machines to the market. These machines attempt to provide high power by linking a number of processing units working together in an efficient manner. These machines can often be justified in relation to supercomputers on a performance/cost analysis basis. This chapter will briefly review the parallel processing field and examine the implications which this area has for the real-time control and simulation industries.

11.2 PARALLEL ARCHITECTURES AND MACHINES

In recent years a number of parallel processing computers have become commercially available. These machines are characterized by the interconnection architecture used to connect the individual processors and memories. The most common architectures are the bus, the crossbar, the hypercube and the multistage switch. Here, these architectures are briefly reviewed, while in each case a commercially available machine having that architecture is described. Because of our interest here in time-critical applications we focus where appropriate on machines which can be loaded with real-time operating system executives. We also cover the transputer, a building block for the development of multiple processor machines of flexible structure.

11.2.1 Bus Based Connections

Bus based multiprocessors are a logical extension of the traditional minicomputer architecture. Individual processors, memory modules and input–output devices are connected by one or more high-speed data buses. A simplified representation of this scheme is shown in Figure 11.1. Since all memory modules are addressable by all

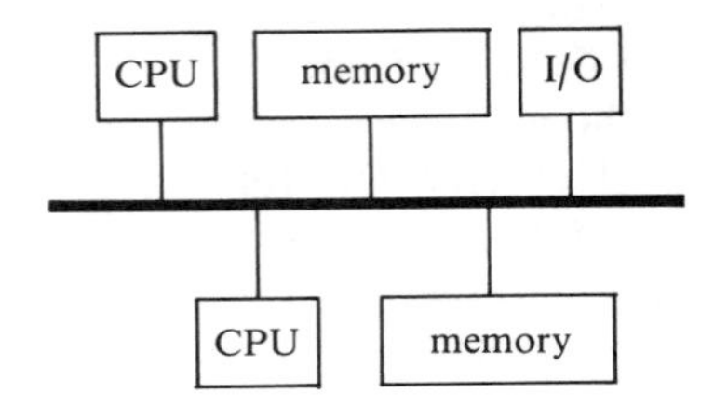

Figure 11.1 Bus-based connections

processors, bus based systems are useful in applications where shared memory is important. Conceptually, the bus based system is the simplest interconnection architecture.

The FX computer series from Alliant Computer Systems is based upon high-speed bus interconnection. The recently announced FX/2800 machines use up to 28 64-bit RISC (reduced instruction set computing) processors, each capable of both scalar and vector computations. The processors are Intel's recently introduced i860 chip. These processors are configured as 14 super computational elements (SCEs) for major computational tasks, and 14 super interactive processors (SIPs) for smaller interactive and systems tasks. Each processor delivers 33 MIPS (million instructions per second) of throughput, which means that a fully configured system has the potential of a peak power of 924 MIPS.

The shared physical memory capacity of the FX/2800 is 1 Gb, with an intermediate cache memory of up to 4 Mb. The actual interconnection architecture of the FX/2800 is somewhat more complex than the simplified representation in Figure 11.1. The processors are in fact connected to the cache memory via a high-speed crossbar interconnect (see below) having a bandwidth of some 1.28 Gb s^{-1}. A memory bus with a bandwidth of 640 Mb s^{-1} then connects the cache to main memory.

The FX series runs under the Concentrix operating system. This is Alliant's version of Unix, enhanced for parallel tasks. In fact, many parallel processing vendors are now converging on Unix System V as an operating system standard (Russell and Waterman, 1987). The operating system allows the creation of up to six clusters of processors, which can each work on different tasks. The series also supports the FX/RT real-time executive for time-critical applications, and a number of standard interfaces. Alliant machines are commonly used for complex dynamic simulation, and real-time data acquisition and control. An example of such an application is given in Section 11.3.2. Alliant says that its entry into the real-time market was driven purely by customer demand and now estimates that around 20% of its business is in this area.

Of particular importance in the realm of real-time industrial control systems is the recent announcement by Alliant of a rack-mounted version of the FX/2800. This machine is essentially a scaled-down version of the 2800.

11.2.2 Crossbar Connections

In a crossbar architecture separate connections exist between each processor and each memory module (see Figure 11.2). This design permits all processors to access all memory modules, provided no two processors attempt to access the same one. The crossbar switch has no theoretical upper limit on the number of processor modules which can be added, and our simplified representation shows an eight processor system.

The C series of machines from Convex Computer Corporation utilizes a

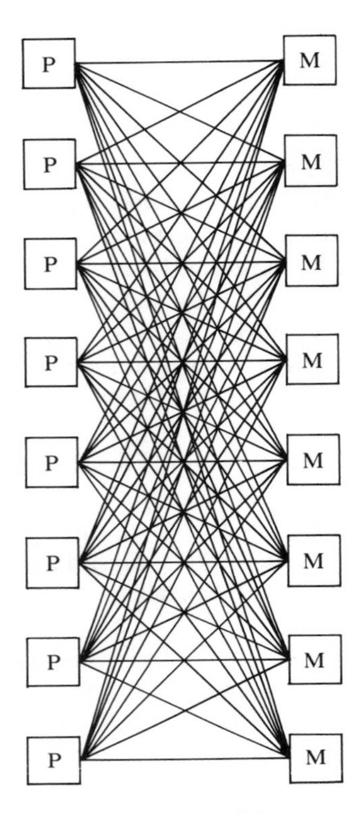

Figure 11.2 Crossbar connections: M = memory, P = processor

crossbar interconnect. The C240, with four processors, is the most powerful system in the C series. Each processor is a 64-bit unit capable of both scalar and vector operations, and has a throughput of 25 MIPS. The machine therefore has a potential 100 MIPS of compute resource.

The maximum shared physical memory in the C240 is 2 GB, accessed through the nonblocking crossbar. Each processor has a dedicated 200 $\mathrm{MB\,s^{-1}}$ port to memory.

The Convex Unit operating system supports a range of standard interfaces together with a very high-volume parallel interface. Convex have also explored the development of a real-time executive for time-critical applications. Convex says that with this development it is gearing up for entry into the real-time market-place, with particular emphasis on very-high-performance simulation, telemetry systems and real-time data capture and analysis.

11.2.3 Hypercube Connections

The hypercube provides multistep communication paths by connecting each processor to a subset of the other processors. In a hypercube of order n, each processor is directly connected to n other processors and their associated memories (see Figure 11.3, where $n = 4$). Two non-adjacent processors can only communicate by passing data via intermediate nodes. The hypercube architecture can support large numbers of processors (see also Chapter 12).

Intel's iPSC range of computers is based upon a hypercube architecture. The latest machine in the series is the iPSC/860. Like Alliant's FX/2800 this machine uses the advanced i860 RISC chip. The iPSC/860 can be configured with from 8 to 128 processors but, as a result of the hypercube architecture, intermediate configurations must use a number of processors which is a power of 2 (i.e. 16, 32 or 64). An iPSC machine with 128 i860 processors has the impressive potential to deliver a peak throughout in excess of 4000 MIPS. Some nodes in the iPSC/860 can alternatively use the 80386 processor. Although less powerful, this processor supports additional languages.

Interprocessor communication is achieved in the iPSC machines using the Direct Connect system. This system dynamically creates dedicated communications links between two processors, with a bandwidth of 2.8 Mb s^{-1}. For two non-adjacent processors this link is via intermediate nodes, although the system does not require message passing. When the communication is complete the link is released. Each node in the system has local memory of either 8 or 16 Mb. A fully configured machine therefore has up to 2 Gb of local memory.

The operating system NX/2, optimized for interprocessor communications, resides on each node. One 80386 node, the System Resource Manager, is dedicated to system work and software development. This node runs under Unix. The iPSC range supports a range of standard interface hardware.

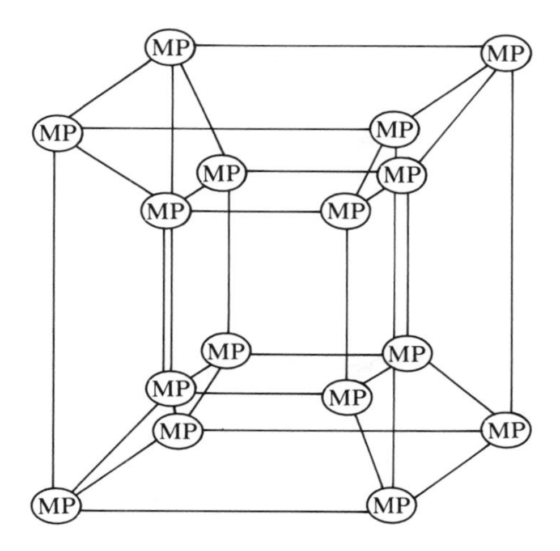

Figure 11.3 Hypercube connections of order $n = 4$: M = memory, P = processor

11.2.4 Multistage Switch Connections

Multistage switch architectures connect every processor card to every other processor card. Unlike the crossbar interconnection, however, dedicated connection paths between all processors are not required (see Figure 11.4).

BBN Advanced Computers has developed a range of machines utilizing the multistage switch interconnection. In its latest machine, the TC2000, any number of processors between 3 and 63 can be configured. Each processor is based upon the Motorola 88000 RISC processor series with 13 MIPS of throughput. A fully configured machine therefore has an aggregate capacity of 819 MIPS.

The processors in a TC2000 system are connected through a network interconnect system, the Butterfly multistage switch. The switch provides around 40 Mb s^{-1} communications bandwidth for each processor. All processors are able to access the potential 2 Gb of shared, physical, memory provided by a fully configured machine.

The TC2000 runs under the nX operating system, BBN's multiprocessor version of Unix. For time-critical applications the machine also supports the $\text{pSOS}^{+\text{m}}$ real-time executive. The TC2000 allows clusters of processors to be dedicated to particular tasks. One cluster running under $\text{pSOS}^{+\text{m}}$ could be dedicated to time-critical operations, another under nX to complex computations, while another might be used for interactive work. The machine supports standard interfaces. An overview of parallel programming tools may be found in Waterman (1988).

BBN Advanced Computers is aiming the TC2000 primarily at the real-time market, and in the past has delivered machines for high-performance simulation, telemetry and data analysis applications. A process control application of the BBN technology is described in Section 11.3.1.

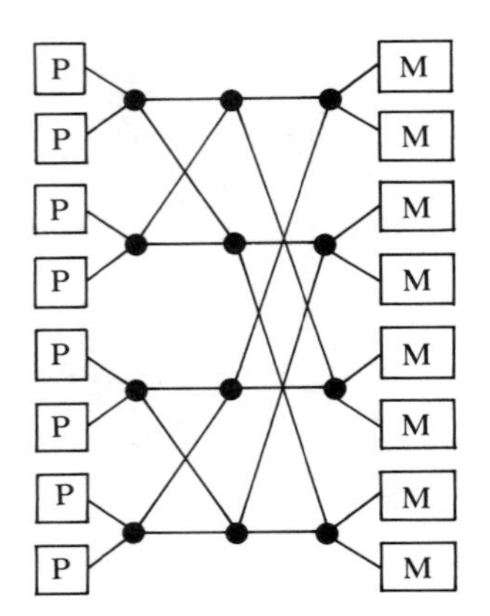

Figure 11.4 Multistage switch connection: M = memory, P = processor

11.2.5 Transputer Based Systems

In the United Kingdom, the transputer is undoubtedly the best known building block for parallel processing systems and no review of this subject would be complete without mentioning it. The transputer, developed by Inmos, is a VLSI microprocessor utilizing a RISC architecture. It has on-chip memory and, most importantly, links for point-to point communications with other transputers. The transputer was designed with concurrency in mind, and any number of transputers can be joined together using the communications links to create a multiprocessor system. The T800 version of the transputer has a peak throughput of 10 MIPS. The transputer has four full-duplex communications links each with a bandwidth of 20 Mbits s^{-1} for connection to other transputers. Transputers may be connected in any topology, including those outlined above.

Meiko has produced an integrated multiprocessing machine, the Computing Surface, based upon multiple transputers. The interconnection of processors is flexible and is done in software. This software configuration means that interprocessor communication is slower relative to the dedicated hardware architectures found in the machines described above. However, the Computing Surface can be configured with a very large number of processors. Currently, the largest machine has more than 1000 processors. Although the operating system does not have any true real-time capability, the Computing Surface does have the ability to dedicate specific processors as data port elements with a bandwidth of 80 MB s^{-1}.

In addition, much of the real-time control research work in the United Kingdom has been done on transputer based systems. For example, the control systems research group in the Department of Mechanical Engineering at the University of Glasgow is currently using a ten-processor Sension system for the real-time control of a robot arm. The team is currently working to link the Sension to a 32-processor Computing Surface to achieve higher throughput.

11.3 CASE STUDIES

11.3.1 CASE STUDY 1: BBN (PROCESS CONTROL)

As an illustration of the application of a parallel processing machine in the process industries the FALCON system will be described. This system was developed by BBN Systems and Technologies using BBN's Butterfly machine. The Butterfly is the predecessor of the TC2000 described above, and has the same architecture. FALCON (fault location and control optimization planner) is the prototype of a system designed for a chemical process plant. FALCON aims to monitor and optimize the production of paracetamol as it proceeds through a four stage chemical reaction. The FALCON system combines a number of advanced technologies: parallel processing, AI rule-based diagnostic reasoning, real-time process simulation, object-oriented programming and colour graphics, and Ethernet (TCP/IP) communications.

The system is intended for deployment and use in the following ways:

1. As a stand-alone simulator for training plant operators.
2. For on-line process monitoring and simulation.

Using FALCON the operators are continuously able to improve the production quality and efficiency of the plant.

Process Overview

The manufacture of paracetamol is a batch process with four primary reaction stages. In addition, production includes subfunctions such as the centrifuge, filtering and drying stages. FALCON includes a detailed simulation of the production process. The system monitors batches to detect imperfections in production and to diagnose faults.

FALCON Architecture

The FALCON system has four main functional modules: simulation, rule-based diagnostics, communications and graphics. The simulation, diagnostic rules and communications modules are implemented as parallel subsystems on the Butterfly. The graphics module is implemented on a Symbolics AI workstation which communicates with the Butterfly (see Figure 11.5).

Simulation

The process is stimulated using object-oriented techniques in Lisp. The simulation contains both deterministic and stochastic representations of the manufacturing

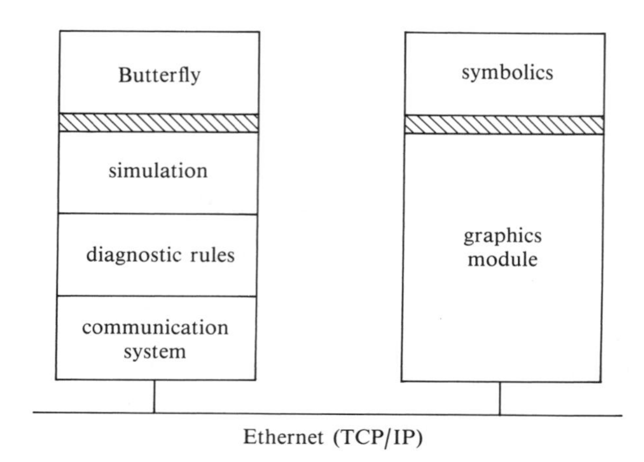

Figure 11.5 FALCON architecture

process. The Butterfly architecture allows the user to run multiple simulations simultaneously on separate nodes. This facility allows the user to conduct multiple 'what if?' scenarios predicting future quality and faults. Comparing results in real-time the user can take anticipative control actions, thereby optimizing production quality.

AI Rule-based Diagnostics

The diagnostic rules in FALCON were implemented using BBN's Butterfly AI tools package (Allen and Sridharan, 1987). The diagnostic rule system is designed to monitor both real and simulated variables. The rules identify problems which could potentially affect the quality and efficiency of the production process.

Communications

The communications between the Butterfly and Symbolics machines is implemented using the TCP/IP protocol running over Ethernet. The communication network provides a link between the simulation and diagnostic rules system (running on the Butterfly) and the colour graphics (running on the Symbolics workstation).

Graphics

The colour graphics displays used in FALCON are based upon the Steamer graphics package. The Steamer graphics system was developed previously by BBN for an AI-based steam plant simulation and training system. Steamer is an object-oriented graphics package which runs on Symbolics workstations. The graphics in FALCON animate the chemical flows in the paracetamol production process.

Using the System

In a typical use of FALCON the intelligent monitoring and display of actual plant parameters takes place on one cluster of processor nodes. In parallel, on another cluster, multiple batch production simulations can be run. This allows the operator to anticipate faults, to take corrective actions and to vary process parameters to optimize production.

Animated process graphics are produced by FALCON. In simulation and training mode a pop-up menu allows the operator to introduce a typical process fault. The operator may also focus on a particular part of the process simulation.

The manufacturing process for the production of paracetamol takes 12 hours. The FALCON system running on the Butterfly completes a simulated batch production in under 12 minutes. The flexibility of the parallel machine allows the simultaneous integration of a number of advanced technologies. Thus, the power and flexibility of the Butterfly allows the implementation of an advanced system which would not be feasible using conventional computer systems.

■ 11.3.2 CASE STUDY 2: ALLIANT (VEHICLE SIMULATION|CONTROL)

The following is an example of very high-fidelity dynamics simulation; the simulator incorporates real-time control of a full-scale mock-up of the real system. Thus we have a full motion, person-in-the-loop automotive simulator. The Dynamic Driving Simulator was developed by Mazda using Alliant's FX/80 technology (the predecessor of the FX/2800 range outlined above). Automotive manufacturers around the world are increasingly using advanced simulators of this kind for the development of new designs and technologies.

The Dynamic Driving simulator utilizes a range of advanced simulation techniques. The aim is to provide the simulator operator with an experience as close as possible to driving a real car. The simulator has a cockpit mock-up identical to that of the real car.

The driver's visual experience is created by projecting the scenery using real-time computer graphics generation. A high-fidelity synthesizer system generates a full spectrum of noises experienced during driving. Finally, a real-time control system drives large-capacity electric actuators to produce motions in all directions. Lateral forces of up to 0.8*G* are produced using a linear motor and rails.

In the FX/80 system used to implement the simulator two processors are dedicated to the real-time control tasks associated with the actuators and sensors. These processors run under the FX/RT real-time executive described above. Two further processors are dedicated to dynamics computation tasks. Other processors handle complex simulations related to aerodynamics (utilizes computational fluid dynamics techniques) and combustion analysis. An array of graphics processors (an integrated extension to the FX/80 system) handles the real-time graphics requirements.

This application demonstrates the technologies of complex dynamics simulation, 3D graphics and real-time control all implemented in an integrated parallel computing machine.

11.4 CONCLUSIONS

The real-time control industries have very high compute resource requirements. In addition, safety and economic benefits can be gained from the use of high-fidelity real-time simulation and advanced computing techniques such as knowledge based systems. This leads to the demand for higher performance, cost-effective, computing engines.

We have seen that a number of parallel processing machines are now commercially available. Some of these machines are equipped with real-time operating system executives and are targeted at the time-critical market. Other vendors currently have firm plans to move into this area. The particular areas likely to benefit most from these developments are: high-fidelity simulation for operator training and process optimization (the case study described above is a good example of this), real-time data acquisition, analysis and control, the application of knowledge based systems, and high volume remote telemetry systems.

Parallel processing machines provide the basis for the cost-effective application of these advanced techniques. At present the introduction of these machines in real-time control situations is at an early stage. However, the technology is now available and this trend looks set to accelerate in the near future.

REFERENCES

ALLEN, D. C. and SRIDHARAN, N. S. (1987) 'Applications of the Butterfly parallel processor in artificial intelligence', in *Parallel Computation and Computers for Artificial Intelligence* (ed. J. S. Kowalik), London: Kluwer.

HUNT, K. J. (1991) 'Parallel processing applications in AI', in *Applications of Artificial Intelligence* (ed. K. Warwick), London: Peter Peregrinus.

RUSSELL, C. H. and WATERMAN, P. J. (1987) 'Variations on UNIX parallel processing computers, *Commun. ACM*, **30**, 1048–55.

WATERMAN, P. J. (1988) 'Programming tools for parallel processing', *Signal Magazine*, April.

12

Using Parallel Algorithms in the Design of Control Systems

E. Rogers and Y. Li

12.1 INTRODUCTION

The past 30 years, in particular, have seen major and far-reaching developments in the design and implementation of control systems. For example, consider the case of linear time-invariant systems described in state-space or transfer-function terms. Then for such systems most of the basic theoretic questions have been answered and a large volume of research is now directed towards robust control, i.e. accounting for the effect of unmodelled (or unknown) plant dynamics.

Computing is, of course, an integral part of the design and implementation of control systems. Further, the rapid developments in this essentially enabling technology, particularly in the last decade, have been exploited to great effect by the general control systems community. In the case of off-line design, this has produced efficient, reliable and portable techniques, generally in the form of numerical algorithms and software, for a large number of commonly used linear time-invariant design procedures.

Currently, there is an active commercial market in interactive 'user friendly' software packages which (typically) contain a cross-section of these techniques. Generally, the latter work in a routine and reliable manner on problems whose state dimension is of the 'order of hundreds'. Further, this statement does not assume the presence of a special structure, such as bandedness or symmetry in the matrices of the defining state-space model. If such a structure is present, and can be appropriately exploited, then clearly such larger problems become routine.

These developments in the general subject area have naturally led to a reappraisal or realistic consideration of systems with greatly increased size and complexity. For example, there is increasing interest in the control of complex distributed systems such as large space structures or interconnected power systems networks. The dynamics of these systems typically extend over 'wide' frequency ranges or spatial domains and hence model order reduction techniques are often not appropriate.

Successful application of established design algorithms to such systems will only be (effectively) achieved, if it is possible, by suitable reformulation or 'algorithmic engineering', to allow for an increase in model dimension by at least one order of magnitude with negligible reduction in reliability. Further, assuming that the underlying theoretical questions can be satisfactorily resolved, it is clear from the work to date that the computation loads generated may well approach or exceed the effective operating ranges of conventional sequentially based machines.

One possible solution to problems such as those outlined above is to make appropriate use of parallel processing. Research has already begun on the design and exploitation of novel parallel processing algorithms (and associated computer architectures) for present and future problems – numerical and otherwise. As a result, it is known that many existing control algorithms can be restructured to exploit (ideally to the maximum possible extent) the capabilities of parallel machines. It is also clear, however, that completely new algorithms will have to be developed to solve problems previously considered impractical or impossible. Obviously, such algorithms will have to exploit fully any generic structural properties present in the application area under consideration – such as sparse or otherwise structured matrices in the defining models of large space structures.

This chapter gives a tutorial-level survey of the current 'state of the art', plus some open research problems, in this general area using, as an illustrative basis, some algorithms and programming techniques for use with hypercube multiprocessors – a particular class of parallel computer. The algorithms detailed are for the computation of a multivariable frequency response matrix from the plant state-space description and the solution of an algebraic Riccati equation. The need for such algorithms is motivated by a brief review of an application area for each of them but their final form is sufficiently generic to permit application to a wide range of such areas.

The main body of this chapter begins in Section 12.2 with a brief description of the two applications areas which motivate the need to treat matrices of very high dimension in certain applications of modern control and systems theory. Section 12.3 reviews hypercube computing systems with particular emphasis on their architecture and programming structures. Detailed algorithm development, plus software considerations, for the two sample problems are then given in Section 12.4. Finally, some open research problems, of both specific and general interest, are discussed in the concluding section.

12.2 APPLICATIONS AREAS

12.2.1 2D Systems

Basically, two-dimensional, or 2D, systems propagate information in two separate directions, termed horizontal and vertical respectively. Applications areas include picture processing, image processing in geophysical systems and processes with repetitive dynamics. The development of a control and systems theory for such systems has been an active research topic over the last two decades with particular emphasis in the discrete case on the so-called Roesser state-space model. This general-purpose model, together with numerous extensions and refinements, can be used to describe examples which are recursive in the positive quadrant. The text by Lim (1990) is one source of a comprehensive overview of the theoretical and practical aspects of the modelling and control of (discrete) 2D linear systems.

This chapter only considers repetitive, or multipass, processes which are a class of dynamic systems characterized by a recursive action consisting of a number of sweeps, or passes, of finite duration – the pass length – through a set of dynamics. On each pass an output, or pass profile, is produced which acts as a forcing function on, and hence contributes to, the new pass profile. In a more general setting, it is the previous $M > 1$ pass profiles which contribute to the current one. The integer M is termed the memory length; such processes are simply termed non-unit memory, and if $M = 1$ the special case of a unit memory process is obtained.

Industrial examples of repetitive processes include long-wall coal cutting systems and certain classes of metal rolling. Further, an important systems theoretical role for these processes has recently been established in the analysis and control of certain classes of learning systems with application to, for example, robot manipulators. Rogers and Owens (1992a) and the relevant references therein describe these and other examples in considerable detail.

Two parameters are required to specify a variable in a repetitive process. In particular, it is necessary to specify the 'position' (or time) along a pass and the pass index or number. Here the notation used is of the form $Y_k(t)$, where Y denotes the variable under consideration, k denotes the pass number, and $0 \leqslant t \leqslant \alpha$, denotes the position along a pass of (assumed) constant finite length a. Using this notation, the following formal definition can now be given.

Definition 12.1

A non-unit memory repetitive process is a dynamic system where the previous pass profiles $Y_{k+1-j}(t)$, $0 \leqslant t \leqslant \alpha$, $1 \leqslant j \leqslant M$, act as forcing functions on, and hence contribute to, the new pass profile $Y_{k+1}(t)$, $0 \leqslant t \leqslant \alpha$, $k \geqslant 0$.

Consider now the case of linear dynamics where this is a valid assumption, at least for initial simulation and control related studies, in a significant number of cases. Then in this case an important subclass is so-called differential non-unit memory linear repetitive processes with state-space model:

$$\begin{aligned} \dot{X}_{k+1}(t) &= AX_{k+1}(t) + BU_{k+1}(t) + \sum_{j=1}^{M} B_{j-1}Y_{k+1-j}(t) \\ Y_{k+1}(t) &= CX_{k+1}(t), 0 \leqslant t \leqslant \alpha, k \geqslant 0 \end{aligned} \tag{12.1}$$

Here $Y_{k+1}(t)$ is the $m \times 1$ current pass profile, $X_{k+1}(t)$ is the $n \times 1$ current pass state vector and $U_{k+1}(t)$ is the $l \times 1$ current pass input vector. Further, no loss of generality arises in assuming the special form of the output equation used in (12.1), i.e. deleting 'direct feedthrough' terms arising from the current pass inputs and the previous pass profiles. Similarly, the following initial conditions can be imposed without loss of generality:

$$d_{k+1} = 0, k \geqslant 0$$
$$Y_{1-j}(t) = 0, 0 \leqslant t \leqslant \alpha, 1 \leqslant j \leqslant M \tag{12.2}$$

i.e. zero state initial conditions on each pass and zero initial pass profiles.

As an alternative to (12.1), a 2D transfer-function matrix description can be used; see Rogers and Owens (1992a) for complete technical details. This is defined in terms of the standard Laplace variable (s) and another complex variable (z) which, in effect, takes account of the basic interpass interaction. This has a number of equivalent forms and the one which is required here is

$$Y(s,z) = G(s,z)U(s,z) \tag{12.3}$$

where $G(s,z)$ is the $m \times l$ 2D transfer-function matrix defined by

$$G(s,z) = (I_m - \sum_{j=1}^{M} G_j(s)z^{-j})^{-1} G_0(s) \tag{12.4}$$

where

$$G_0(s) = C(sI_n - A)^{-1}B \tag{12.5}$$

and

$$G_j(s) = C(sI_n - A)^{-1}B_{j-1}, 1 \leqslant j \leqslant M \tag{12.6}$$

One immediate use of $G(s,z)$ is to provide the block diagram interpretation of the process dynamics shown in Figure 12.1 or, equivalently, Figure 12.2. This shows that these processes can be regarded as a dynamic precompensator followed by a positive feedback loop with unity gain in the forward path and dynamic elements in the feedback loop. These elements are the basic interpass interaction.

The basic unique control problem for repetitive processes is the possible presence in the output sequence $\{Y_k\}_{k \geqslant 1}$ of oscillations which increase in amplitude from pass to pass. Such cases clearly require strong control action and it is known

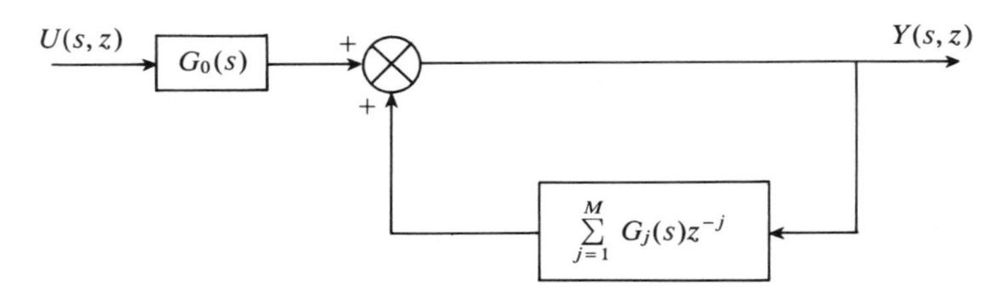

Figure 12.1 Block diagram interpretation of $G(s,z)$

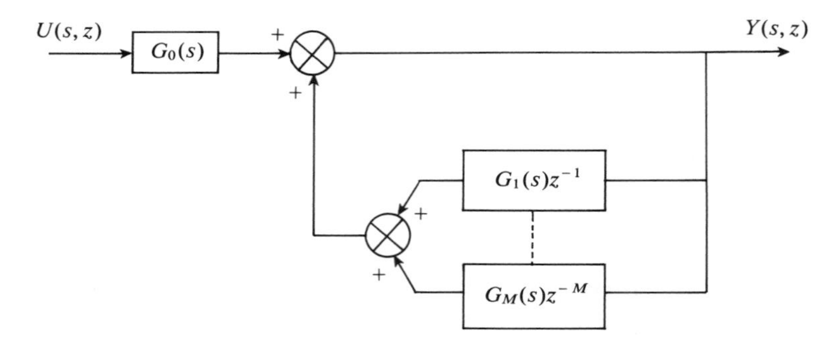

Figure 12.2 Alternative version of Figure 12.1

that direct application of standard techniques will almost always end in failure. This, in turn, has led to the development of a rigorous stability theory as the essential starting point for a comprehensive control theory, based on a general abstract model formulated in Banach space terms.

In effect, this stability theory consists of the distinct concepts of asymptotic stability and stability along the pass respectively (see Rogers and Owens, 1992a, for complete details). Further, the former is a necessary condition for the latter which is required in all practical applications. To provide a basic explanation of this fact, note that asymptotic stability ensures that $\{Y_k\}_{k \geqslant 1}$ converges to a steady, or so-called limit, profile Y_∞ as a function $\alpha < +\infty$ and stability along the pass is independent of this parameter. This, in turn, means that asymptotic stability alone would permit 'growth' terms in the dynamics along a pass – an obviously totally unacceptable situation.

Asymptotic stability of (12.1) holds since this property is only dependent on the presence of 'direct feedthrough' terms from the previous passes to the current one, i.e. the output equation contains terms of the form $D_j Y_{k+1-j}(t)$, $1 \leqslant j \leqslant M$, which have been deleted here without loss of generality. To introduce a set of necessary and sufficient conditions for stability along the pass, first define the so-called interpass transfer-function matrix as

$$G(s) = \begin{bmatrix} 0 & I_m & 0 \\ & & I_m \\ G_M(s) & & G_1(s) \end{bmatrix} \tag{12.7}$$

where, in effect, this matrix describes the combined effects of the previous pass profiles. Then the following is the required result.

Theorem 12.1

Under certain well-defined controllability and observability assumptions, the process of (12.1) is stable along the pass if, and only if,

1. all eigenvalues of the matrix A have strictly negative real parts; and
2. all eigenvalues of $G(s)$ have modulus strictly less than unity, $s = i\omega$, $\forall\, \omega \geq 0$.

Other sets of necessary and sufficient conditions for stability along the pass exist and are detailed in Rogers and Owens (1992a) and the relevant references therein. One major advantage of Theorem 12.1 is that its conditions and, in particular, 2 can also be used as an effective basis for computer aided control systems design. For example, with a little extra work, this result can be used to provide relative stability/performance indicators in the spirit of (say) gain and phase margins for standard linear systems. This and related subjects are detailed in Owens and Rogers (1992).

In terms of applying Theorem 12.1, the major computational load will clearly arise with the condition listed under 2, i.e. in the context of this chapter, with the construction of $G(i\omega)$, $\forall\, \omega \geqslant 0$ and the subsequent calculation of its eigenvalues. The problem which will often arise here is that the resulting computational loads required to complete these tasks, using any one of numerous well-known algorithms, on conventional sequentially based computers will exceed their effective operating range. This can arise in a number of ways as result of the effects of combinations of the following:

1. The values of m, n and l (the output, state and input vector dimensions respectively) coupled with the (effective) frequency range to be covered.
2. The memory length M, typically in the range $50 \leqslant M \leqslant 100$ for industrial applications such as bench mining systems.

As the first step to a 'complete parallel solution', Section 12.4 develops basic hypercube algorithms for the construction of $G(i\omega)$, $\forall\, \omega \geqslant 0$. In this respect, the following structural properties of $G(s)$ of (12.7) should be noted:

1. The block companion structure.
2. The link between the constituent elements $G_j(s)$, $1 \leqslant j \leqslant M$, of (12.6). In particular, these transfer-function matrices differ only in the driving matrix B_{j-1}.

These properties will play a major role in the algorithm development studies.

12.2.2 Large Space Structures

Balas (1982) and Mackay (1983) are two possible sources of a comprehensive treatment of large space structures – with particular emphasis on modelling and control problems. Here, however, it is only necessary to summarize the main points. First note, therefore, that these structures are physically 'quite large' but mass limitations mean that they are very flexible and often have a 'large' number of closely spaced resonances at low frequencies. Hence model order reduction techniques, with obvious advantages in terms of the computational loads generated by simulation and control studies, are (typically) not appropriate.

This situation is often compounded by the requirements imposed for 'acceptable control'. Typically, stringent specifications are demanded on pointing accuracy, vibration suppression and shape maintenance. These require active control with a high controller bandwidth which leads to dynamic interaction between the controller and the structural vibration nodes.

Generally, these structures are modelled by partial differential equations and hence an infinite dimensional state vector. Usually, however, the sensors and actuators are regarded as point devices. Consequently the control and measurement vectors, and hence the controller, are finite dimensional.

In terms of controller design and, in particular, the associated computations, two approaches exist. The first of these is to use finite element methods to produce a 'large' finite dimensional model from the partial differential equation(s). Alternatively, the design studies could be undertaken on the partial differential equation(s) model and the result discretized. Both of these approaches have their advantages and disadvantages, whose implications are almost a separate subject area. Greene and Stein (1979) give an extensive treatment of this subject which is not considered further here. Instead, second-order models are (briefly) considered since they are commonly used to provide (realistic) approximate finite dimensional models for controller design purposes.

The generic form of the second-order or so-called Rayleigh model is the matrix differential equation:

$$\begin{aligned} &M\ddot{X} + C\dot{X} + KX = DU \\ &Y = PX + S\dot{X} + DU \end{aligned} \tag{12.8}$$

where Y is the $m \times 1$ output vector, X is the $n_1 \times 1$ state vector, and U is the $l \times 1$ input vector. In general terms, (12.8) is the matrix generalization of the familiar single-input single-output second-order model used to describe simple arrangements such as spring–mass–damper systems. Often it is possible to impose the following structure on the defining matrices:

$$\begin{aligned} &M = M^{\mathrm{T}} > 0 \\ &K = K^{\mathrm{T}} \geqslant 0 \\ &X = C_1 + C_2, C_1 = C_1^{\mathrm{T}} \geqslant 0, C_2 = -C_2^{\mathrm{T}} \end{aligned} \tag{12.9}$$

where $>$ and $\geqslant$ denote positive definiteness and positive semi-definiteness respectively.

This Rayleigh model has many application areas and can, for example, arise as a model of the system itself derived by applying Newton's laws or, as here, from applying finite element methods to a dynamic continuum problem. Generally, M is termed the mass matrix, C the Rayleigh (damping) matrix and K the Hooke or stiffness matrix. In the decomposition of the Rayleigh matrix, C_1 is termed the dissipation matrix and represents structural damping forces, and C_2 is the gyroscopic matrix which represents such forces.

The following should be noted:

1. A more general stiffness matrix can be used to take account of circulatory, as opposed to only conservative, force fields.
2. Additional assumptions are often imposed on the matrices of (12.9) with the general objective of easing analytical analysis.
3. Arnold (1984), for example, gives an in-depth treatment of this model (and its limitations).

The standard approach to the analysis of (12.8) in a control systems context is to augment the state vector and hence obtain an equivalent first-order model. As one obvious choice, define the augmented state vector $W = [X^{\mathrm{T}}, \dot{X}^{\mathrm{T}}]^{\mathrm{T}} \varepsilon R^n n = 2n_1$ to obtain

$$\dot{W} = \begin{bmatrix} 0 & I_n \\ -M^{-1}K & -M^{-1}C \end{bmatrix} W + \begin{bmatrix} 0 \\ M^{-1}D \end{bmatrix} U$$

$$Y = [P \;\; S]\, W + DU \tag{12.10}$$

Alternatively, the generalized realization

$$\begin{bmatrix} I_n & 0 \\ 0 & M \end{bmatrix} \dot{W} = \begin{bmatrix} 0 & I_n \\ -K & -C \end{bmatrix} W + \begin{bmatrix} 0 \\ D \end{bmatrix} U \tag{12.11}$$

could be used. One advantage of this realization is that it avoids the need to construct the inverse of M which is often desirable for numerical reasons.

Both of these state-space realizations and, in particular, the defining matrices have a special structure which is very often ignored in subsequent calculations using standard algorithms. This has prompted research on developing realizations (of the 'standard' first-order form, i.e. $\dot{X} = AX + BU$ or $E\dot{X} = AX + BU$) which attempt to exploit, for example, any special structure associated with the damping matrix C. Further, there has been research on developing certain essential elements of a systems and control theory for direct application to the second-order model of (12.8).

The basic objective of this work is to exploit structural properties, such as symmetry, in a more effective manner. Bender and Laub (1985), for example, detail work in this general area. One (potentially) limiting factor of this approach, however, is the absence to date of numerically reliable canonical forms for (12.8).

Suppose now that (12.8) is available and rewritten in first-order form

$$\dot{W} = AW + BU,\; W \in R^n \tag{12.12}$$

Then a wide range of well-established techniques can be used to complete the basic controller design exercise. Here, however, the interest is in the following so-called generalized algebraic Riccati equation which arises in the solution of linear quadratic optimal control and filtering problems:

$$A^{\mathrm{T}}XE + E^{\mathrm{T}}XA - E^{\mathrm{T}}XRXE + Q = 0 \tag{12.13}$$

All entries here are $n \times n$ matrices, X is the unknown, $Q = Q^{\mathrm{T}}$, $R = R^{\mathrm{T}}$ and nonnegative definite, and $|E| \neq 0$. Further, $Q \geqslant 0$ is required in a number of cases of practical interest, such as the linear quadratic regulator.

Note that setting $E = I_n$ in (12.12) generates the standard algebraic Riccati equation which is central to the linear quadratic optimal control problem for (12.12).

The basic question of the existence of a unique stabilizing solution of (12.13) satisfying $X = X^{\mathrm{T}}$ has been extensively studied by, for example, Molinari (1977). This has led to the conclusion that there are three sets of conditions which cover a large number of cases of interest. In the context of this chapter, however, it suffices to assume that one of these sets holds.

Computing the solution, X, of (12.13) can be undertaken using any one of numerous well-tested and reliable algorithms, with supporting software, whose effective operating range is generally accepted to be problems whose state dimension does not exceed the range 100 to 300. Note, however, that these algorithms (typically) have computational times proportional to n^3 and storage requirements of the order of n^2. This is an immediate source of difficulty for applications such as large space structures, where it is by no means unusual for n to be of the order of 10^3 or above, and the resulting computational load is outside the effective operating range of sequentially based computers. Section 12.4 therefore develops basic hypercube algorithms for solving (12.13).

12.3 HYPERCUBES

This section summarizes the required background on hypercube multiprocessors with particular emphasis on architectural and programming aspects. A more detailed treatment can be found in Wiley (1987), Karp (1987) and the proceedings of three hypercube conferences edited by Heath (1986, 1987) and Fox (1988). Included in these references are examples of their use in nonscientific computing problems, such as those arising in artificial intelligence studies, and comparisons with alternatives based, for example, on systolic arrays. Further, it is clear that hypercube machines are already a useful contributor to research-related activities.

In terms of the problems considered in this chapter, it is so-called distributed memory hypercubes which are of use. Basically this means that the processors execute programs independently and the only means of communication between them is by message passing. As such, therefore, they are an example of a MIMD computer with no shared memory.

At this stage, note that a fundamentally desirable property of parallel algorithms for problems from the general area of linear control systems is so-called scaleability. Basically, this means that an increase in the problem size (resulting from an increase in the state dimension n in the case of the Riccati equation of (12.13), for example) can be dealt with by adding processors, i.e. without recourse to the (possibly nontrivial) task of reformulating the algorithm. This requirement is met by a distributed, as opposed to a shared, memory architecture. Note that hypercubes (typically) cost less for the same computing power as alternatives, such as Cray machines. Further, the algorithms of Section 12.4.2, for example, are essentially independent of the hypercube topology and could be used on any distributed memory multiprocessor equipped with a message passing facility and an 'appropriately large' number of interconnections. These factors justify the use of the hypercube multiprocessor (at least for initial studies), despite the fact that it can prove more difficult to use effectively than some of its competitors.

Basically, the hypercube topology is a particular pattern of interconnections between the processors of a computer which can be visualized, see Figure 12.3, as a cube of dimension d. As noted above, the interest here is in a hypercube multiprocessor with no shared memory. Further, the term 'hypercube' only applies to the topology of the communications network linking the processors. Hence most of the remainder of this section also applies to distributed memory multiprocessors based on other topologies.

In Figure 12.3, each vertex of the cube is a processor, more commonly termed a node, and the edges are the communications lines or links. Clearly, the number of nodes is always given by $p = 2^d$ and these are only nearest-neighbour connected (a highly desirable feature in parallel processing architectures). As a result, messages intended for more distant nodes must be forwarded through one or more intermediate nodes. Further, of all the interconnection schemes investigated to date, the hypercube is (arguably) the most successful.

The maximum distance between any two nodes in Figure 12.3 is the dimension d and the total number of interconnections is $pd/2$. In contrast, a fully connected system would allow direct communication between all its processors. This, in turn, means that the number of communications links required would be proportional to p^2 which is (prohibitively) expensive if p is 'large'. At the opposite end of the scale, a ring is one of the simplest possible topologies which requires only p links but has

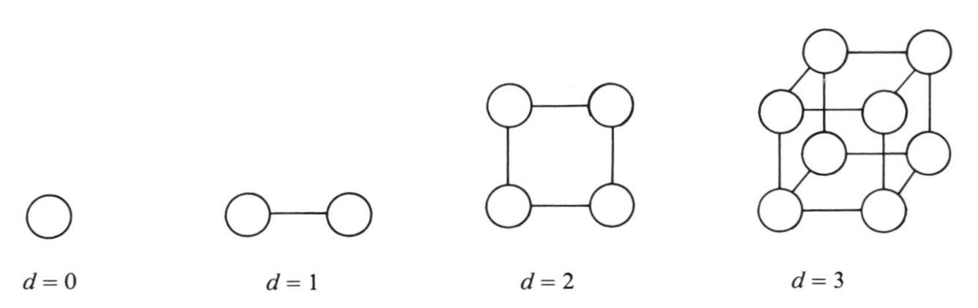

Figure 12.3 Hypercube topology for $d = 0, 1, 2, 3$

a maximum internode distance of $p/2$ and hence 'slow' communications for many tasks. Consequently, the hypercube is a 'good' compromise between these two extremes.

One other major advantage of the hypercube of note here is its flexibility which can be exploited to simulate other connection schemes. For example, by using only an appropriate subset of the available communications links, a hypercube can be considered as a ring, a two or three-dimensional mesh, or a tree. Note also that optimal topology selection is very heavily dependent on the application being considered.

The following is a summary of the main features of the nodes:

1. Each processing node can be made as powerful as required.
2. Each node is equipped with its own processor and memory, a copy of the operating system, and copies on any applications programs to be used. Consequently each of them can be regarded as an independent computer.
3. Nodes may run different programs or operate copies of the same program on different data sets.

In operation, the processing nodes are usually supplemented by another processor termed the host or cube manager. This controls access to the cube itself and is used for the following general functions:

1. Program development.
2. Input and output facilities.
3. User interface.

Further, it is not recommended practice to use the host itself either for computational purposes or to coordinate this activity in the cube. The reason is that such use creates a bottleneck which (usually) slows program execution. Note also that host/node communication is often much slower than that between two nodes.

A high-level language, plus special subroutines for message passing, is usually employed to program a hypercube. For example, Fortran and C have been very popular in the scientific applications considered to date. Administrative functions such as data input, presentation of results, and starting the node programs are undertaken by a program running on the host computer. All productive (real) work is done by the node programs and very often scientific applications use a single-node program duplicated in all of them.

Often applications require specialized processing. To provide this, a node can determine its identity, as a node number indicator from 0 to $p-1$. For example, suppose that matrix computations are being undertaken. Then in such cases the node can identify the columns of the matrix it is working on.

Unless the application under consideration is totally parallel (rarely the case), the node programs must communicate with each other during problem solving. Consequently, the manufacturers (usually) provide a node operating system and utility subroutines for this purpose which, in effect, complement the basic message passing functions of 'send' and 'receive'. In a more general setting, communication between the nodes, or segments, of an algorithm is a crucial factor in the successful application of parallel processing.

Typically, internode communication is slower than computation. This, coupled with 'high' start-up times on messages, makes it highly desirable to send a few 'long' messages rather than many 'short' messages which, in control-related computations, will often consist of a vector or a column of a matrix. Hence, for efficiency, communications should be between adjacent processors, or nearest-neighbour based, if at all possible. To deal with cases where this is not possible, most hypercube operating systems have an automatic message passing system between non-adjacent nodes.

Problem decomposition, i.e. the generation of algorithms, is obviously crucial to the successful application of parallel processing. In some cases, there is an obvious parallel structure which can be exploited, such as when the problem domain is a physical plate or rod where each processor can be allocated to a particular region to yield nearest-neighbour connections. Consider also matrix computations which are of interest here. Then one extensively used method of splitting such computations is termed data parallelism, i.e. each node is assigned certain columns of the matrix which it works on. The basic essential objectives of this are the following:

1. To retain the orientation of the columns in the original algorithm(s).
2. To establish and maintain a balance of the computational load amongst the processors.

Clearly the first of these objectives is required for theoretical reasons and if the second is not achieved, to within at least acceptable bounds, then the processors present will not be fully used. This, in turn, will lead to a degraded performance from parallel algorithms since the processors not fully employed will be idle during the time taken by the overloaded ones to complete their workload. Consequently the objective of so-called load balancing is for each processor to perform an equal share (within 'acceptable' bounds) of the total workload.

Load balancing is another key factor governing the successful application of parallel processing and is, in general, a nontrivial problem. This general problem area can be decomposed into two parts, termed static and dynamic respectively. In the case of the former, an a priori estimate of the work distribution is available, as in solving sets of dense linear equations, and hence load balancing can be immediately incorporated into a specific applications program. Dynamic load balancing is the case where no a priori estimate is available and the quantity of work being assigned to individual processors only becomes known during the actual program execution. Specifically, as the computations evolve, different processors can end up being responsible for unequal quantities of work. One example of dynamic load balancing is solving partial differential equations using adaptively generated grids.

Cybenko (1988) and Cybenko and Allen (1991), together with the relevant references therein, detail progress to date and open research problems in the general

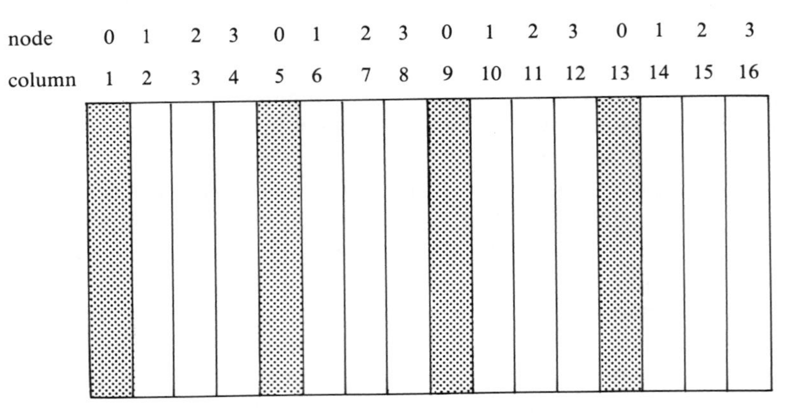

Figure 12.4 Column-wrapped distribution for a hypercube with four nodes

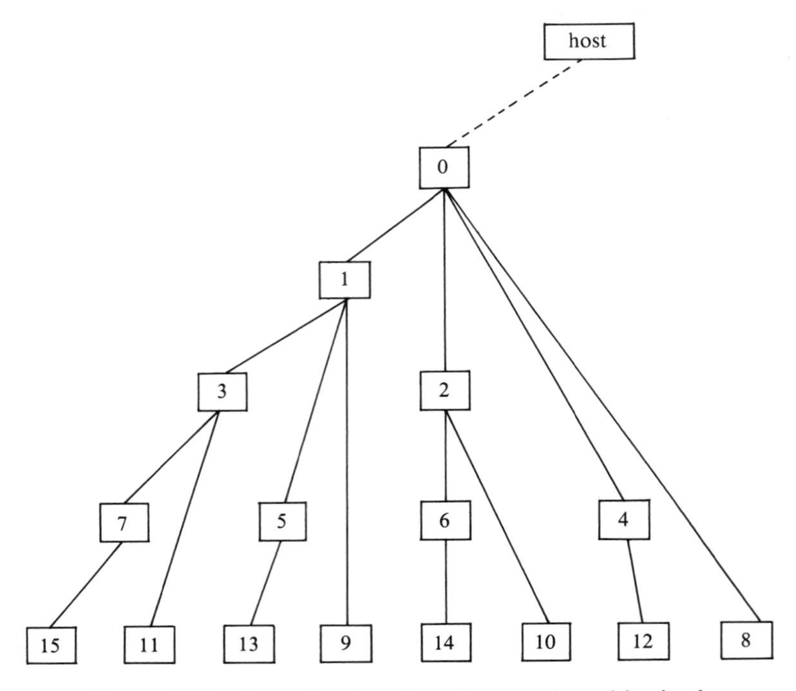

Figure 12.5 Spanning tree for a hypercube with $d = 4$

area of dynamic load balancing. In this work, however, it is matrix computations which are to be executed and one approach to load balancing is to use a so-called column-wrapped distribution. This approach has already been used for other research work on a hypercube and in the case of an $n \times n$ matrix it means that each node stores and processes approximately n/p of its columns. Effectively, it stores these columns in its memory and performs any required computations on them. Figure 12.4 illustrates this approach for a four node hypercube with $n = 16$ and column j is in node $(j-1)$ mod p(i.e. the remainder on dividing $j-1$ by p).

Using this column-wrapped approach, communications between nodes are both nearest neighbour and quite regular in structure. This latter property is also a very desirable feature of parallel architectures in general since it typically makes expansion of a given case much easier in construction or fabrication terms. Typically, internode communication is undertaken using broadcasts and tree-oriented operations where, for example, Figure 12.5 illustrates a spanning tree for the case of $d = 4$. To illustrate the basic mechanism, suppose that a global maximum must be found after each node has found its local maximum – a problem which is common in control-oriented computations. Here, values are passed up the tree until the so-called root node O has the maximum value which is then passed to the host for display to the user.

12.4 ALGORITHMS

This section develops algorithms for the problems of Section 12.2, i.e. the construction of $G(s)$, $s = i\omega$, $\forall \omega \geq 0$, of (12.7) and the solution of the Riccati equation of (12.13). Software development and expected performance from these algorithms is also discussed. The analysis of this section is based on Laub and Gardiner (1988) and Rogers and Owens (1992b) for the multivariable frequency response, and Gardiner and Laub (1991) for (12.13).

12.4.1 Multivariable Frequency Response Matrix

Return to the differential non-unit memory linear repetitive process of (12.1) and define the so-called augmented previous pass driving matrix as

$$D = (B_0, B_1, \ldots, B_{M-1}) \tag{12.14}$$

Consider also the construction of the frequency response of the interpass transfer-function matrix $G(s)$ of (12.7), i.e. the complex valued matrix resulting from setting $s = i\omega$ for N values of ω where typically $N \gg n$. Then it follows immediately that the nontrivial problem here is the construction of the frequency response matrix

$$\tilde{G}(i\omega) = C(i\omega I_n - A)^{-1} D \tag{12.15}$$

over the same range of ω. This is just the frequency response matrix for a linear time-invariant system parameterized by the state-space triple (A, D, C). Hence the analysis which follows, based on the original work of Laub and Gardiner (1988) with refinements by Rogers and Owens (1992b), can be applied to any such system.

Basically, the development of algorithms for hypercube computation of $\tilde{G}(i\omega)$ can be split into two distinct cases. The first of these is the case where the model (i.e. the state-space triple (A, D, C)) is 'small enough' to enable the response at a given value of ω to be computed on a single processor. In this case a parallel program can be created from a sequential one, based on an appropriate choice from numerous existing algorithms for this problem, with virtually no extra development work. Consequently a very significant speedup can be achieved by using, say, 32 processors and completing all computations using unmodified existing software. This exploits the natural parallelism of this case in the sense that the response at each value of ω can be computed independently. Laub and Gardiner (1988) give a detailed treatment, including code for a specific implementation on an Intel iPSC/d5 hypercube, of this case which is not considered further here.

The second (more interesting) case is the, by no means unusual, situation where the total memory available in the nodes is 'sufficiently' large to store (A, D, C), plus some workspace but the data need not necessarily fit into a single node. Then clearly the above approach cannot be directly used. One option in such cases is to use the following algorithm which is known to be reliable and efficient:

1. Transform A to upper Hessenberg form using orthogonal similarity transformations and, simultaneously, appropriately transform C and D, i.e. transform (A, D, C) to, say, (A^1, D^1, C^1). Hence at the end of this step
$$\tilde{G}(i\omega) = C^1(i\omega I_n - A^1)^{-1}D^1 \tag{12.16}$$
2. For each value of ω solve (12.16) where, under the assumption that $N \gg n$, this step consumes most of the computation time.

Consider now the details of each step. Then the Hessenberg reduction at step 1 is undertaken using Householder similarity transformations, i.e.

$$\left.\begin{array}{l} A^1 \leftarrow T_{n-2}\ldots T_2T_1AT_1T_2\ldots T_{n-2} \\ D^1 \leftarrow T_{n-2}\ldots T_2T_1D \\ C^1 \leftarrow CT_1T_2\ldots T_{n-2} \end{array}\right\} \tag{12.17}$$

where

$$T_j = I_n - \sigma_j u_j u_j^{\mathrm{T}}, \sigma_{\mathrm{j}} = \frac{u_j^{\mathrm{T}} u_j}{2} \tag{12.18}$$

Here each u_j is selected to null simultaneously the elements of column j below the subdiagonal and leave the previously processed columns invariant. Further, dropping all subscripts for ease of notation, introduce

$$v = A^{\mathrm{T}}u, w = \tilde{A}u, \tilde{A} = A - \sigma u v^{\mathrm{T}} \tag{12.19}$$

and hence (with obvious notation)

$$w = [\tilde{a}_1 \dots \tilde{a}_n] \begin{bmatrix} \mu_1 \\ \vdots \\ \mu_n \end{bmatrix} = \sum_i \mu_i \tilde{a}_i \tag{12.20}$$

Using this framework,

$$PAP = A - \sigma u v^{\mathrm{T}} - \sigma w u^{\mathrm{T}} \tag{12.21}$$

and the following is a parallel algorithm for this key computation, $1 \leqslant j \leqslant n-2$.

(i) Node holding a_j: compute u and σ from a_j as in sequential case. Others: wait.
(ii) Broadcast: u, σ.
(iii) All nodes: compute $v_i = a_i^{\mathrm{T}} u$ for each local a_i.
(iv) All nodes: compute $a_i \leftarrow a_i - \sigma v_i u$ for local a_i. At this stage $A - \sigma u v^{\mathrm{T}}$ is available.
(v) All nodes: compute the partial sum $w = \Sigma \mu_i a_i$ for local a_i.
(vi) Global add; add up the components of w from all nodes.
(vii) Broadcast w.
(viii) All nodes: compute $a_i \leftarrow a_i - \sigma \mu_i w$ for local a_i.

At this stage, the following points are relevant:

1. Updating D is similar to steps (iii) and (iv). Hence the details are omitted as are those for updating C which is similar to steps (v)–(viii).
2. Steps (i), (ii), (vi) and (vii) involve idle time for some of the nodes due to either communication delays or a sequential portion of the algorithm.
3. Steps (iii), (iv), (v) and (viii) exhibit higher levels of parallelism which could be exploited by including vector processors in the nodes.

In this chapter, the implications of (2) and (3) will not be considered further for ease of presentation. Instead, reference can be made to Rogers and Owens (1992b) for an in-depth treatment of these areas.

Consider now step 2 of the basic algorithm given above and, in particular, the computation of

$$\tilde{G}(i\omega) = C^1 (i\omega I_n - A^1)^{-1} D^1 \tag{12.22}$$

for a single value of ω. Further, suppose that LU factorization of the complex Hessenberg matrix $i\omega I_n - A^1$ is performed by either Gaussian elimination with partial pivoting or QR factorization. Then

$$\begin{aligned} \tilde{G} &= C^1 (LU)^{-1} D^1 \\ &= (C^1 U^{-1})(L^{-1} D^1) \\ &= YX \end{aligned} \tag{12.23}$$

where

$$YU = C^1 \tag{12.24}$$

$$LX = D^1 \tag{12.25}$$

At this stage, the basic problem has been replaced, in effect, by the computation of X and Y which can be undertaken in parallel (an example of so-called high-level functional parallelism). Further, set $H = i\omega I_n - A^1$ with elements h_{ij}. Then the following steps represent the essential 'mechanics' of this factorization:

1. Use h_{jj} and h_{j+1j} to compute a transformation L_j to eliminate h_{j+1j}.
2. Apply L_j to the $(n+1-j)\times(n+1-j)$ trailing submatrix of H.
3. At this stage the final forms of L_j and the jth row of U are available. The last row of U is available as $\mu_{nn} := h_{nn}$.
4. Write the triangular system $YU = C^1$ in the form

$$Y_i = \frac{1}{\mu_{ii}}\left[c_i - \sum_{j=1}^{i-1} \mu_{ji} Y_j\right] \tag{12.26}$$

Then defining

$$Y_i^{(0)} = c_i, i = 1, \ldots, n \tag{12.27}$$

permits at step j the computation

$$Y_j = \frac{1}{\mu_{jj}} Y_j^{(j-1)} \tag{12.28}$$

$$Y_i^{(j-1)} = Y_i^{j-1} - \mu_{ji} Y_j, i = j+1, \ldots, n \tag{12.29}$$

In this scheme each factor L_j can be applied to D^1 immediately it is produced and it alters only rows j and $j+1$ of this matrix to produce the jth row of X. Consequently there is no need to store L_j or L. Similarly, the rows of U are used immediately they are produced and hence there is also no need to store this matrix.

Using this approach, the matrix Y emerges by columns and the matrix X by rows. Hence the natural way to compute G is an outer (or dyadic) product. Storage is not required for X but is for Y since its columns are computed recursively. A data flow diagram for the complete procedure is shown in Figure 12.6 and this can be regarded as data pipelining at a high level.

Under ideal conditions, each box of Figure 12.6 would execute immediately it has input available and completely ignore the internal operations of all other boxes. Laub and Gardiner (1988) have implemented this system with synchronous processing in each node where the overlap is still important in order to minimize wasted time. Details of the parallel algorithms, given H, are omitted here for brevity except to note that this also contains idle time and higher levels of parallelism. Rogers and Owens (1992b) again consider how these can be exploited to increase efficiency.

Proceeding to consider hypercube implementation of the algorithm developed above in detail, a study of the two basic steps shows that a high degree of parallelism has been introduced at several levels. In particular, after Hessenberg reduction the computations at the N different frequencies are independent. Hence, if the coefficient matrices are 'small enough', the processing nodes can work independently on different values of ω and the problem is again almost perfectly

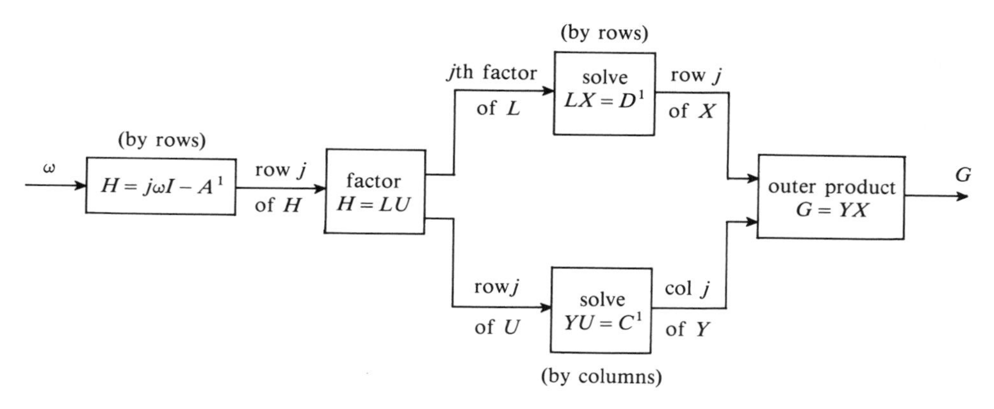

Figure 12.6 Data flow diagram for frequency response calculation

parallel. (Recall the case discussed earlier where it is possible to compute the response at a single value of ω on one processor.)

Of more general interest is the case where the matrices (A, B, D) must be split amongst multiple nodes. This is an obvious application for the column-wrapped distribution detailed in Section 12.3. In particular, each processor receives approximately n/p columns of A, mM/p columns of D, etc. and produces mM/p columns of $\tilde{G}$.

By appropriately combining the two approaches outlined above, it is possible to allow for a problem of any size. In particular, let d be the dimension of the smallest subcube capable of computing $\tilde{G}(i\omega)$ for a single value of ω and denote the dimension of the entire cube by q. Then the essential basis of this approach is to treat the overall cube as 2^{q-d} independent subcubes of dimension d. Further, these subcubes receive identical matrices but different frequency ranges and the columns are distributed amongst the nodes of each of them.

Using this approach means that step 1, the Hessenberg reduction, is performed redundantly by all of the subcubes. This work duplication could, for example, be avoided by first having all nodes working together on the Hessenberg reduction and then redistributing the resulting matrices to the appropriate subcube configuration. Note, however, that this would increase both the program complexity and the communications burden between the nodes. Further, this step is not really required if a 'large' number of frequencies is to be considered (almost always the case) since the duplicated work is then insignificant.

The code given in Laub and Gardiner (1988) consists of a host program and a node program where the latter runs on all nodes of the cube which are used. Normally all nodes are used but in every case the actual number is always a power of 2. Further, the host program reads the input matrices, i.e. the triple (A, D, C), and determines the size of the smallest subcube $(p = 2^d)$ capable of computing $\tilde{G}(i\omega)$ at a single value of ω. On completion of this, the node programs are loaded and configured to operate as 2^{q-d} independent subcubes of dimension d. Each subcube receives (A, D, C), where each matrix is distributed by columns to the nodes within it, and the Hessenberg transformation is performed by all subcubes.

Proceeding to discuss the execution of step 2, the host requests the input of the frequency range to be considered and the number of points. Each subcube computes the frequency response at logarithmically spaced points within its frequency range and returns the results to the host for onward transmission. An open research question at this stage is the 'optimal' choice of the frequencies.

Communication within the cube is restricted to broadcasts and so-called 'dual operations' which are the dual of them. These are achieved by a spanning tree where each subcube has its own version and hence has a totally independent communications structure from each of its contemporaries. Further, this approach pipelines data through the program and does not estimate condition numbers. Hence, see also the earlier discussion, the need to store the complex matrix H and its factors L and U is avoided. Note again that a column of Y and a row of X are used directly to form G and hence no storage is required for X.

In terms of storage requirements, this is easily concluded to be approximately $(n+3m+4)(n_{\text{loc}}+m_{\text{loc}})+4(n+m)$ double words per node. Here the terms n_{loc} and m_{loc} refer to the number of columns of A and D owned by each node where, for example, $n_{\text{loc}} \simeq n/p$. Laub and Gardiner (1988) have completed some preliminary 'benchmark' tests for this algorithm based on an iPSC/d5 (i.e. 32 processors) and a VAX 11/780 with floating point accelerator. All computations were completed in Fortran using a f77 compiler on the VAX and ftn 286 on the iPSC together with double precision or double precision complex arithmetic.

A representative sample of the timing results obtained is detailed in Table 12.1. In this preliminary analysis, the comparative performance measure used is speedup S_p – defined as the ratio of the time taken to solve a problem on a single processor to that for solving the same problem on $p(>1)$ processors. This shows that problems 'large enough' to require more than one node on the hypercube are beyond

Table 12.1 Timing results (in seconds) for frequency response computation

					Hessenburg reduction			*Frequency response (per freq.)*		
n	*m*	*l*	*Subcube dimension*	*Number of subcubes*	*iPSC time*	*VAX time*	S_p	*iPSC time*	*VAX time*	S_p
10	3	2	0	32	<1	0.1	–	0.146	0.05	3.4
50	10	10	0	32	16	8.2	0.5	0.367	2.70	7.4
100	15	15	0	32	118	56	0.5	1.820	13	7.1
180	20	20	1	16	334	[a]	[a]	7.130	[a]	[a]
300	32	32	3	4	444	[a]	[a]	63	[a]	[a]
520	32	32	4	2	1175	[a]	[a]	91	[a]	[a]

[a] Too large for VAX.

the practical operating limit of a VAX. These timings also provide 'benchmarks' against which to compare this basic algorithm with proposed modifications or extensions – such as those proposed by Rogers and Owens (1992b). The analysis and results of this section are discussed again in Section 12.5, with possible areas for further research.

12.4.2 Generalized Algebraic Riccati Equation

The development of solution procedures for (12.13), i.e. systematic algorithms for computing the $n \times n$ matrix X, is a well-researched area and numerous alternatives exist. In the current context, therefore, it is appropriate to begin with an overview of these methods with particular emphasis on their potential for parallel computation. Complete theoretical details of these methods can, for example, be found in Gardiner and Laub (1991) and the relevant references therein.

Consider first the standard algebraic Riccati equation which results from setting $E = I_n$ in (12.13). Then a reliable class of methods for solving this equation is based on constructing a particular invariant subspace of an associated Hamiltonian matrix. This, in turn, has led to the development of algorithms for solving the generalized version based on finding a stable deflating subspace (i.e. its basis vectors span the eigenspace associated with the stable eigenvalues) of the following $2n \times 2n$ matrix pencil which has a 'Hamiltonian-like' structure

$$P - \lambda L := \begin{bmatrix} A & -R \\ -Q & -A^{\mathrm{T}} \end{bmatrix} - \lambda \begin{bmatrix} E & 0 \\ 0 & E^{\mathrm{T}} \end{bmatrix} \tag{12.30}$$

The eigenvalues of this pencil occur in symmetric pairs about the imaginary axis of the complex plane. Further, introduce the matrix

$$J = \begin{bmatrix} 0 & I_n \\ -I_n & 0 \end{bmatrix} \tag{12.31}$$

where $J^{-1} = J^{\mathrm{T}} = -J$. Then the following symmetry properties will be of use here:

$$(JP)^{\mathrm{T}} = JP, (JL)^{\mathrm{T}} = -JL \tag{12.32}$$

Suppose now that $[Z_{11}{}^{\mathrm{T}}, Z_{21}{}^{\mathrm{T}}]^{\mathrm{T}}$ denotes a basis for the stable deflating subspace of (12.30). Then it can be shown that

$$X = Z_{21} Z_{11}^{-1} E^{-1} \tag{12.33}$$

is a symmetric stabilizing solution of (12.13). Further, consider the case when (12.13) is to be solved to construct the feedback gain matrix is an optimal control problem. Then this matrix, K, has the structure

$$K - S^{-1} B^{\mathrm{T}} XE \tag{12.34}$$

when R is of the form $BS^{-1}B^{\mathrm{T}}$ and hence it is only necessary to compute the matrix $XE = Z_{21} Z_{11}^{-1}$.

A basic approach to constructing the stable deflating subspace is to solve the eigenvalue problem it represents. Further, orthogonal transformation methods which make use of QR and QZ algorithms currently rank amongst the most reliable and efficient. These yield an orthonormal basis for the required subspace where, in effect, a QZ-like algorithm is used for the generalized case and a QR-like algorithm for $E = I_n$.

In order to use this method as the basis of a parallel algorithm, it is clearly necessary to develop parallel implementations of the QR and QZ algorithms. Attempts reported to date at developing parallel versions of the QR algorithm (which is structurally similar to the QZ algorithm) have proved largely disappointing. Consequently this approach is not considered further here.

Jacobi's method is an alternative orthogonal transform based method for solving the standard eigenvalue problem and hence the standard algebraic Riccati equation. This approach is particularly attractive for parallel computation purposes despite the fact that it has been largely replaced by the QR algorithm for sequentially based computation. Further, parallel forms of Jacobi's method are available which have been targeted to Hamiltonian matrices. The major disadvantage of unsymmetric Jacobi iteration is the fact that the operation count is in excess of $O(n^3)$ and hence this approach is also discarded here.

Another approach to computing the stable deflating subspace is to use the matrix sign function. Basically, this is a generalization of the well-known sign or signum function for a scalar quantity. Its major advantage in this context is that it can be computed using simple matrix operations such as multiplication by a scalar, multiplication and the solution of linear equations which are easily converted to run in parallel.

Background on the matrix sign function is given in the appendix. In terms of the generalized version, use of the matrix sign function, in effect, separates the deflating subspaces corresponding to left and right-half plane eigenvalues of the pencil (12.30) without computing them. This, together with the algorithm detailed below, allows the solution of (12.13) to be computed.

Define

$$Z_0 = P \tag{12.35}$$

and introduce the iteration

$$Z_{k+1} = \frac{1}{2}\left[\frac{1}{c_k} Z_k + c_k L Z_k^{-1} L\right], k \geqslant 0 \tag{12.36}$$

where

$$c_k = \left[\left|\frac{\det Z_k}{\det L}\right|\right]^{1/2n} \tag{12.37}$$

is a scalar factor used to increase convergence. Then it can be shown that the sequence $\{z_k\}_{k \geqslant 0}$ converges if (12.13) has a solution, and denote this limit by S.

In which case the required solution is obtained by solving, with an obvious partitioning of S,

$$\begin{bmatrix} S_{12} \\ S_{22} + E^{\mathrm{T}} \end{bmatrix} XE = - \begin{bmatrix} S_{11} + E \\ S_{21} \end{bmatrix} \tag{12.38}$$

Noting again the special structure of P and L, the sign function can be written in the alternative symmetric form

$$Z_0 = JP \tag{12.39}$$

$$Z_{k+1} = \frac{1}{2}\left[\frac{1}{c_k} Z_k - c_k(JL)Z_k^{-1}(JL)\right], k \geqslant 0 \tag{12.40}$$

where c_k is defined by (12.37). Then $\{Z_k\}_{k \geqslant 0}$ converges to JS where S is defined as above. This form of the iteration has a number of advantages in terms of parallel implementation(s) which are exploited below.

In common with numerous matrix algorithms, the sign function algorithm offers the potential of parallelism on several levels. Here the column-wrapped distribution (Figure 12.4) is used to exploit parallelism in the data by assigning the matrices such that each processor has approximately the same number of columns to store and perform operations on. If the symmetric form of (12.39) and (12.40) is used, the storage requirements can be reduced by storing only the upper triangular portion of all iterations of the symmetric matrix Z. This is a major reason for using the symmetric form as the basis of the implementation studies described below.

A secondary level of parallelism exists here in the form of coarser-grained functional parallelism. In particular, most of the overhead (loss of efficiency) in the type of parallel algorithm used here results from either processor idle time to allow sequential portions of the algorithms to be completed or from communication time. Using functional parallelism, the dead-time inherent in one function can be used by some other operation. For example, that arising during matrix factorization could be used to complete part of a matrix multiplication operation. This feature has been exploited in the implementations described here by overlapping different functions where possible to give the greatest efficiency.

Gardiner and Laub have used a parallel form of the LINPACK routines (see Dongarra *et al.*, 1979, for the necessary background on this standard linear algebra package) for solving indefinite linear systems to develop a parallel algorithm for computing the sign function. This is detailed below together with two variations to serve as the basis for a critical comparative study in terms of computational and communication complexity – so-called algorithm complexity – and measured running times.

The programs used by Gardiner and Laub have been written in Fortran supplemented by special subroutines for sending and receiving messages. Further, all processors execute the same code, and the 'node identity facility', see Section 12.3, has been used to provide specialized node processing where required, i.e. each node 'recognizes' which columns of a matrix it is processing. A virtually regular communications structure consisting, essentially, of broadcasts and global operations (also termed 'fan-out' and 'fan-in' respectively) has been employed.

Basically, see also Section 12.3, in a broadcast one node sends a message, such as the columns of a matrix, to all other nodes. In the case of the hypercube topology, this can be completed in d – the cube dimension – steps. Conversely, consider the case when all nodes have some information which must be merged: for example, the computation of a global sum. This is an example of a global operation which also requires d steps on a hypercube. Note, however, that the actual implementation of these operations is totally independent of the algorithms used and this, in turn, means that they are independent of topology. Hence they can be implemented on alternatives to the hypercube for which Rogers (1991) details some highly promising initial results.

The basic algorithm developed by Gardiner and Laub, the so-called diagonal pivoting algorithm, is based on rewriting the sign function in suitably factored form. To introduce this, first note that the scale factor, c_k, can be deleted for ease of presentation in what follows and set $F = JL$. Further, factor Z_k as

$$Z_k = UDU^{\mathrm{T}} \tag{12.41}$$

where U and D have a special form, such as triangular or diagonal, which is easily invertible. Then (12.40) can be written as

$$Z_{k+1} = \frac{1}{2}(Z_k - YD^{-1}Y^{\mathrm{T}}) \tag{12.42}$$

where

$$UY^{\mathrm{T}} = F \tag{12.43}$$

Further, what follows concentrates on the construction of $YD^{-1}Y^{\mathrm{T}}$ since this operation requires the vast majority of the computational work expended.

In Gardiner and Laub's work, the decomposition (12.41) has been performed using a parallel version of the LINPACK algorithm DSIFA for factoring symmetric indefinite matrices. Essentially, this algorithm, also termed diagonal pivoting factorization, is a symmetric form of Gaussian elimination with partial pivoting.

Use of this algorithm to compute (12.41) involves three steps, detailed in turn below, and the required data flow is shown schematically in Figure 12.7.

Step 1: Factor Z_k as

$$Z_k = PUDU^{\mathrm{T}}P^{\mathrm{T}} \tag{12.44}$$

where D is a block diagonal matrix where each diagonal block is either 1×1 or 2×2, U is unit upper triangular, and P now denotes a permutation matrix used for pivoting. (For notational convenience Figure 12.7 refers to the case when P is the identity matrix). The reason for the 2×2 diagonal blocks in D is to guarantee that the factorization exists (it is also reasonably stable) which may not be the case if it was demanded that D be diagonal.

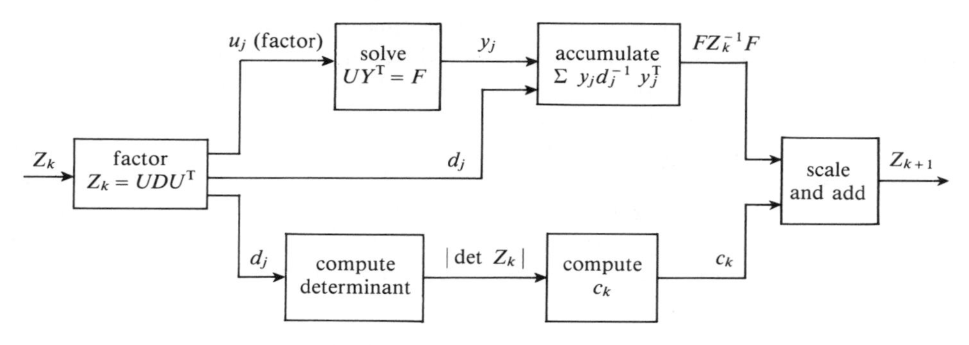

Figure 12.7 Data flow diagram for sign function iteration

Step 2: Solve

$$PUY^T = F \tag{12.45}$$

for Y^T.

Step 3: Construct $F^T Z_k^{-1} F$ as an outer (dyadic) product using

$$F^T Z_k^{-1} F = YD^{-1} Y^T \tag{12.46}$$

In particular, write

$$D = \text{diag}\,\{d_i\}_{1 \leqslant i \leqslant r} \tag{12.47}$$

where d_i is either a 1×1 or a 2×2 matrix and $r \leqslant 2n$ is the number of blocks. Further, partition Y conformally with D, i.e. write this matrix as $[y_i, \ldots, y_r]$ where each entry is either a one or two column matrix. Then

$$YD^{-1}Y^T = \sum_{i=1}^{r} y_i\, d_i^{-1} y_i^T \tag{12.48}$$

In terms of data flow, Figure 12.7 (with c_k included) can be regarded as pipelining at a high level. Further, in the ideal case, each box should execute immediately it has input available and independently of the internal operations of the other boxes. (Compare with Figure 12.6 for the frequency response matrix computation.) The currently available implementation has two interacting processes in each node – one for the factorization and the other for the solve and accumulate steps. Consequently processors which would be idle during communication portions of the factor operation can be employed to complete other operations.

Using the column-oriented approach, coupled with column-wrapped distribution of the matrices Z, F and Y, means that all functions (which operate on their inputs) produce output column by column. Further, U is generated in factored form and each of these can be applied to F as soon as they are available. The matrix Y emerges by columns and the dyadic product of (12.48) is accumulated.

Define $\|\,.\,\|$ as the sum of the absolute values of all elements on or above the diagonal of a matrix. Then convergence monitoring is included here by asking each node to compute its contribution to the total difference $\| Z_k - Z_{k+1} \|$. This is then followed by a global addition to all nodes and the iteration terminates when the difference is 'small enough', i.e. formally $\| Z_k - Z_{k+1} \| \leqslant \varepsilon$ for given ε.

Suppose now that the sign function iteration has been completed. Then it remains to compute the required solution *XE*. This can be achieved by straightforward *LU* factorization followed by forward and backward substitution.

Algorithm complexity and measured running times are the performance (comparative) measures used here. In the case of the former, this consists of the following:

1. The number of floating point operations required – the flop count – where one flop is approximately equivalent to the scalar operation $ax + b$.
2. The number of broadcasts and global operations.
3. The number of random distribution messages.

Note also that no actual computation is completed sequentially in this algorithm and the following is a summary of these measures for each of the four main steps (as above plus the final sum and convergence monitoring).

Step 1 ('Factor'): $4n^3/3$ flops, $6n$ broadcasts or global operations of average length n, and $2n+1$ local messages sent to another task in the same processor with a total length of $8n^3$.
Step 2 ('Solve'): $4n^3$ flops, $2n$ broadcasts each of length $2n$, and p random messages of average length n where $p \leqslant 2n$ is the number of times pivoting is required.
Step 3 ('Accumulate'): $4n^3$ flops and no communication.
Step 4 ('Final Sum and Convergence Monitoring'): $12n^2$ flops and 2 broadcasts or global operations of length 2.

A basic weakness of the above algorithm is that both row and column pivoting must be performed in order to provide stability and preserve symmetry. Further, the first class of operations introduce no problem since the elements to be exchanged reside in the same processor. This is not true of column exchanges, however, which, together with the selection of the pivot, require extra communication in an irregular pattern. Consequently the removal of this pivoting problem is an obvious starting point of any attempt at improving the basic efficiency of this algorithm.

One obvious approach to attempt here is to use orthogonal rather than elementary transformations. In particular, at least one possible parallel subroutine for reducing a symmetric matrix to tridiagonal form (the upper and lower bandwidth is 1) using orthogonal similarity transformations is known. Further, tridiagonal matrices are easily factored and this approach leads to an alternative algorithm which has the same three functions as before.

The details of these steps are omitted here for brevity and the following is a summary of the main points from a comparative complexity analysis:

1. An irregular communication pattern for pivoting has been replaced by a

more regular but at least as time-consuming global operation in the form of 'fan-in'.

2. The ratio of orthogonal to diagonal pivoting factorization times is 'four'.

On the basis of these, therefore, it is to be expected that its performance – see also below – will be inferior to that of the diagonal pivoting algorithm.

Work by Charlier and Van Dooren (1989) is the basis of a third algorithm which has a somewhat different structure from that of the previous two. The essential difference is that instead of performing repeated factorizations, either of the two methods described above is used to perform an initial factorization and the factors are then updated directly. Further, when a 'sufficient' degree of convergence has been achieved these factors are recombined to yield the sign matrix JS. This update is one-sided – taken to mean that it only requires row operations – and hence it can be expected to require less communication amongst the processors with a consequent increase in efficiency.

The details of the steps involved in this algorithm are also omitted for brevity. Instead, it suffices to note that a key one of these is analogous to QR factorization which, on the assumption that it exists, can be completed with a combination of orthogonal (also termed Givens or plane) and hyperbolic rotations. Further, this factorization may not exist and this is usually the case for the problems of interest here. In fact, the required factorization exists, if and only if, the product matrix Z_{k+1} can be factored as UDU^T, with U^T upper triangular and D diagonal, without pivoting or 2×2 diagonal blocks for D. It is also known that a necessary condition for the existence of this factorization is positive definiteness of either R or Q in (12.13).

A factorization is guaranteed to exist if a certain triangular matrix can be generalized to block upper triangular with diagonal blocks of dimensions 1×1 and 2×2. Further, it is necessary to perform column pivoting which is what this algorithm set out to avoid. Hence this is not a particularly feasible option and the practical choice of pivot columns is still an open research question.

Gardiner and Laub have used a form of preconditioning on the matrix Z_0 to avoid this problem. This has worked on all problems attempted to date but is not guaranteed to work in general without pivoting. Note also that the basic form of this update algorithm (without pivoting or preconditioning) was originally developed for a fine-grained processing environment based on systolic arrays. The work under consideration here, however, shows that it performs 'very poorly' (see also the discussion immediately following) in the coarse-grained processing environment of the hypercube multiprocessor.

Discounting the initial factorization (or preconditioning), the operation count for each iteration of the update is $\frac{32}{3}\ n^3$ and the communications requirement consist of $2n$ broadcasts of average length $3n$ per iteration. Further, since it requires the same number of iterations as the other two methods, this algorithm actually uses eight times as much computation as the first (diagonal pivoting) algorithm and two times that for the second (orthogonal tridiagonalization) algorithm. Note also that the flop count could be halved by replacing the Givens rotations with a form of Householder reflection.

This reflection approach would still require four times the number of floating point operations as the diagonal pivoting algorithm. Further, the update process has a 'large' sequential component which is required for computing the transformations. These factors are the major contributors to the 'poor' performance of this algorithm which is clearly highlighted in the timing results discussed next.

At this stage, the matrix sign function has been used to develop three candidate parallel algorithms for solving the generalized algebraic Riccati equation of (12.13). Further, in terms of algorithm complexity the one based on diagonal pivoting is clearly superior. In terms of measured running times, the following summarizes tests conducted by Gardiner and Laub which (a) compare the three parallel algorithms relative to a single node; and (b) compare all three parallel algorithms relative to a sequential one run on a VAX 11/780.

The basic data used consisted of matrices with randomly generated coefficients (i.e. A, E, R and Q) for $20 \leqslant n \leqslant 100$ and the timing results obtained are given in Table 12.2. This work used an iPSC/d5 (32 nodes) hypercube with the standard configuration (i.e. no vector board or memory) and iPSC software version 3.2 together with the latest operating system. Those results relating to the VAX were obtained using an 11/780 with floating point accelerator running bsd 4.2 UNIX. Further, the Schur algorithm (an orthogonal transformation method) was used for tests on the single node of the VAX and figures with superscript 'a' in Table 12.2 denote times for the Schur algorithm extrapolated from other data.

Table 12.2 shows that all the parallel algorithms offer some speedup over the sequential approach. Further, in comparative terms, the VAX is approximately three to four times faster than a single node of the iPSC and for $n = 20$ (the smallest case considered) the VAX is faster. This advantage disappears as n increases until at $n = 100$ the multiple processor hypercube easily solves a problem which is not feasible on the single processor VAX.

Clearly, the diagonal pivoting algorithm is the best (by a considerable margin) of the parallel group. Note also that the behaviour of the parallel algorithms is very

Table 12.2 Timing results (in seconds) for the Riccati equation

	n				
	20	*40*	*60*	*80*	*100*
Diagonal pivoting	12.2	36.3	94.6	166	323
Tridiagonalization	31.9	117	266	518	977
Update algorithm	39.9	163	391	758	1465
Schur algorithm: 1 node	77.6	578	2000[a]	4600[a]	9100[a]

[a] Times for Schur algorithm extrapolated from other data.

much less than cubic for 'small' n. In particular, for 'small' n the communication overhead significantly degrades the performance of the parallel algorithms, and for 'larger' n the computation dominates.

Given the fact that the diagonal pivoting algorithm is the 'best' of the parallel group, an obvious next stage is to undertake a detailed assessment of its performance for 'benchmarking' and related purposes. This is clearly a wide-ranging problem for which the only currently available output is again due to Gardiner and Laub. They have run this algorithm on problems of many sizes using different numbers of processors. Again, the basic data consisted of matrices with randomly generated coefficients and the solution times averaged over five problems of the same size.

Using these results, graphs of solution times, speedup and efficiency against the problem size, n, can be plotted as shown in Figures 12.8, 12.9 and 12.10 respectively. These graphs represent data measurements for problem sizes which are multiples of ten with linear interpolation between points. The graph for $p = 1$ is for the sequential sign function algorithm run on a single node with n bounded above by 60. Further, efficiency, denoted E_p, is a normalized form of speedup where $E_p = S_p/p$ with S_p bounded above by p and by E_p by unity.

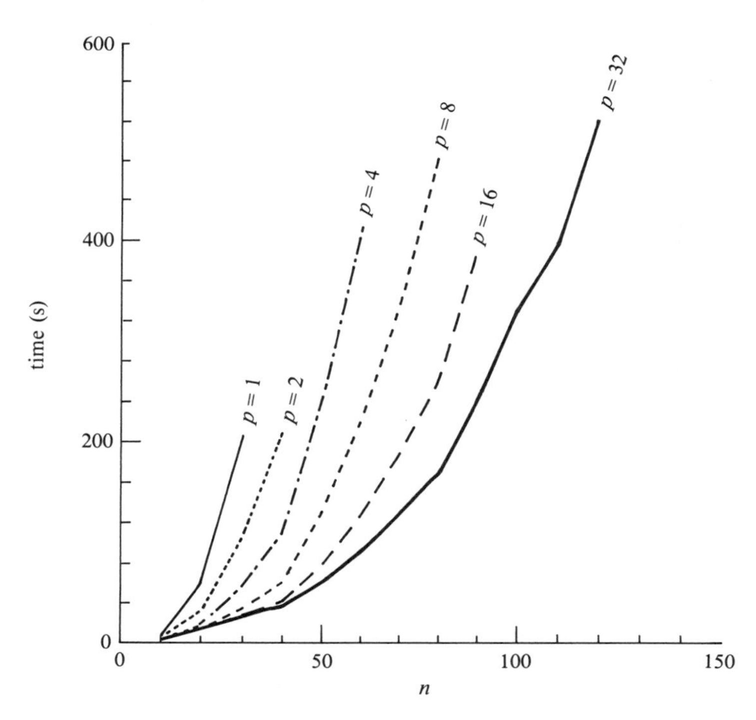

Figure 12.8 Run times for the diagonal pivoting algorithm

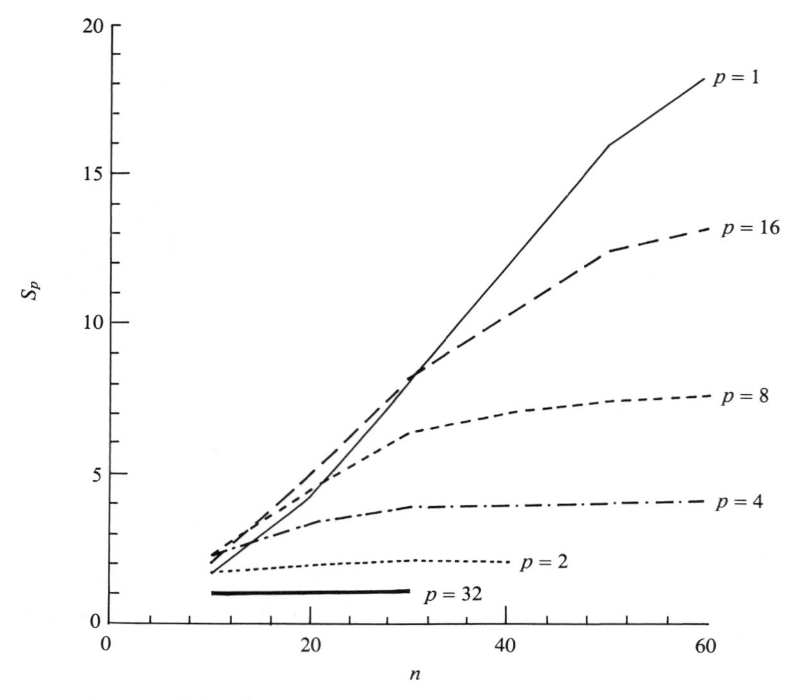

Figure 12.9 S_p plots for the diagonal pivoting algorithm

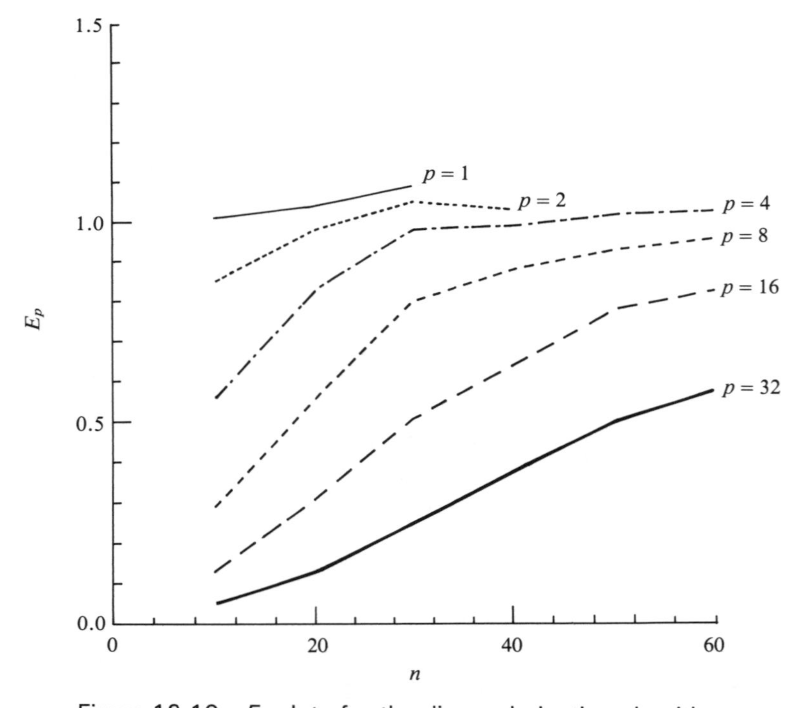

Figure 12.10 E_p plots for the diagonal pivoting algorithm

The important point concerning the graphs of Figures 12.8–12.10 is that the trends shown are valid. In particular, the discrepancy in Figures 12.9 and 12.10, i.e. a small number of measured efficiencies greater than unity, can be explained by the fact that the sequential program makes extensive use of standardized routines (mainly from LINPACK) but the parallel code is special purpose and hence more highly optimized. Further, S_p and E_p for the sign function algorithm can be estimated theoretically using a model developed from the complexity figures presented earlier. The details are omitted here for brevity except to note that the predicted versions of Figures 12.9 and 12.10 have the same shape as their experimental counterparts, but lower values overall. These can be explained by factors such as neglecting any overlapping of communication and computation which occur in practice and serve to increase the apparent communication overhead.

12.5 CONCLUSIONS

Results already available in the open literature have been used as the basis for a tutorial style survey of the current 'state of the art' on the development of parallel algorithms for (linear) control systems design. Two potential application areas have been briefly discussed to motivate the need for such algorithms which, in effect, arises from an incurred computational load which approaches or exceeds the effective operating range of sequentially based machines. For example, the dynamics of large space structures typically extend over wide frequency ranges or spatial domains and hence model order reduction techniques (with consequent benefits in terms of a reduced computational load for analysis and design) are often not appropriate.

This chapter started with a brief review of two possible application areas – 2D systems and large space structures. Further, it has been concluded that the successful application of established design techniques, or algorithms, to such areas will only be effectively achieved if it is possible to allow for an increase in model dimension of at least one order of magnitude with negligible reduction in reliability. Two (reasonably) generic problems have been used, i.e. the computation of a multivariable frequency response matrix from the plant state-space description and the solution of a generalized algebraic Riccati equation.

The algorithms detailed in this chapter are based on use of a hypercube multiprocessor. One immediate general point arising from this work is that the development of parallel algorithms for standard extensively used approaches to control systems design is not simply a case of modifying existing software. This point has been clearly demonstrated by the Riccati equation example where it has been found that a number of well-known solution algorithms are not suitable for parallel implementation.

In effect, the results presented consist of basic feasible algorithms plus some test results which compares their performance with sequentially implemented algorithms on a VAX 11/780. One point here is the difference in memory size between the two machines where the hypercube has a 'fairly large' memory (the sum of that in each node) and the VAX uses virtual memory. Consequently when the VAX is running a program, the overhead incurred by data transfer between the main memory and the disk can overwhelm the time required for computation.

This last fact partly explains the large speedup factor of the hypercube over the VAX and suggests that the comparison is not really valid. In actual fact, however, this argument really shows that the VAX was never intended to solve problems of the general size under consideration here. Further, this discrepancy will become even more pronounced as the need arises to consider even larger problems coupled with continuing developments in the hypercube infrastructure (more processors, faster communications, etc.).

To discuss the test results given here in terms of the currently known and potential roles for parallel processing in linear control systems design (and beyond), it clearly suffices to restrict attention to one of the two cases analyzed. In the case of the algebraic Riccati equation, therefore, the diagonal pivoting factorization form of the matrix sign function algorithm is very clearly the best available. Further, it is the most direct translation of a sequential algorithm. This is a highly desirable property which, unfortunately, is not likely to be present in all cases.

The results given in Section 12.4.2 (i.e. algorithm complexity and measured running times) clearly show that for the essentially linear algebra based problem type under consideration here, the computation intensity (or amount) is more important than communication. Further, this communication tends to be very 'small' and a major factor limiting potential speedup is, of course, the amount of inherent sequential computation.

These results also indicate that the algorithm used is practical with regard to the fact that the iPSC/d5 is, in relative terms, a simple machine. In particular, currently available hypercubes with more processors, more powerful computing power in the processors at each node, and enhanced interprocessor communications are capable of producing computing times at least three orders of magnitude faster with, essentially, the original code used by Gardiner and Laub. This compares with 25 minutes of CPU time, and well in excess of one hour of elapsed time, for the Schur algorithm with $n = 80$ run on a VAX 11/780 with floating point accelerator.

Another highly attractive feature of the algorithms detailed in this chapter is their flexibility in terms of exploiting recent and potential future developments in parallel (or concurrent) processing architectures. For example, there is potential for significantly higher speedup using a hypercube with vector processors in the nodes. This point is discussed further in Rogers and Owens (1992b), for the multivariable frequency response case. Further, many of the inner loops in the diagonal pivoting algorithm for the Riccati equation have a structure which is very amenable to vector processing. A complete appraisal of this aspect must, however, await the results of appropriately targeted future research.

One note of caution concerning the use of faster processors must be introduced at this stage. In particular, faster processors will increase overall performance but processor use, and hence speedup, will decrease if there is not also an increase in communication speeds. The 'trade-off' here is, therefore, to accept some decrease in speedup, since for 'large' problems the speedup and efficiency found to date are 'quite high'.

Summarizing, therefore, extremely encouraging results for both problems considered are available with relatively slow processing nodes. Use of a high-performance hypercube (or an alternative multiprocessor) should yield greatly improved results but lower efficiencies. Further, additional algorithm enhancement should offer yet more scope for solving larger problems even faster. These are obvious general target areas for short to medium term research effort. Coupled with this, research is obviously desirable in terms of extending the underlying results of this chapter to related problems, such as computing the H_∞ norm of $G(s)$ of Sections 12.2.1 and 12.4.1.

One very important aspect which has not been directly considered here is that of choosing the 'best' architecture. This is an obvious area to address after work, such as that reported here, on developing basic feasible algorithms has been 'satisfactorily' completed. In terms of the hypercube approach, the following points are relevant:

1. It is a distributed memory architecture and therefore meets the requirement of scaleability. This means that an increase in problem size can be dealt with by adding processors, i.e. without recourse to reformulation of the algorithm, which has obvious benefits in the control systems design application area.
2. It is typically lower priced for the same computational power in comparison to alternatives such as Cray machines. Further, in a number of cases considered to date, such as the Riccati equation of Section 12.4, the algorithms are essentially independent of the hypercube topology. This, in turn, means that they could be used on any distributed memory multiprocessor with a message-passing facility and an 'appropriately large' number of interconnections.

APPENDIX

The matrix sign function is a generalization of the well-known sign (or signum) function for a scalar. Suppose that the $q \times q$ matrix M has no eigenvalues on the imaginary axis of the complex plane and assume the Jordon decomposition

$$M = T(D + N)T^{-1} \tag{12.A1}$$

where $D = \operatorname{diag}\{\lambda_i\}_{1 \leqslant i \leqslant q}$, N is nilpotent and commutes with D. Then the matrix sign of M can be defined as

$$Z = \operatorname{sgn}(M) = T \operatorname{diag}\{\operatorname{sgn}(\operatorname{Re} \lambda_i)\}_{1 \leqslant i \leqslant q} T^{-1} \tag{12.A2}$$

One of a number of simple methods for computing sgn(M) is the following iteration

$$Z_{k+1} = \frac{1}{2}(Z_k + Z_k^{-1}); \; Z_0 = M \tag{12.A3}$$

and then

$$Z = \lim_{k \to +\infty} Z_k = \text{sgn}(M) \tag{12.A4}$$

In effect, this is Newton's method for solving $Z^2 - I = 0$ and usually gives excellent performance. High-order formulas are, however, of interest for two reasons. The most obvious of these is to provide higher convergence rates where, for example, Halley's method uses second-order derivative information to provide third-order convergence.

The second reason why higher-order methods are of interest here is that they result in algorithms which are more amenable to parallel implementation. Consider, therefore, rational iterative methods with the general form

$$X_{n+1} = P_k(X_n)Q_m^{-1}(X_n) \tag{12.A5}$$

where P and Q are polynomials of order k and m respectively. These methods are of special interest here because their partial fraction forms lead naturally to parallel algorithms. In particular, sequential matrix multiplications which arise in evaluating P_k and Q_k are avoided by using a partial fraction expansion

$$\frac{P_k(z)}{Q_m(z)} = \sum_{i=1}^{m} \frac{w_i}{z - z_i} \tag{12.A6}$$

where $k \leqslant m$ and z_i, $1 \leqslant i \leqslant m$, are the roots of Q_m which are assumed to be distinct and nonrepeated.

Use of this partial fraction expansion enables algorithm-level parallelism to be employed since each fraction can be evaluated in parallel on a different processor. Note, however, that this expansion does not always produce numerically reliable algorithms since, in general, the parameters w_i and z_i must be computed numerically. This can cause inaccuracy if $Q_m(z)$ is ill-conditioned, but this problem does not arise here.

Rational function approximations of sgn(M) can be developed by exploiting a link between this function and the hypergeometric function

$$\text{f}(z) := \frac{1}{\sqrt{(1-z)}} \tag{12.A7}$$

In particular, considering the real case,

$$\text{sgn}(x) = \frac{x}{|x|} = \frac{x}{\sqrt{(1-(1-x^2))}}, \quad \forall x \neq 0 \tag{12.A8}$$

and hence $\text{sgn}(x) = x\text{f}(z)$ results on setting $z = 1 - x^2$. Then the required approximations follow on use of Pade approximations of f, where the main diagonal and first subdiagonal ones (obtained by setting $k = m$ and $k = m - 1$ respectively) can be shown to result in iterations which are globally convergent. Further, they include Newton's and Halley's methods as special cases.

An in-depth analysis of this link yields the approximation

$$\operatorname{sgn}(x) \simeq x \sum_{i=1}^{m} \frac{1}{mx_i}\left(\frac{1}{\alpha_i^2 + x_i^2}\right) \tag{12.A9}$$

where $\alpha_i^2 = 1/x_i - 1 > 0$ and x_i are the roots of shifted Chebyshev polynomials. Hence rational approximations to $\operatorname{sgn}(x)$ in partial fraction form result with the crucial feature that w_i and z_i of (12.A6) are available with a high degree of accuracy.

A straightforward generalization of the scalar results summarized above to the matrix case is possible. This yields the following iterative formula for computing the matrix sign function:

$$X_{n+1} = X_n \sum_{i=1}^{m} \frac{1}{mx_i} (X_n^2 + \alpha_i^2 I)^{-1}$$

$$X_0 = X \tag{12.A10}$$

REFERENCES

ARNOLD, N. F. (1984) 'Numerical solution of algebraic matrix Riccati equations', PhD thesis, University of Southern California.

BALAS, M. J. (1982) 'Trends in large space structure control theory: fondest hopes, wildest dreams, *IEEE Trans Auto-Control*, **AC-27**, 522–35.

BENDER, D. J. and LAUB A. J. (1985) 'Controllability and observability at infinity of multivariable second order lags', *IEEE Trans Auto-Control*, **AC-29**, 163–5.

CHARLIER, J. P. and VAN DOOREN, P. (1989) 'A systolic algorithm for Riccati and Lyapunov equations', *Math. Control. Sig. Syst*, **2**, 109–36.

CYBENKO, G. (1988) 'Load balancing and partitioning for parallel signal processing and control algorithms, in *Advanced Computing Concepts and Techniques in Control Engineering* (eds M. J. Denham and A. J. Laub), Berlin: Springer Verlag, pp. 391–408.

CYBENKO, G. and ALLEN, T. G. (1991) 'Multidimensional binary partitions: Distributed data structures for spatial partitioning', *Int. J. Control,* **51**(6), 1335–52.

DONGARRA, J. J., BUNCH, J., MOLER, C. B. and STEWART, G. W. (1979) *LINPACK Users Guide,* Philadelphia: Society for Industrial and Applied Mathematics.

FOX, G. (ed.) (1988) *Proceedings of the Third Conference on Hypercube Concurrent Computers and Applications.*

GARDINER, J. D. and LAUB, A. J. (1991). 'Parallel algorithms for alegbraic Riccati equations', *Int. J. Control*, **51**(6), 1317–34.

GREENE, C. S. and STEIN, G. (1979) 'Inherent damping, solvability conditions and solutions for structural vibration control, in *Proceedings of the 18th IEEE International Conference on Decision and Control, Fort Lauderdale*, pp. 230–2.

HEATH, M. T. (ed.) (1986) *Hypercube Multiprocessors*, Philadelphia: Society for Industrial and Applied Mathematics.

HEATH, M. T. (ed.) (1987) *Hypercube Multiprocessors*, Philadelphia: Society for Industrial and Applied Mathematics.

KARP, A. H. (1987) 'Programming for parallelism', *Computer,* **20**, 43–56.

LAUB, A. J. and GARDINER, J. D. (1988) 'Hypercube implementation of some parallel algorithms in control', in *Advanced Computing Concepts and Techniques in Control Engineering* (eds. M. J. Denham and A. J. Laub). Berlin: Springer Verlag, pp. 361–390.

LIM, J. S. (1990) *Two-dimensional Signal and Image Processing*, Englewood Cliffs, NJ: Prentice Hall.

MACKAY, M. K. (1983) 'Active control of large flexible space structures', PhD Thesis, University of Southern California.

MOLINARI, R. P. (1977) 'The time invariant linear quadratic optimal control problem', *Automatica,* **13**, 347–57.

OWENS, D. H. and ROGERS, E. (1992) 'H_∞ norm minimisation and the stabilisation of systems with repetitive behaviour', *Trans. Inst. M. C.*, **14**(3), 126–9.

ROGERS, E. and OWENS, D. H. (1992a) *Stability Analysis for Linear Repetitive Processes*, Lecture Notes in Control and Information Sciences Series, Vol. 175. Berlin: Springer Verlag.

ROGERS, E. and OWENS, D. H. (1992b) 'Enhanced parallel algorithms for computing the multivariable frequency response', in preparation.

ROGERS, E. (1991) *Parallel Algorithms for Controller Design: Alternatives to the hypercube*, Research Report, Department of Aeronautics and Astronautics, University of Southampton.

WILEY, P. (1987) 'A parallel architecture comes of age at last', *IEEE Spectrum,* **24**, 46–50.

Index

UNIVERSITY OF STRATHCLYDE
30125 00447552 0